RESIDUE REVIEWS

VOLUME 25

SPECIAL VOLUME – SEMINAR ON

EXPERIMENTAL APPROACHES TO PESTICIDE METABOLISM, DEGRADATION, AND MODE OF ACTION

UNITED STATES – JAPAN SEMINAR

August 16-19, 1967

Nikko, Japan

SPRINGER-VERLAG

BERLIN · HEIDELBERG · NEW YORK

1969

RESIDUE REVIEWS

Residues of Pesticides and Other
Foreign Chemicals in Foods and Feeds

RÜCKSTANDS-BERICHTE

Rückstände von Pesticiden und anderen
Fremdstoffen in Nahrungs- und Futtermitteln

Edited by

FRANCIS A. GUNTHER

Riverside, California

VOLUME 25

SPRINGER-VERLAG
BERLIN · HEIDELBERG · NEW YORK
1969

ISBN-13: 978-1-4615-8445-2   e-ISBN-13: 978-1-4615-8443-8
DOI: 10.1007/978-1-4615-8443-8

© 1969 by Springer-Verlag New York Inc.
Softcover reprint of the hardcover 1st edition 1969
Library of Congress Catalog Card Number 62–18595.

Title No. 6627

Preface

That residues of pesticide and other "foreign" chemicals in foodstuffs are of concern to everyone everywhere is amply attested by the reception accorded previous volumes of "Residue Reviews" and by the gratifying enthusiasm, sincerity, and efforts shown by all the individuals from whom manuscripts have been solicited. Despite much propaganda to the contrary, there can never be any serious question that pest-control chemicals and food-additive chemicals are essential to adequate food production, manufacture, marketing, and storage, yet without continuing surveillance and intelligent control some of those that persist in our foodstuffs could at times conceivably endanger the public health. Ensuring safety-in-use of these many chemicals is a dynamic challenge, for established ones are continually being displaced by newly developed ones more acceptable to food technologists, pharmacologists, toxicologists, and changing pest-control requirements in progressive food-producing economies.

These matters are of genuine concern to increasing numbers of governmental agencies and legislative bodies around the world, for some of these chemicals have resulted in a few mishaps from improper use. Adequate safety-in-use evaluations of any of these chemicals persisting into our foodstuffs are not simple matters, and they incorporate the considered judgments of many individuals highly trained in a variety of complex biological, chemical, food technological, medical, pharmacological, and toxicological disciplines.

It is hoped that "Residue Reviews" will continue to serve as an integrating factor both in focusing attention upon those many residue matters requiring further attention and in collating for variously trained readers present knowledge in specific important areas of residue and related endeavors; no other single publication attempts to serve these broad purposes. The contents of this and previous volumes of "Residue Reviews" illustrate these objectives. Since manuscripts are published in the order in which they are received in final form, it may seem that some important aspects of residue analytical chemistry, biochemistry, human and animal medicine, legislation, pharmacology, physiology, regulation, and toxicology are being neglected; to the contrary, these apparent omissions are recognized, and some pertinent manuscripts are in preparation. However, the field is so large and the interests in it are so varied that the editor and the Advisory Board earnestly solicit suggestions of topics and authors to help make this international book-series even more useful and informative.

"Residue Reviews" attempts to provide concise, critical reviews of timely advances, philosophy, and significant areas of accomplished or needed endeavor in the total field of residues of these chemicals in foods, in feeds, and in transformed food products. These reviews are either general or specific, but properly they may lie in the domains of analytical chemistry and its methodology, biochemistry, human and animal medicine, legislation, pharmacology, physiology, regulation, and toxicology; certain affairs in the realm of food technology concerned specifically with pesticide and other food-additive problems are also appropriate subject matter. The justification for the preparation of any review for this book-series is that it deals with some aspect of the many real problems arising from the presence of residues of "foreign" chemicals in foodstuffs. Thus, manuscripts may encompass those matters, in any country, which are involved in allowing pesticide and other plant-protecting chemicals to be used safely in producing, storing, and shipping crops. Added plant or animal pest-control chemicals or their metabolites that may persist into meat and other edible animal products (milk and milk products, eggs, etc.) are also residues and are within this scope. The so-called food additives (substances deliberately added to foods for flavor, odor, appearance, etc., as well as those inadvertently added during manufacture, packaging, distribution, storage, etc.) are also considered suitable review material.

Manuscripts are normally contributed by invitation, and may be in English, French, or German. Preliminary communication with the editor is necessary before volunteered reviews are submitted in manuscript form.

The Editor is grateful to DR. JOHN E. CASIDA who very generously assumed responsibility for collecting these manuscripts, for checking galley and page proofs, and for preparation of the index of this special volume. This assistance is gratefully acknowledged.

Department of Entomology   F.A.G.
University of California
Riverside, California
December 6, 1968

Foreword

This volume of Residue Reviews presents the collected papers of a Seminar on "Pesticide Metabolism, Degradation, and Mode of Action" held in Nikko, Japan from 16-19 August 1967. This was the second in a series of three seminars on pest control being held as a part of the United States-Japan Cooperative Science Program under the Joint Sponsorship of the National Science Foundation and the Japan Society for the Promotion of Science. The first seminar was held in April 1967, in Fukuoka, Japan; it dealt with "Microbiological Control of Pests." The third seminar, held in Honolulu in January 1968, concerned "New Biochemical Approaches to Pest Control." In each case, an approximately equal number of participants from Japan and the United States shared in the presentations and discussions.

The continued development of safe and efficient pesticides is critical to the maintenance of an adequate food and fiber supply and of a high standard of health for man and domestic animals. An increasing proportion of the research work that supports this need is conducted in Japan and the United States, the two nations participating in the Seminar. While the total amount of pesticides used in Japan is second only to the United States, the largest amount of pesticide usage per unit area is in Japan. Both Japan and the United States are in the vanguard of the nations that recognize certain serious problems which result from the adverse effects on animals and plants associated with the rapidly increasing use of pesticides. Cooperation between the two countries can accelerate a solution to problems arising from pesticide misuse, environmental contamination, potentially hazardous residues, failure to control resistant pest populations, and related problems.

The papers published in this volume encompass the following general areas of research: pesticide photodecomposition, herbicide metabolism and mode of action, fungicide mode of action, insecticide metabolism and mode of action, and physico-chemical approaches to structure-activity relationships. The two aspects considered in most detail are 1) the biochemistry of pesticide action in target and non-target organisms with particular reference to the mode of action and metabolism, and 2) the progress being made in development of selective and biodegradable pesticides. The emphasis on compounds useful in rice culture is justified because of the importance of this crop in the agricultural economy of the orient. The discussions following each paper are not included here but their content is reflected to some degree in the updated versions of the papers published in this volume, especially in regard to future areas of useful research.

The co-organizers and participants of the August, 1967 Seminar are pleased that the information exchanged at Nikko is now more broadly available as a result of this publication of the collected papers.

JOHN E. CASIDA
(Co-organizer of Seminar)
Division of Entomology
University of California
Berkeley, California

KAZUO FUKUNAGA
(Co-organizer of Seminar)
National Institute of
Agricultural Sciences and
Institute of Physical and
Chemical Research, Tokyo

Table of Contents

Experimental approaches to
pesticide photodecomposition

by
Donald G. Crosby [*]

Contents

I. Introduction

The past few years have witnessed a rapidly increasing interest in the decomposition ("metabolism") of pesticides by micro-organisms, animals, and higher plants. Although the influence of non-living environmental forces on pesticides has been observed or suspected for decades, the significance of heat, light, air, surfaces, and moisture has received relatively little scientific attention. In fact, such processes as the conversion of P—S to P—O compounds, the hydrolysis of organophosphorus insecticides, the decomposition of dithiocarbamate fungicides to isothiocyanates in soil, and the dehydrohalogenation of diphenylmethane insecticides, generally thought to be restricted to the domain of living organisms, are carried out efficiently by physical forces alone (Crosby 1968 a).

The sum total of the effects of these environmental factors on pesticides is termed "weathering" (Gunther and Blinn 1955). While it is apparent that each may bring about chemical transformations, the

---

[*] Agricultural Toxicology and Residue Research Laboratory, University of California, Davis, California.

influence of light appears to be perhaps the most important and wide-spread of all. Although plant photosynthesis and human sunburn represent well-recognized examples of photochemical reactions in biological systems, the bleaching of clothing and color photography demonstrate the power of sunlight in bringing about extensive chemical changes outside of living things.

Many pesticides are decomposed by ultraviolet (UV) light in the laboratory (MITCHELL 1961). In most instances, the UV component of sunlight likewise is responsible for the environmental effects; little influence generally is to be expected at wavelengths longer than 450 $m\mu$ (CROSBY 1968 b). On the other hand, the atmosphere effectively absorbs UV light of short wavelengths; 287 $m\mu$ is the minimum ever recorded on the earth's surface, and 300 $m\mu$ probably represents the practical UV limit at most locations (KOLLER 1965). Intensity is strongly affected by season, climate, latitude, elevation, and angle of incidence. However, shading is not as important a factor as might be suspected; the intensity of UV reflection from open sky often exceeds that from direct sunlight, and the total UV energy received each summer day at ground level in Alaska exceeds that in Hawaii in the potential for pesticide photolysis (*U. S. Department of Commerce* 1966).

While the environmental transformations of pesticides are obviously of great practical importance, the wide variety of chemical structures represented by these compounds offers many photochemical reactions which are themselves of basic scientific interest. Almost without exception, the photolytic reactions of pesticides investigated in our laboratory have not been described previously. It is the intent of the present review to provide the reader with a brief operational introduction to this fascinating field of study and, by examples from our own recent work, to afford a glimpse of what the new investigator can expect.

II. Experimental approaches

a) Light sources

Pesticides generally are exposed to sunlight during field application and use. Therefore, the sun assumes particular experimental importance as a source of radiation. However, the hourly, daily, and seasonal variation in intensity and wavelength distribution, difficulty of temperature regulation, and extended periods of cloudiness all lead to poor reproducibility, inconvenience, and lack of experimental control.

Various laboratory devices have been designed to reproduce the broad ranges of wavelengths and intensities which comprise natural sunlight (Fig. 1 E) (KOLLER 1965). They generally are expensive and

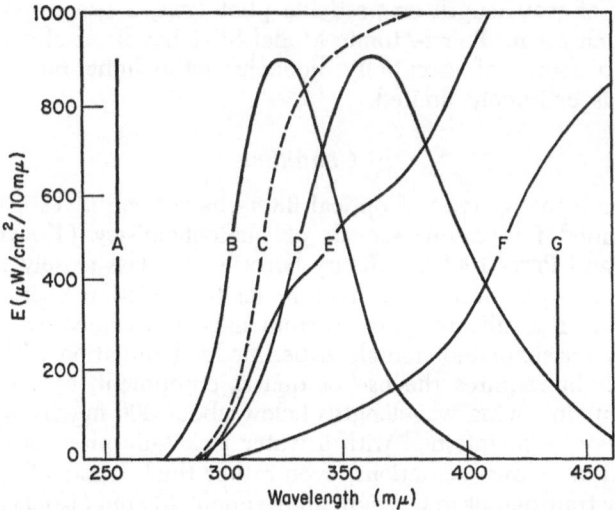

Fig. 1. Emission spectra of a mercury arc (*A. G*); fluorescent sunlamp (*B*) and black-light (*D*); normal sunlight (*E*); and daylight fluorescent lamp (*F*); *C* represents the transmission of borosilicate glass.

approximate natural conditions only very roughly. Actually, only the wavelength region between 285 and 450 mμ appears to be of wide importance to pesticide photolysis in sunlight; commercially-available fluorescent lamps emitting in this limited wavelength range are proving to be of great value in our laboratory. They are inexpensive, cool, efficient, and approach sunlight in their intensity in this spectral region (Fig. 1 D). Their effective range is limited, of course, and the existence of several intense mercury lines (Fig. 1 G) may give rise to photochemical transformations not observed under sunlight irradiation. Fluorescent "sunlamps" (Fig. 1 B) also may be useful in some applications, and while ordinary white fluorescent lamps (Fig. 1 F) are too weak for most purposes, it is apparent that extended exposure of some pesticide samples under normal conditions in the laboratory could lead to photodecomposition.

Unshielded mercury discharge lamps have been widely employed in experimental photochemistry (CALVERT and PITTS 1966). Low- and medium-pressure lamps are convenient, of moderate expense, and concentrate the major part of their UV energy at 254 mμ (Fig. 1 A) with smaller proportions at 265, 302, 313, and 365 mμ. While relatively high intensities may be obtained, most of the emission is at energies much greater than are found in sunlight and generally results in a variety of photochemical transformations which are not observed in nature. The "germicidal" lamps commonly employed in laboratory fluoresence detection emit the same radiation but are much less intense than the unshielded arcs and also much cooler. For precise measurement of the effect of wavelength on pesticide photolysis, a spectrophotofluor-ometer such as the Baird-Atomic Model SF-1 has been of exceptional value as a source of essentially monochromatic light, but again the intensity is extremely limited.

b) Conditions

Although many types of optical filters have been investigated and recommended for various aspects of photochemistry (KOLLER 1965, CALVERT and PITTS 1966), ordinary borosilicate glass closely simulates the emission spectrum of sunlight in its transmission characteristics (Fig. 1 c). Borosilicate glass reactors have performed well in our laboratory for simulated sunlight experiments. Irradiation with 254 mμ mercury light requires the use of quartz equipment, as do sunlight experiments involving wavelengths below about 300 mμ. Temperature control may be maintained with a water jacket–despite both popular belief and one's own sensations, even rather thick layers of water are essentially transparent to UV light above about 260 mμ (KOLLER 1965)– and either air or inert gas may be introduced. The useful reactor shown in Figure 2 includes a condenser outlet (A), fluorescent tube (B), gas inlet (C), reaction chamber (D), and thermowell (E).

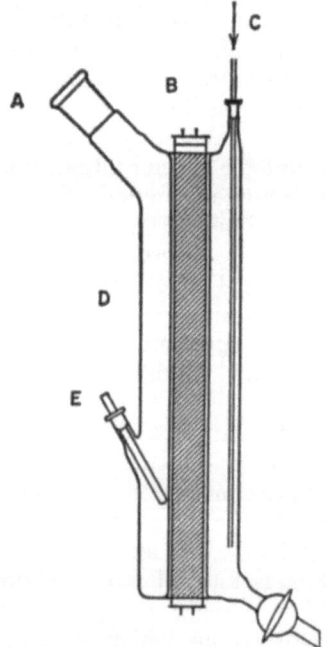

Fig. 2. Vertical photoreactor. See text for details

For laboratory irradiation by sunlight, fluorescent lamp, or mercury arc, the pesticide generally is present in the solid state, usually as a thin film on a borosilicate plate; as a solid or in solution on the surface of filter paper; or in solution or suspension in water (to simulate an aquatic environment) or in a neutral organic solvent (to simulate an environment at a leaf surface). Most of our own experiments have been conducted in solution, either in open dishes or in reactors in the presence of air. Light intensity, concentration, irradiation time, and pH all effect the photo-products and their rates of formation. Irradiation in solution is convenient, but it should not be considered necessarily to provide close simulation of field conditions.

c) Detection, isolation, and identification

The means of detection, isolation, and identification of pesticide photolysis products will vary with the particular compound and conditions. Application of the many tools available to the organic chemist will be shown in later examples. Thin-layer chromatography (TLC) has provided the generally most useful technique for isolation despite the frequent difficulty of overlapping bands. Gas chromatography (GLC) often may be satisfactory, but it is only applicable to volatile compounds and may lead to thermal decomposition at high temperatures. Reliance should not be placed on any single technique; crystallization, sublimation, column chromatography, and other methods provide useful alternatives.

A number of organic chemical, enzymatic, and physical methods have been used to detect photodecomposition products. Radioactive compounds, although expensive in quantity, may become of great importance for field experiments on photolysis; they presently are limited in applicability because they usually provide so little help in structure identification. Unless great care is taken in interpretation of results, the large number of radioactive products frequently formed in the field also will include products of the metabolism of the parent pesticide, its other weathering products, and any impurities normally present. However, if authentic specimens of the photolysis products are available or if an unknown can be shown unequivocally to result from photolysis, isotopic tracers can be of unrivaled value (TUTASS 1966), particularly where multiple labels are employed (CASIDA 1968).

Elucidation of the chemical structures of photolysis products is encumbered by the very small amounts being handled in most experiments. Consequently, the powerful physical tools of microscale infrared spectroscopy (IR), mass spectrometry, and, where quantities permit, nuclear magnetic resonance (NMR) spectrometry offer a decided and often necessary advantage. While the classical techniques of organic chemistry obviously cannot be disregarded, they frequently require modification for use with minute amounts of possibly unstable compounds. Under these circumstances, the presence of artifacts is always to be suspected.

d) Photochemical transformations

A few examples will illustrate both type reactions to be expected and the laboratory methods which may be employed.

Irradiated solutions of N-methylcarbamate insecticides could be resolved into individual constituents by TLC (CROSBY et al. 1965). Those which inhibited cholinesterases were detected colormetrically upon spraying with enzyme and indicator by the absence of the enzymatic liberation of acetic acid from acetylcholine. Other spray reagents indicated the presence of phenolic decomposition products including 1-naphthol from carbaryl (I), but unstable 1-naphthol (II) could only be isolated as its complex with 2,4,7-trinitro-9-fluorenone (TNF) formed directly on the TLC plate. N-Methylcarbamates are unstable to heat, and gas chromatography produced methyl isocyanate (III) and the substituted phenol; the same products were formed at room temperature in the photochemical reactor (CROSBY and LEITIS 1966), demonstrating that the detected phenols indeed are derived by the action of light rather than simple hydrolysis.

The initial application of TLC and GLC separations to photolyzed dieldrin (IV) indicated that only unchanged starting material was present. Eventually, GLC conditions were found which permitted the isolation of a relatively pure photolysis product having physical and chemical properties essentially identical with those of dieldrin (HENDERSON and CROSBY 1967). The mass spectrum indicated the loss of one chlorine atom and the addition of one hydrogen. Although the IR spectra of the unknown and dieldrin were almost the same, the NMR spectrum showed that a chlorine adjacent to the dieldrin double bond had been replaced by hydrogen abstracted from the solvent to give compound V. Irradiation under different conditions (ROBINSON *et al.* 1966, Rosen *et al.* 1966), produced a cyclic isomer of dieldrin (VI). In both instances, NMR spectra were of vital importance in structure elucidation.

Guthion (VII) likewise appeared to abstract hydrogen from its solvent in UV light (Kurihara and Crosby 1966). Irradiation in hexane solution produced benzazimid (VIII) and trimethyl phosphorodithioate (IX), together with a small amount of anthranilic acid and much cholinesterase-inhibiting polymer. However, irradiation of VII in methanol also produced a large proportion of X and demonstrates the importance of solvent interaction during photolysis.

Such solvent interaction becomes especially important in the presence of water. 2,4-Dichlorophenoxyacetic acid (2,4-D) was photolyzed in aqueous solution with cleavage of the ether group to give 2,4-dichlorophenol and replacement of ring chlorines by hydroxyl to provide the two possible chlorohydroxyphenoxyacetic acids (Crosby and Tutass 1967). Eventually, 1,2,4-benzenetriol was formed which was rapidly air-oxidized to a polyquinoid humic acid. Neither TLC nor GLC provided satisfactory isolations in this case; fractional vacuum sublimation permitted recovery of various key intermediates for identification, and the last, oxidative step was inhibited by bisulfite.

Attempts to extend these aqueous photolytic reactions to the herbicide fenac (principally 2,3,6-trichlorophenylacetic acid) resulted in an extremely complex mixture, inseparable by GLC and TLC, which IR spectra indicated to contain lactones, benzyl alcohols, and phenols. Fenac, itself, is a mixture of halogenated phenylacetic acids; to simplify the picture, *o-*, *m-*, and *p*-chlorophenylacetic acids used as models revealed the replacement of chlorine by hydroxyl to give the phenols, replacement of chlorine by hydrogen to give phenylacetic acid, decarboxylation to give benzyl alcohol, and subsequent oxidation to benzaldehyde (Crosby 1966). The consistent reactions of the models allow a straightforward prediction of the probable photolysis products from fenac; without them, such prediction would not have been possible.

Aqueous solutions of the salts of chlorobenzoic acids behaved similarly, producing hydroxybenzoic acids and benzoic acid itself (Crosby 1966). The products were well separated by GLC as their methyl esters, together with an acidic unknown which crystallized well, melted sharply, and gave a single sharp GLC peak. However, mass spectrometry revealed this unknown to be a mixture of the completely unexpected *p*-acetylbenzoic and terephthalic acids, (eventually resolved by TLC); IR spectra and elemental analysis had been completely misleading. The mechanism by which these acids are formed is still unknown, but their confirmed presence shows that we have a lot to learn about pesticide photochemistry.

From these examples, it may be seen that isomerization (dieldrin), reduction (chlorinated compounds and Guthion), replacement of aromatic halogens by hydroxyl (chlorinated herbicides), elimination (carbamates), polymerization, and a variety of other more specific reactions take place when pesticides are irradiated with UV light. We are only beginning to understand these processes and their significance to the utility and environmental fate of agricultural chemicals.

III. Discussion

It is apparent that most pesticides may be decomposed by UV light. Yet so complex and little understood are the physical and chemical nature of the environment that perhaps in only a few simplified instances has photolysis been demonstrated in the field. Photolysis is a dynamic process leading, in one direction, to the rapid formation of stable polymers such as humic acids; in another direction, the photolytic intermediates may be dispersed by volatility (dichlorophenol), reaction (methyl isocyanate), and absorption and metabolism by plants and microorganisms. The nature of the pesticide deposit or its solution in plant cuticle, reactivity of the substrate with photolysis products, and the amount of moisture at the reacting surface all may be expected to be of extreme importance to photodecomposition.

Unfortunately, we doubtless do not yet appreciate all of the environmental factors which may influence photochemical reactions. For instance, the mechanism by which solar energy becomes available for many pesticide transformations remains unknown. Dieldrin and 2,4-D exhibit appreciable UV absorption only at wavelengths far below those available at the earth's surface (215 mμ and 230 mμ, respectively), yet they rapidly form the same photolysis products in sunlight that are observed upon irradiation at their UV absorption maxima. These may represent sensitized free-radical reactions; the possible importance of hydroxyl and hydrogen radicals generated from water certainly would indicate sensitization, although peroxidation and hydrogen abstraction offer more likely alternatives.

Regardless of mechanism, it has been demonstrated that light provides a means for the initial destabilization of many pesticides. This leads to the possibility of structurally introducing a desired degree of stability or lability into a pesticide, or of intentionally degrading pesticide residues on crops to less toxic photoproducts (CROSBY 1965). The demonstrated generality of pesticide photodecomposition raises intriguing questions about the toxicity of the intermediate and end products, their relationship to residue analysis, their importance to the metabolism of pesticides under field conditions, and their significance and use in the study of basic photochemical processes.

Summary

Although past scientific attention has been principally directed toward the metabolic fate of pesticides in living organisms, the influence of other environmental factors—light, air, water, heat, etc.— may be of greater importance. The ultraviolet (UV) component of sunlight (290 to 400 mμ) is responsible for most pesticide photolysis in the environment, but the variation in intensity with time of day, season, and atmospheric conditions lends particular importance to controlled laboratory experiments.

Most previous laboratory work has utilized light sources emitting at 254 or 366 mμ, but newer equipment and methods permit practical investigation in the more critical 290 to 350 mμ range. Physical state, solvents, and chemical structure also are of crucial importance. Although exploratory experiments employ organic solvents, glass equipment, etc., field experiments require the use of plant surfaces and other environmental substrates. Radio tracers may provide important information under such circumstances, but eventual isolation and identification of photolysis products still are of utmost practical and scientific importance.

As knowledge of the chemical nature of these products increases, such reactions as aromatic substitution, reduction, elimination, cyclization, and polymerization emerge as common modes of photodecomposition. However, the reactions of any particular pesticide may be influenced by light intensity, concentration, pH, and other factors, and the relationship between photolysis products and the observed absorption of UV light still is obscure.

Pesticide photodecomposition is a relatively new area of study, but it holds great significance for safety, application, and residue analysis as well as for advances in fundamental photochemistry.

Résumé *

Approches expérimentales relatives à la photodécomposition des pesticides

Bien que, dans le passé, l'attention scientifique ait été principalement orientée vers le métabolisme des pesticides dans les organismes vivants, l'influence d'autres facteurs ambiants, tels que la lumière, l'air, l'eau, la chaleur, etc., peut être de plus grande importance. Les rayons ultraviolets (UV) de la lumière solaire (290-400 mµ) sont, le plus souvent, responsables de la photolyse des pesticides, mais la variation de leur intensité selon le moment de la journée, la saison et les conditions atmosphériques confère une importance particulière aux expériences contrôlées en laboratoire.

La plupart des travaux expérimentaux antérieurs ont utilisé des sources lumineuses émettant à 254 mµ ou 366 mµ, mais l'instrumentation et les méthodes plus récentes permettent l'investigation pratique dans la région critique de 290 à 350 mµ L'état physique, les solvants et la structure chimique sont aussi décisives. Bien que les essais d'orientation utilisent des solvants organiques, une instrumentation de verre, etc., les expériences de plein champ nécessitent l'usage de surfaces végétales et d'autres substrats du milieu. Les traceurs radioactifs peuvent fournir des renseignements importants dans ces conditions, mais la séparation éventuelle et l'identification des produits de photolyse sont de la plus haute importance pratique et scientifique.

Etant donné que la connaissance de la nature chimique de ces produits s'améliore, des réactions telles que la substitution aromatique, la réduction, l'élimination, la cyclisation et la polymérisation apparaissent comme étant des modes courants de photodécomposition. Cependant, les réactions de chaque pesticide peuvent être influencées par

---

* Traduit par S. DORMAL-VAN DEN BRUEL.

l'intensité lumineuse, la concentration, le pH et d'autres facteurs, de sorte que la relation entre les produits de photolyse et l'absorption observée de lumière UV reste encore obscure.

La photodécomposition des pesticides est un domaine d'étude relativement neuf, mais de la plus haute importance pour la sécurité, l'application et l'analyse des résidus, ainsi que pour les progrès de la photochimie fondamentale.

Zusammenfassung *

Experimentelle Betrachtungen zum Pestizidabbau durch Licht

Obwohl die wissenschaftliche Aufmerksamkeit in der Vergangenheit hauptsächlich auf das metabolische Schicksal der Pestizide in lebenden Organismen gerichtet worden ist, können der Einfluss von andern Umweltfaktoren wie Licht, Luft, Wasser, Hitze usw. von grösserer Wichtigkeit sein. Der Ultraviolettbestandteil (UV) des Sonnenlichtes (290-400 mμ) ist für die meisten Pestizidphotolysen in der Umwelt verantwortlich, aber der Wechsel in Intensität mit der Tageszeit, Jahreszeit und den atmosphärischen Gegebenheiten verleiht kontrollierten Laborexperimenten besondere Bedeutung.

In den meisten früheren Laborarbeiten wurden Lichtquellen benutzt, welche bei 254 und 366 mμ ausstrahlten, aber neuere Einrichtungen und Methoden erlauben zweckmässige Untersuchungen in dem kritischeren Bereich von 290 bis 354 mμ. Physikalischer Zustand, Lösungsmittel und chemische Struktur sind auch von entscheidender Bedeutung. Obwohl Forschungsexperimente organische Lösungsmittel, Glasapparaturen usw. verwenden, erfordern Feldexperimente die Benutzung von Pflanzenoberflächen und anderen Umweltsubstraten. Radio-tracers können unter solchen Umständen wichtige Auskunft geben, aber eventuelle Isolierung und Identifizierung von Photolyseprodukten sind immer noch von äusserster praktischer und wissenschaftlicher Wichtigkeit.

Indem das Wissen um die chemische Natur dieser Produkte zunimmt, kommen solche Reaktionen wie aromatische Substitution, Reduktion, Eliminierung, Ringbildung und Polymerisation als übliche Formen der Zersetzung durch Licht zum Vorschein. Jedoch können die Reaktionen jedes einzelnen Pestizids durch Lichtintensität, Konzentration, pH und andere Faktoren beeinflusst werden, und die Beziehung zwischen den Photolyseprodukten und der beobachteten Absorption des UV-Lichtes ist immer noch undeutlich. Pestizidabbau durch Licht ist ein relativ neues Arbeitsgebiet, aber es besitzt grosse Bedeutung für Sicherheit, Applizierung und Rückstandsanalyse als auch für Fortschritte in grundlegender Photochemie.

---

* Übersetzt von A. Schumann.

References

CALVERT, J. C., and J. N. PITTS, JR.: Photochemistry. New York: Wiley (1966).

CASIDA, J. E.: Radiotracer studies on metabolism, degradation, and mode of action of insecticide chemicals. Residue Reviews **25**, (1968).

CROSBY, D. G.: The intentional removal of pesticide residues. In: Research in pesticides (C. O. CHICHESTER, ed.), p. 213. New York: Academic Press (1965).

—— Photodecomposition of herbicides. Abstr. 152nd National Meeting, Amer. Chem. Soc., New York, sec. A, p. 32 (1966).

—— The non-metabolic decomposition of pesticides. Ann. N.Y. Acad. Sciences. In press (1968 a).

—— The photochemistry of herbicides. In: Degradation of the herbicides (P. C. Kearney and D. D. Kaufman, eds.). In press. New York: Dekker (1968 b).

——, and E. LEITIS: Unpublished research (1966).

——, ——, and W. L. WINTERLIN: The photodecomposition of carbamate insecticides. J. Agr. Food Chem. **13**, 204 (1965).

——, and H. O. TUTASS: Photodecomposition of 2,4-dichloro-phenoxyacetic acid. J. Agr. Food Chem. **14**, 596 (1966).

GUNTHER, F. A., and R. C. BLINN: The analysis of insecticides and acaricides. New York: Interscience (1955).

HENDERSON, G. L., and D. G. CROSBY: Photodecomposition of dieldrin and aldrin. J. Agr. Food Chem. **15**, 888 (1967).

KOLLER, L. R.: Ultraviolet radiation, 2 ed. New York: Wiley (1965).

KURIHARA, N. H., D. G. CROSBY, and H. F. BECKMAN: Photodecomposition of Guthion. Abstr. 152nd National Meeting, Amer. Chem. Soc., New York, sec. A, p. 47 (1966).

MITCHELL, L. C.: The effect of ultraviolet light (2537Å) on 141 pesticide chemicals by paper chromatography. J. Assoc. Official Agr. Chemists **44**, 643 (1961).

ROBINSON, J., A. RICHARDSON, G. BUSH, and K. E. ELGAR: A photoisomerization product of dieldrin. Bull. Environ. Contamination Toxicol. **1**, 127 (1966).

ROSEN, J. D., D. J. SUTHERLAND, and G. R. LIPTON. The photochemical isomerization of dieldrin and endrin and effects on toxicity. Bull. Environ. Contamination Toxicol. **1**, 133 (1966).

TUTASS, H. O.: The contribution of light to the decomposition of 2, 4-D on the corn plant. Thesis, University of California, Davis (1966).

U. S. *Department of Commerce:* Climatological data **17**, 306 (1966).

Photochemical degradation
products of pentachlorophenol

by
Katsura Munakata * and Masao Kuwahara *

Contents

I. Introduction

For the control of barnyard grass, *Panicum crusgalli* L., on the paddy fields in Japan, farmers utilize the herbicide, PCP-Na (sodium pentachlorophenoxide) on about the fifth day after transplanting of rice plant seedlings. On the day of the treatment the paddy field water is highly toxic to fish and many injuries by PCP herbicides to fish cultures occur during the transplantation season of rice plants, but the fish-killing activity in the water disappears several days after treatment. This fish toxicity in the water was prolonged by covering over the surface of field water by sheets and shutting out the sunshine. These phenomena taught the authors that sunshine promoted the degradation of PCP-Na in the water, just as in the case of the windowside PCP-Na solution turning from colourless to purple, after about ten days.

Our experiments to obtain information on the chemical structures of the degradation products of PCP-Na and reaction mechanisms were undertaken with the scale of one kg. of PCP-Na in 50 l. of water irradiated with sunshine for ten days until the solution turned purple and the content of PCP-Na decreased to 50 percent of the added amount, through quantitative chemical analysis.

---

* Fàculty of Agriculture, Nagoya University, Nagoya, Japan.

13

The reaction mixture was acidified with hydrochloric acid (1:1) and extracted by ether; the ether extracts were chromatographed on a silica gel column. After elution with ether for recovering of PCP (*ca.* 500 g.), eluation by chloroform gave three zones in the column: purple, yellow, and violet. The purple zone gave three kinds of crystals with large amounts of dark red resinous materials, a red crystal $C_{12}HO_4Cl_7$ (I) (0.16 percent), tetrachlororesorcinol (II) (0.10 percent), and an orange-red crystal $C_{12}H_2O_5Cl_6$ (III) (0.08 percent). From the yellow zone two crystalline materials came out, $C_{12}HO_4Cl_7$ (IV) (Ys – 2) (0.5 percent) and $C_{18}H_2O_6Cl_{10}$ (V) (Ys – 1) (0.2 percent), and the violet zone gave chloranilic acid (VI) (1.8 percent). The average yields of these degradation products were calculated from starting PCP-Na. The compounds I, III, IV, and V were identified to be new compounds (KUWAHARA *et al.* 1966). Regarding the action of sunshine on Na-PCP, HAITT *et al.* (1960) reported the degradation followed first-order kinetics. However, there had been no knowledge about the photochemical decomposition products. On oxidation of PCP by various oxidants, chloranilic acid, chloranil, tetrachloro-*o*-benzoquinone, and 2,3,4,5,6-pentachloro-4-pentachlorophenoxy-2,5-cyclohexadienone were identified.

II. Chemical structures of degradation products

a) The structure of the red compound (I)

Compound I was red prisms, $C_{12}HO_4Cl_7$, which had following properties: m.p. 258° to 259° C., λ_{max} 320 mμ (log ϵ 4.38) and 430 mμ (log ϵ 2.61) ($CHCl_3$), ν_{max} 3430 (OH) and 1682 (C = O) cm.$^{-1}$ ($CHCl_3$). The ultraviolet, visible, and infrared spectra of I which were similar to those of chloranilic acid, λ_{max} 304 mμ (log ϵ 4.25) and 447 mμ (log ϵ 2.74) ($CHCl_3$), ν_{max} 3429 and 1681 cm.$^{-1}$ ($CHCl_3$) (HANCOCK *et al.* 1962), indicated the existence of a *p*-quinoid structure.

On the hydrogenation in the presence of 10 percent palladium charcoal, I absorbed quantitatively one mole of hydrogen to yield the corresponding quinol, $C_{12}H_3O_4Cl_7$, m.p. 199° to 201° C., which was reconverted to I by air oxidation in an ethereal solution. The quinol was characterized as a trimethyl ether, $C_{15}H_9O_4Cl_7$, mp. 175° to 176° C. The derivatives of I were as follows: monoacetate, $C_{14}H_3O_5Cl_7$, m.p. 188° to 190° C., λ_{max} 291 mμ (log ϵ 4.38) and 402 mμ (log ϵ 2.84) ($CHCl_3$), ν_{max} 1787 (CH_3CO) and 1680 (C = O) cm.$^{-1}$ (KBr), and reductive triacetate, $C_{18}H_9O_7Cl_7$, molecular weight 567.5 by the Rast method, m.p. 191° to 192° C., ν_{max} 1780 (CH_3CO) cm.$^{-1}$ (KBr). On oxidation with chromium trioxide in acetic acid at 70° C. I gave chloranil, and further on partial oxidation chloranil and pentachlorophenol. Acid fission and alkaline reductive cleavage with sodium dithionite I afforded pentachlorophenol. Alkaline cleavage in 10 percent sodium hydroxide at 80° C. gave one mole of chloranilic acid (yield 84.9 percent) and pentachlorophenol (yield 81.8 percent), respectively. Thus I was conclusively established to be 2,5-dichloro-3-hydroxy-6-pentachlorophenoxy-p-benzoquinone.

b) The structure of the orange-red compound (III)

Compound III was orange-red needles, $C_{12}H_2O_5Cl_6$, m.p. 245.0° to 245.5° C., λ_{max} 304 mμ (log ϵ 4.34) and 420 mμ (log ϵ 2.67) (CHCl$_3$), ν_{max} 3530 (OH), 3410 (OH), and 1679 (C = O) cm.$^{-1}$ (CHCl$_3$). These spectral data indicated the existence of a p-quinoid structure as in the case of I.

The compound III dissolved in alkali to give a violet solution, indicating the same behavior as that observed for I. Reduction of III with zinc dust in acetic acid gave the corresponding quinol, $C_{12}H_4O_5Cl_6$, m.p. 109° to 111° C., recovering the starting material by air oxidation, which was characterized as a tetramethyl ether, $C_{16}H_{12}O_5Cl_6$, m.p. 106° to 107° C., with diazomethane. III also gave the following derivatives: with acetic anhydride and sodium acetate monoacetate, $C_{14}H_4O_6Cl_6$, m.p. 223° to 224° C., λ_{max} 302 mμ (log ϵ 4.39) and 431 mμ (log ϵ 2.73) (CHCl$_3$), ν_{max} 3210 (OH), 1758 (CH$_3$CO) and 1678 (C = O) cm.$^{-1}$ (KBr); with acetic anhydride and concentrated sulfuric acid diacetate, $C_{16}H_6O_7Cl_6$, m.p. 157.0° to 158.5° C., λ_{max} 292 mμ (log ϵ 4.34 and 403 mμ (log ϵ 2.87) (CHCl$_3$), λ_{max} 1787 (CH$_3$CO) and 1688 (C = O) cm.$^{-1}$ (KBr); and with reductive acetylation by zinc and acetic anhydride tetraacetate, $C_{20}H_{12}O_9Cl_6$, molecular weight 619.6 by the Rast method, m.p. 150° to 151° C., ν_{max} 1781 (CH$_3$CO) cm.$^{-1}$ (KBr). Alkaline cleavage of III with 10 percent sodium hydroxide yielded chloroanilic acid and tetrachlororesorcinol in yields of 91.2 and 94.0 percent respectively. Thus the structure for III was elucidated as 2,5-dichloro-3-hydroxy-6-(2',4',5',6'-tetrachloro-3'-hydroxyphenoxy)-p-benzoquinone.

c) The structure of the yellow compound (Ys – 2) (IV)

Compound IV was yellow prisms, $C_{12}HO_4Cl_7$, m.p. 218° to 219° C. (dec.), λ_{max} mμ (log ϵ): 253 (4.53), 295 (3.39), 302 (3.60), and 330 (3.26) (CHCl$_3$); λ_{max} mμ (log ϵ): 235 (4.57), 295 (3.56), 305 (3.63), and 401 (3.62) (MeOH); λ_{max} 3580 (OH), 3480 (OH), 1680 (C = O) and 1640 (C = O) cm.$^{-1}$ (KBr). The spectra of IV exhibited the absorption curves which were similar to those of tetra-chloro-o-benzoquinone, λ_{max} 262 mμ (log ϵ 3.71 and 462 mμ (log ϵ 3.36) (CHCl$_3$), λ_{max} 243 mμ (log ϵ 3.76), 353 mμ (log ϵ 3.31) and 420 mμ (log ϵ 3.08) (MeOH), and thus indicative of an o-quinoid structure for IV. Compound IV was stable to hydrochloric acid and dissolved in aqueous alcoholic alkali to give a deep yellow color. On hydrogenation over 20 percent palladium-on-carbon, IV absorbed quantitatively one mole of hydrogen to yield the corresponding quinol, $C_{12}H_3O_4Cl_7$, m.p. 172° C., which was reconverted to IV by air oxidation in an ethereal solution. Reductive acetylation of IV with zinc, acetic anhydride, and concentrated sulfuric acid gave a triacetate, $C_{18}H_9O_7Cl_7$, m.p. 171° to 172° C., molecular weight 585.4 by mass, ν_{max} 1685 (C = O) cm.$^{-1}$ (KBr), and a colorless methylene dioxide, $C_{14}H_5O_4Cl_7$, m.p. 184° to 185° C., respectively. Alkaline reductive cleavage of IV with sodium dithionite in an alcoholic solution at 70° C. afforded tetrachloropyrocatechol. On the basis of these data, the structure for IV should be either 3,4,5-trichloro-6-(2′,3′,4′,5′-tetrachloro-6′-hydroxyphenoxy)-o-benzoquinone or 3,4,6-trichloro-5-(2′,3′,4′,5′-tetrachloro-6′-hydroxyphenoxy)-o-benzoquinone (XVIII) (JACKSON and CARLETON 1908). The physical and chemical properties of synthetic models were quite different from those of IV, that is, XVIII; m.p. 214° C., λ_{max} mμ (log ϵ): 263 (4.26), 296 (3.60), 302 (3.60), and 3.55 (3.12) (CHCl$_3$); λ_{max} mμ (log ϵ): 248 (4.33), 289 (3.54), 299 (3.41), and 390 (3.82) (MeOH); ν_{max} 3500, 3450, and 1695 cm.$^{-1}$ (KBr). Therefore XVIII was abandoned and the structure for IV was concluded to be 3,4,5-trichloro-6-(2′,3′,4′, 5′-tetrachloro-6′-hydroxyphenoxy)-o-benzoquinone.

d) The structure of the yellow compound (Ys – 1) (V)

Compound V was yellow prisms, $C_{18}H_2O_6Cl_{10}$, m.p. 250° C. (dec.), λ_{max} 242 mμ (ϵ 37000), 305 mμ (ϵ 8300), 398 mμ (ϵ 3700) in EtOH, ν_{max} 3350, 1670, 1655, and 1611 cm.$^{-1}$ (KBr disk). It gave following derivatives: reductive acetate $C_{26}H_{12}O_{10}Cl_{10}$ m.p. 216° to 218° C., colorless prisms, ν_{max} 1778 cm.$^{-1}$; acetate $C_{20}H_6O_8Cl_{10}$ m.p. 252° to 255° C. yellow prisms, λ_{max} 240 mμ (ϵ 44200), 303 mμ (ϵ 9120), 400 mμ (ϵ 6450) in EtOH, ν_{max} 1790, 1780, 1690, and 1625 cm.$^{-1}$ (KBr); and methyl ether $C_{20}H_6O_6Cl_{10}$, m.p. 251° to 254° C., yellow prisms, λ_{max} 254 mμ (ϵ 27800), 291 mμ (ϵ 8500), 300 mμ (ϵ 9320), 334 mμ (ϵ 2050), 392 mμ (ϵ 1400) in EtOH, ν_{max} 1690, and 1624 cm.$^{-1}$. These spectral data showed V has the o-benzoquinoid structure. The reduction of V with lithium aluminum hydride yielded three identified substances: the quinol substance of V, M.P. 153° to 4° C., max 3450 cm.$^{-1}$, a material easily converted to the reductive acetate of V; tetrachlorohydroquinone; tetrachloropyrocatechol. V was oxidized by hydrogen peroxide-acetic acid into oxidized substance $C_{18}H_6O_7Cl_{11}$ (by mass), m.p. 235° to 7° C. (dec.), λ_{max} 269 mμ (ϵ 12000) 302 mμ (ϵ 8300) (EtOH), ν_{max} 3350, 1738, and 1627 cm.$^{-1}$ (KBr), which could be reconverted to V by potassium iodide in acetone. The UV spectrum of the oxidized substance of V was quite similar with that of the oxidized substance of IV. The acetate of V was also given from the oxidized substance of V through its acetate, m.p. 138° to 40° C. (dec.) ν_{max} 1778, 1779, and 1732 cm.$^{-1}$ (KBr) and reduction with potassium iodide in acetic anhydride. These evidences, together with the data of mass spectroscopy, suggested the chemical structure V for this yellow substance.

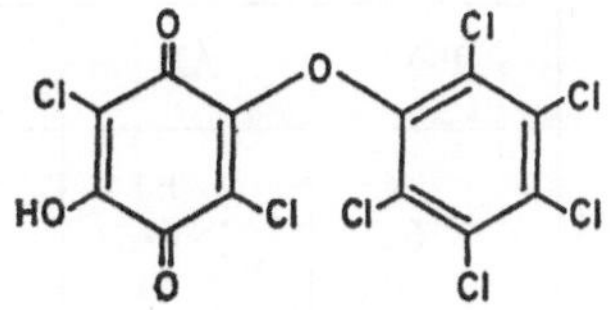

(I)
Red substance

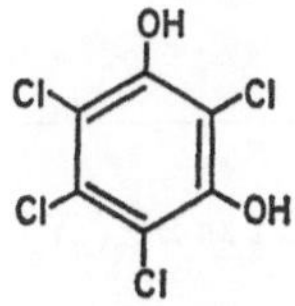

(II)
2,4,5,6-Tetrachlororesorcinol

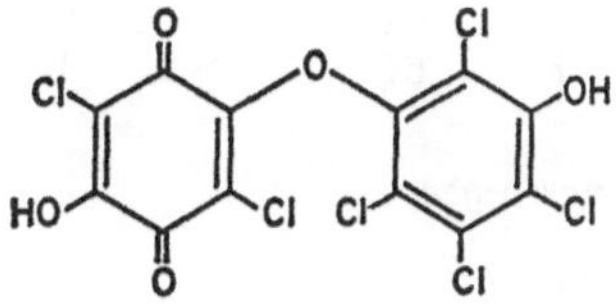

(III)
Orange-red substance

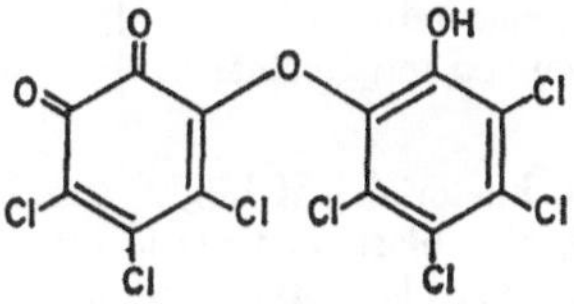

(IV)
Ys-2

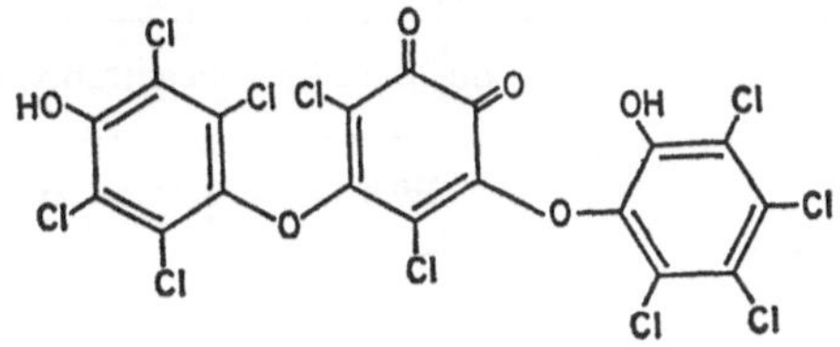

(V)
Ys-1

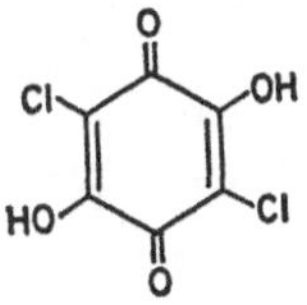

(VI)
Chloranilic acid

III. Physiological activities of the degradation products of PCP-Na

Several preliminary physiological tests of the photochemical de-gradation products of PCP-Na were undertaken as shown in the following tables using several remaining samples after chemical experiments. Fungicidal activities were examined using four kinds of fungus and minimum lethal concentrations were designated as in Table I. The compounds of IV-oxide and I and III-monoacetate showed rather same fungicidal activities with PCP, but others could not show any stronger activities.

Seed-killing activities were tested in Petri dishes using seeds of rice and rape for only two kinds of degradation products, and the activity became very much weaker than PCP-Na (Table II); at the same time fish-killing activities were examined using Japanese killifishes, *Oryzias latipes*; half-killing concentrations (TLm) at 24, 48, and 120 hours after treatments were compared in Table III showing the samples

Table I. *Fungicidal activities of the degradation products of PCP-Na*

Product	Fungicidal [a] concentration (p.p.m.)			
	XO	PO	AK	GC
PCP	16	4	64	64
Yellow subst. (Ys-2) (IV)	64	64	> 64	> 64
IV-Oxide	16	4	64	64
Red subst. (I)	16	4	64	64
Red subst.-tri Me.	> 64	> 64	> 64	> 64
Orange red subst. (III)	64	64	> 64	⩾ 64
III-Monoacetate	16	4	. 64	. 64
III-Reduced tetraacetate	> 64	64	> 64	> 64

[a] XO, bacterial leaf blight, *Zanthomonas oryzae.*
PO, rice blast, *Piricularia oryzae.*
AK, pear black spot, *Alternaria kikuchiana.*
GC, grape ripe rot, *Glomerella cingulata.*

Table II. *Seed killing activities of the degradation products of PCP-Na*

Product	Seed-killing concentration (%)	
	Rice	Rape
PCP	0.01	0.01
Chloranilic acid	0.5	0.5
Yellow substance (Ys-2) (IV)	0.25	0.25

Table III. *Fish toxicities of the degradation products of PCP-Na*

Product	TLm (p.p.m.)		
	24 hrs	48 hrs	120 hrs
PCP	0.63	0.51	0.47
Chloranilic acid (VI)	400.0	400.0	400.0
IV	23.5	23.5	20.2

tested to be weakened by photochemical degradation.

IV. Reaction mechanisms of the products

In the aqueous solution of PCP-Na irradiated by sunshine, several activation reactions might occur and higher energy radical pairs would be formed (JOSCHEK and MILLER 1966).

(1)

$$C_6Cl_5OH \xrightarrow{h\nu} \dot{C}_6Cl_5O\cdot + \cdot H$$

$$\cdot H + C_6Cl_5OH \longrightarrow C_6Cl_5O\cdot + \cdot H_2$$

$$\cdot H + H_2O \longrightarrow \cdot OH + H_2$$

$$\cdot OH + C_6Cl_5OH \longrightarrow C_6Cl_5O\cdot + H_2O$$

$$\cdot H + O_2 \longrightarrow \cdot O_2H$$

(2)

(a)　(b)　(c)　(d)

Chloranilic acid
(VI)

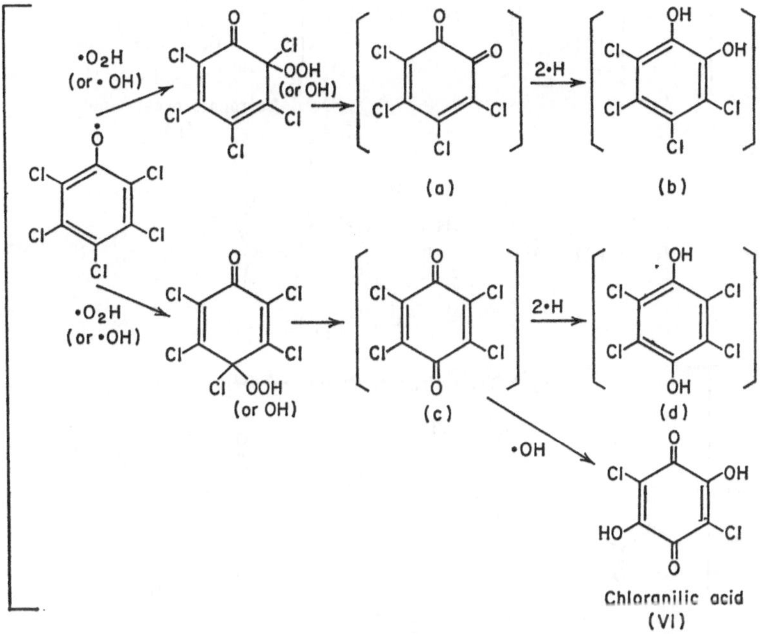

At first, the activation reaction of equation (1) may produce penta-chlorophenoxy radical, hydroxy radical, etc., and subsequent reaction to form the intermediates (a),(b),(c),(d), and chloranilic acid (VI) as in equation (2). Tetrachlororesorcinol (II) would be formed by the activation of the *meta*-position of pentachlorophenol in the triplet state irradiated by sunshine as in equation (3). The red substance (I) would be formed with the intermediate (c) and PCP as in equation (4); the intermediate (c) may also give the orange-red compound (III) with tetrachlororesorcinol as in equation (5). The yellow sub-stance (IV) (Ys – 2) would come from intermediate (a) plus (b), and the other yellow substance (V) (Ys – 1) might be produced by intermediate (d) plus (IV) as in the equation (6) and (7), re-spectively. These reaction mechanisms would be only an assumption, but the several photochemical degradation products isolated and identified here promote some considerations for their reaction mechanisms.

(3)

Tetrachlororesorcinol
(II)

(4)   (c) + PCP $\xrightarrow{h\nu}$

(I)

(c) + Tetrachlororesorcinol $\xrightarrow{h\nu}$

(5) $\xrightarrow{\cdot OH}$

(III)

(6)   (a) + (b) $\xrightarrow{h\nu}$

(IV)

(7)   (d) + (IV) →

(V)

V. Conclusions

Through the information of the practical use of PCP-Na for paddy field herbicides and also the observations of colour change of aqueous solutions of PCP-Na at windowside we have undertaken experiments to identify the colouring materials for the first time.

At the point of half break-down time of PCP-Na, the photochemical degradation reaction was stopped and the extraction of degradation products was undertaken; however, the chemical characters of the end products of photodegradation by longer irradiation time remained of interest. Six kinds of products together with large amounts of resinous products were isolated and identified to be oxidized monomers, dimers, and a trimer. It was revealed that the main purple colour came from chloranilic acid formation, but several new chemical structures were elucidated by chemical and physical properties.

Summary

Farmers in Japan utilize the herbicide, PCP-Na (sodium penta-chlorophenoxide) for the control of barn yard grass, *Panicum crusgalli* L., on the paddy field. On the day of treatment with PCP-Na, the paddy field water is highly toxic to fishes, but the fish toxicity in the water disappears several days after treatment. A clear colourless solution of PCP-Na turns purple as an effect of sunshine when placed by the window for about ten days. These evidences attracted our interests to the study of the photochemical degradation products of PCP-Na.

A photochemical reaction mixture of PCP-Na exposed to sunshine was chromatographed on a silica gel column with chloroform and chloroform-methanol systems. $C_{12}HO_4Cl_7(I)$ as red crystals, tetrachlororesorcinol (II), $C_{12}H_2O_5Cl_6$ (III) as orange-red crystals, $C_{12}HO_4Cl_7$ (IV) as yellow crystals, $C_{18}H_2O_5Cl_6$ (V) as yellow crystals, and chloranilic acid (IV) (the origin of the purple colour) were obtained after being eluted in this order. I, III, IV, and V were found to be new compounds, and their chemical structures were elucidated by chemical reactions and spectrometric procedures. The reaction mechanisms involved in the formation of these products after the photochemical degradation of PCP-Na were also discussed.

It was also of interest from the standpoint of developing new pesticides that these new compounds showed stronger fungicidal but weaker phytotoxic and fish-killing activities than PCP-Na.

Résumé *

Produits de dégradation chimique du pentachlorophénol

Au Japon, les fermiers utilisent l'herbicide PCP-Na (pentachloro-phénate de sodium) pour la lutte contre le pied-de-coq, *Panicum*

---

* Traduit par S. DORMAL-VAN DEN BRUEL.

crusgalli L., dans les champs de riz. Le jour du traitement, l'eau des champs de riz est très toxique pour les poissons, mais la toxicité disparaît quelques jours après le traitement. Une solution incolore et limpide de PCP-Na vire au pourpre sous l'effet de la lumière solaire, lorsqu'elle est placée devant une fenêtre pendant dix jours environ. Ces faits ont dirigé l'attention vers l'étude des produits de dégradation photochimique du PCP-Na.

Un mélange provenant de la réaction photochimique du PCP-Na exposé à la lumière solaire a été chromatographié sur une colonne de silica-gel avec du chloroforme et un mélange de chloroforme-méthanol. Les produits suivants ont été obtenus selon l'ordre de l'élution: $C_{12}HO_4Cl_7$ (I) sous forme de cristaux rouges, tetrachlororésorcinol (II), $C_{12}H_2O_5Cl_6$ (III) sous forme de cristaux rouge-orangés, $C_{12}HO_4Cl_7$ (IV) sous forme de cristaux jaunes, $C_{18}H_2O_5Cl_6$ (V) sous forme de cristaux jaunes, et acide chloranilique (VI) (source de la couleur pourpre). Les composés I, III, IV et V sont nouveaux; leur structure chimique a été élucidée par des réactions chimiques et des procédés spectrométriques. Les mécanismes de réaction impliqués dans la formation de ces composés après la dégradation photochimique du PCP-Na ont été discutés sur la base d'expériences permettant d'élucider les mécanisms de réaction.

Il était aussi intéressant de constater, sous l'angle de la découverte de nouveaux pesticides, que ces nouveaux composés présentaient un pouvoir fongicide plus fort que celui du PCP-Na, mais une phytotoxicité et une toxicité à l'égard du poisson plus faibles.

Zusammenfassung *

Photochemische Abbauprodukte von Pentachlorphenol

Bauern in Japan benutzen das Herbizid PCP-Na (Natriumpentachlorphenolat) zur Bekämpfung des "barn yard"-Grases, *Panicum crusgalli L.*, im Reisfeld. Am Tage der Behandlung mit PCP-Na ist das Wasser des Reisfeldes sehr toxisch für Fische, aber die Fischtoxizität des Wassers verschwindet mehrere Tage nach der Behandlung.

Eine klare, farblose Lösung von PCP-Na wird purpur von der Wirkung der Sonnenstrahlung, wenn sie für etwa 10 Tage ans Fenster gestellt wird.

Diese Tatsachen zogen unser Interesse auf die Untersuchung von photochemischen Abbauprodukten des PCP-Na.

Eine photochemische Reaktionsmischung von PCP-Na, welche der Sonnenstrahlung ausgesetzt worden war, wurde auf einer Silicagelsäule mit Chloroform und einem Chloroform-Methanolsystem chromatographiert. $C_{12}HO_4Cl_7$ (I) ist rot, Tetrachlor-resorcinol (II) und

---

* Übersetzt von A. SCHUMANN.

$C_{12}H_2O_5Cl_6$ (III) sind orangerot, $C_{12}HO_4Cl_7$ (IV) und $C_{18}H_2O_5Cl_6$ (V) sind gelb, und Chloranilinsäure ist die Ursache der Purpurfarbe; sie wurden in kristalliner Form erhalten, nachdem sie in dieser Reihenfolge eluiert worden waren. I, III, IV und V wurden als neue Verbindungen erkannt, und ihre chemische Struktur wurde durch chemische Reaktionen und spektrometrische Methoden aufgeklärt. Reaktionsmechanismen zur Bildung dieser Produkte nach dem photochemischen Abbau von PCP-Na wurden auch diskutiert.

Es war auch von Interesse, vom Standpunkt der Entwicklung neuer Pestizide, dass diese Verbindungen stärkere fungizide, aber schwächere phytotoxische und fischtötende Wirksamkeit zeigten als PCP-Na.

References

CROSBY, D. G., and H. O. TUTASS: Photodecomposition of 2,4-dichlorophenoxyacetic acid. J. Agr. Food. Chem. 14, 596 (1966).

HAITT, C. W., W. T. HASKINS, and L. OLIVER: The action of sunshine on sodium pentachlorophenate. Amer. J. Trop. Med. Hyg. 9, 527 (1960).

HANCOCK, J. W., C. E. MORRELL, and D. RHUM: Trichlorohydroxyquinone. Tetrahedron Letters, p. 987 (1962).

JACKSON, C. L., and F. W. CARLETON: Derivatives of tetrachloro-o-benzoquinone. Amer. Chem. J. 39, 493 (1908).

JOSCHEK, H. H., and S. I. MILLER: Photocleavage of phenoxyphenols and bromophenols. J. Amer. Chem. Soc. 88, 3269 (1966).

——, —— Photooxidation of phenol, cresols, and dihydroxybenzenes. 88, 3273 (1966).

KUWAHARA, M., N. KATO, and K. MUNAKATA: The photochemical reaction of pentachlorophenol. Part I. The structure of the yellow compound. Agr. Biol. Chem. Japan 30, 232 (1966).

——, ——, —— The photochemical reaction of pentachlorophenol. Part II. The chemical structures of minor products. Agr. Biol. Chem. Japan 30, 239 (1966).

Reactions of pesticides in soils

by
PHILIP C. KEARNEY [*] and CHARLES S. HELLING [*]

Contents

I. Introduction

Once a pesticide has destroyed the target organism, the compound has served its intended function and may remain as a residue in air, water, and soil. Residual pesticides in soils create two problems: (1) at nonphytotoxic concentrations pesticides or their metabolites may be absorbed into plants and eventually enter the food chain at levels below established tolerances; or (2) at phytotoxic levels (in the case of herbicides) they may injure or destroy subsequent crops. The latter situation is further confounded by recent evidence that other pesticides may interact with herbicides to cause synergistic phytotoxic effects or extend their persistence.

An understanding of the processes that operate on a pesticide in soils is imperative if we are to minimize their undesirable effects and ultimately devise methods for controlling their persistence. The processes which govern the fate and behavior of pesticides include photo-

[*] Crops Research Division, Agricultural Research Service, U. S. Department of Agriculture, Beltsville, Maryland 20705.

25

decomposition, adsorption, leaching, volatilization, plant uptake, chemical reactions, and microbial metabolism. Since the organic herbicides have been investigated so extensively, the outline of the present chapter has been planned to include some of the major classes of reactions undergone by herbicides in soils. Microbiological and chemical reactions (exclusive of photochemical reactions) are active in rendering many herbicides innocuous. Free of the soil environment, microbial reactions can be studied in detail at the cellular and enzymic levels. Studies with sterile soils and model systems have revealed that purely chemical systems also play a role in destroying herbicides. Intensive study of representative compounds within a particular group of herbicides has elucidated certain classes of reactions which have underlying similarities from a chemical and biochemical standpoint. In a few cases, a thorough understanding of these reactions has allowed us to predict the type of decomposition product that could be produced within certain classes of compounds and the ease with which they are mediated.

Until recently far less information existed on reactions of other pesticides in soils. Since some clear trends in mechanisms of degradation are now appearing in the literature, and since they logically fit within the outline used to describe the reaction of herbicides in soils, examples of other pesticide decompositions are included in the text. Finally, as there are now many commercial herbicides and insecticides on the market, we will not cover reactions associated with every pesticide.

II. Chemical reactions

The least understood transformations of pesticides in soils are those reactions catalyzed by the purely chemical systems as opposed to the biochemical reactions mediated by soil microorganisms. The major difficulty in distinguishing chemical from biochemical reactions has been to obtain a sterile soil system that has not undergone extensive physical and chemical destruction due to autoclaving or other methods of sterilization. As a consequence of the difficulty of obtaining intact sterile soil systems in which to study chemical reactions, a large majority of the pesticide transformations have been ascribed to soil microorganisms. Recently, however, several clear examples of chemical hydrolysis have appeared in the literature. One of the simplest cases relates to the solvolysis of *cis-* and *trans-*1,3-dichloropropene (Castro and Belser 1966). These fumigants are hydrolyzed in wet soil to the *cis-* and *trans-*3-chloroallyl alcohols. Solvolysis rates, in the presence of soil, are enhanced up to three-fold as compared to rates in water.

$$\underset{Cl}{\overset{H}{\diagup}}C=C\underset{CH_2Cl}{\overset{H}{\diagdown}} \quad \xrightarrow[\text{Soil}]{H_2O} \quad \underset{Cl}{\overset{H}{\diagup}}C=C\underset{CH_2OH}{\overset{H}{\diagdown}}$$

There is considerable evidence accumulating that certain phosphate insecticides are cleaved by chemical hydrolysis in soils (KONRAD *et al.* 1967, GUNNER 1967, TIEDJE and ALEXANDER 1967). Diazinon [1] in culture solutions of soil microorganisms is apparently first chemically hydrolyzed to 2-isopropyl-4-methyl-6-hydroxypyrimidine and O,O-diethyl-thiophosphoric acid.

$$\text{(diazinon)} \xrightarrow{\text{H}_2\text{O}} \text{(2-isopropyl-4-methyl-6-hydroxypyrimidine)} + \text{(}O,O\text{-diethyl-thiophosphoric acid)}$$

GUNNER has shown that species of *Pseudomonas, Arthrobacter,* and *Streptomyces* are the predominant microbial populations arising in diazinon treated soils. These organisms, however, attack the products of the chemical reaction, *i.e.,* the pyrimidine and diethyl-thiophosphoric acid, rather than intact diazinon. The soil microbes appear to first oxidize the sulfur atoms of the phosphorothioate and subsequently degrade the resultant diethyl phosphoric acid.

Likewise TIEDJE and ALEXANDER (1967) have shown that in soil cultures enriched with malathion, the initial reaction appears to be a non-enzymic hydrolysis to yield a product tentatively identified as a malathion monoacid.

$$\underset{\substack{| \\ \text{OCH}_3}}{\text{CH}_3\text{O–P}}\overset{\text{S}}{\underset{}{\|}}\text{–S–CH–}\overset{\text{O}}{\underset{}{\|}}\text{C–OH} \quad \text{or} \quad \underset{\substack{| \\ \text{OCH}_3}}{\text{CH}_3\text{O–P}}\overset{\text{S}}{\underset{}{\|}}\text{–S–CH–}\overset{\text{O}}{\underset{}{\|}}\text{C–OC}_2\text{H}_5$$

A second degradation product seemed to support limited growth. HINDIN (1963) also has evidence that malathion, demeton, and mevinphos appear to be decomposed primarily by chemical hydrolysis in soils.

Various soil fractions may participate in facilitating the initial catalytic hydrolysis of the phosphate insecticides. KONRAD *et al.* (1967) followed diazinon hydrolysis in acid and soil solutions. In soils, hydrolysis is apparently catalyzed by adsorption on some soil component rather than by acid hydrolysis. In acid solutions at pH 2.0, diazinon hydrolysis is rapid, while at pH 6.0 the insecticide is quite

[1] Pesticide chemicals mentioned in this text and their common or trade names are given in Table I.

stable. By comparison, diazinon hydrolysis in a Poygan silty clay at pH 7.2 was fairly rapid and amounted to 11 percent/day.

Further evidence that soil components may enhance the degradation of the phosphate insecticides is provided by Mortland and Raman (1967). Catalytic hydrolysis of several organophosphates, including diazinon and Dursban, occurred in cupric chloride solutions and in copper-montmorillonite clay suspensions. The mechanism was suggested to be bidentate chelation through the S and the heterocylic N atoms; the resultant exocyclic resonance system decreases activation energy for hydrolysis. Catalysis by Cu(II) was in direct proportion to its activity. Thus, clay minerals which bond metallic cations more strongly than does montmorillonite (e.g., beidellite, vermiculite, nontronite) and organic soils allowed little or no rapid hydrolysis.

Environmental factors influence the persistence of the phosphate insecticides in soils. Menzer and Ditman (1968) have shown that disulfoton and Phorate are degraded in the soil more slowly in cold weather. Phorate is generally dissipated faster than disulfoton under all environment conditions.

A recent study by Hance (1967) was undertaken to determine the rate of decomposition of six herbicides in aqueous slurries of soils and clays maintained at 85°, 95°, and 107° C. Under these conditions, it was assumed that biological activity would not occur. Hance's conclusion, after reacting these herbicides for 164 hours, was that degradation by purely chemical means is unlikely to be an important pathway of degradation for the herbicides atrazine, CIPC, diuron, linuron, paraquat, and picloram, as the half-lives at 20° C. would be in the range of nine to 116 years. As will be discussed in the *Biochemical reactions* section, evidence exists for the microbial transformations of CIPC and paraquat in soils. Much is still unknown about the transformation of picloram by chemical or biochemical means, but most surprising were findings with atrazine.

In contrast to Hance's (1967) finding with atrazine, one of the most studied chemical transformations of a herbicide in soils is the hydrolysis of the 2-chloro-s-triazines to their respective 2-hydroxy analogues. It would be well at this point to review the evidence for chemical hydrolysis of the chlorotriazines in higher plants. A plant constituent identified as a benzoxazinone derivative has been isolated which causes the conversion of the chloro-s-triazines to their respective hydroxy analogues (Roth and Knüsli 1961, Hamilton and Moreland 1962). The benzoxazinone participates in a nucleophilic displacement of the chlorine at C-2 in the s-triazine ring to form an unstable adduct which spontaneously hydrolyzes to form the 2-hydroxy-s-triazine. This was one of the first cases of a clearly defined chemical reaction causing herbicide detoxification in higher plants. It is not the only reaction that the s-triazines undergo in the intact higher plant, but is probably one of the major routes of detoxification. Subsequent to the plant work, several investigations have shown that chloro-s-triazine degradation in soils also includes chemical conversion to their respective hydroxy analogues. Working with sterilized soils, 28 to 47 percent of added simazine, atrazine, or propazine was converted to their respective hydroxy forms within eight weeks at 30° C. (HARRIS 1965). SKIPPER et al. (1967) also concluded that chemical hydrolysis of atrazine to hydroxyatrazine was the major pathway of degradation in soils. Hydroxyatrazine accounted for about 20 percent of the extracted activity from atrazine-^{14}C incubated for two to four weeks in both nonsterile and sterile soils. Hydroxyatrazine has also been detected as the major degradation product in perfused soils varying in organic matter, clay content, and pH (ARMSTRONG et al. 1967). Atrazine hydrolysis in aqueous and in sterilized soil systems is first order with respect to atrazine at constant pH. Contact between soil and atrazine increases the rate of hydrolysis, implying catalysis by some soil component.

An interesting development in atrazine degradation relates to recent work showing the participation of soil fractions (RUSSELL et al. 1968). Chemical hydrolysis of the chloro-s-triazine, following interaction with montmorillonite clay, was facilitated by protonation at the colloid surface. Comparison of the infrared spectra of atrazine-clay and propazine-clay complexes with those of the non-complexed hydroxy analogues suggests that the s-triazine component adsorbed on the clay is not the hydroxy analogue per se. It is rather the hydroxy analogue present in the cationic form, which is tightly held on the clay. The authors claimed, by NMR studies, that protonation of the hydroxy analogues on the ring nitrogen favors the formation of a double-bond external to the ring.

Infrared studies of montmorillonite-adsorbed atrazine and propazine also suggested the formation of an external double-bond between the ring and the alkyl nitrogen. RUSSELL et al. (1968) speculate that formation of an external double-bond favors delocalization of the π electrons, causing a weakening of the C-Cl bond and thus facilitating acid hydrolysis. The surface chemistry of adsorbed pesticides, with respect to mechanisms of adsorption and degradation, has received relatively little consideration. Their approach suggests important new avenues of investigation for future research.

While the dividing line between the purely chemical and biochemical reactions in the case of the chloro-s-triazines is fairly clear, it is less clear in the case of several other classes of organic herbicides.

Decomposition of the N,N-dimethylphenylurea herbicides and amitrole has always been encountered in soils under conditions conducive to microbial degradation, *i.e.*, high temperature, high soil moisture, and organic matter. Although the exact sequential metabolism of the N,N-dimethylphenylureas is fairly well understood and discussed in detail in the *Biochemical reactions* section, metabolites have not been isolated from pure cultures of soil microorganisms in the absence of soil. Somewhat the same situation exists in the case of amitrole. There must exist in soils, therefore, a series of reactions associated with the interaction of the soil, microorganism, and pesticide which we will term for lack of a more descriptive definition *associated reactions*. The term "associated reactions" is interjected to imply that pesticide degradation takes place as a consequence of the normal metabolism of the soil microorganism but not by specifically adaptive or induced processes. One could visualize here that the organism produces some exoenzyme or extracellular product which causes pesticide decomposition. Although the concept is hazy in its present context, recent evidence is highly suggestive of this sequence of events. Amitrole decomposition, in part, appears to be a chemical reaction in soils. The failure of steam-sterilized soil to degrade amitrole has been cited previously as direct evidence for microbial involvement in this process. It was surprising, however, when soils sterilized with azide or ethylene oxide were found to decompose amitrole-5-^{14}C at a rate comparable to non-sterilized soils (Kaufman *et al.* 1968). The heterocyclic ring of amitrole is known to be stable towards a variety of common reagents but is highly susceptible to attack by free radicals. A detailed study was initiated to identify the products of amitrole decomposition in a number of free-radical generating systems. Fenton's reagent, a source of hydroxy radicals, reacts with amitrole-5-^{14}C to give $^{14}CO_2$, unlabeled urea, and cyanamide (Plimmer *et al.* 1967) (Figure 1).

More recent evidence indicates that dealkylation of hydroxyatrazine occurs in one or both of the side chains by the production of hydroxyl radicals from Fenton's reagent (Plimmer and Kearney 1968). There is also some indication that dealkylation occurs with the N,N-dimethylureas, since labeled carbon was trapped in an organic base when methyl-labeled diuron was reacted with Fenton's reagent. The rate of reaction was not rapid. The overall significance of free radical reactions in soils is unclear with regard to pesticide decomposition. Electron paramagnetic resonance studies on several soil components indicated that soil humic acid samples have spin concentrations of about 10^{18} spins/g. In addition to the humic acid fractions, other paramagnetic substances (in addition to metals) in soils include certain lignins, resins, pigments, antibiotics, vitamins, and miscellaneous bacterial products. An excellent review of free radical reactions in soils has recently been published by Steelink and Tollin (1967). It is also possible that one of the major ingredients in Fenton's re-

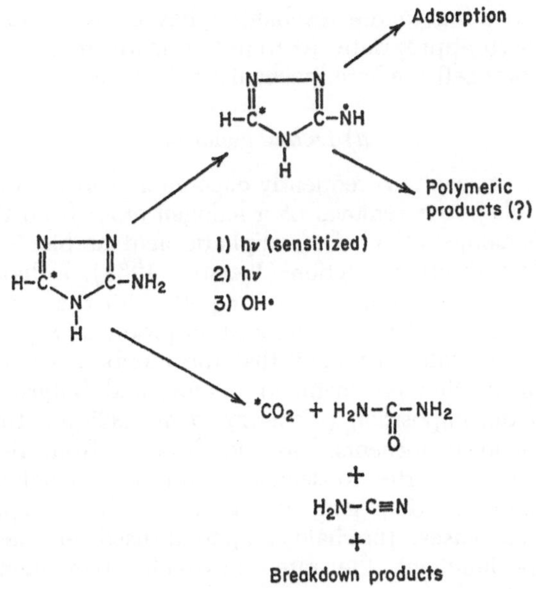

Fig. 1. Proposed reactions of amitrole resulting from irradiation and oxidation with FENTON's reagent (from PLIMMER and KEARNEY 1967)

agent, *i.e.*, hydrogen peroxide, can also be generated as a consequence of the electron transport system in soil microorganisms. Hydrogen peroxide is known to be produced by such enzymes as glucose oxidase, glycolic acid oxidase, and several other oxidase systems. Some of the oxidase enzymes are known to be exoenzymes, excreted into the medium surrounding the microbial culture (KUSAI *et al.* 1960). Therefore, one of the prime elements for the production of free radicals can be produced as a consequence of normal respiration. Experiments are presently underway in this laboratory to determine the role of radicals, produced by exogenous microbial products, in the overall schemes of several important herbicidal associated reactions.

III. Biochemical reactions

The principal microbial reactions associated with pesticide decomposition by soil microorganisms include dehalogenation, dealkylation, amide or ester hydrolysis, oxidation, reduction, ether fission, aromatic ring hydroxylation, and ring cleavage. Some of these reactions and the enzymes, pathways, and intermediate products involved in herbicide decomposition have been recently summarized (KEARNEY 1966, KEARNEY *et al.* 1967). The first five reactions are of considerable importance in soil metabolism and are discussed in the following sections. Although these reactions have been demonstrated to occur in soils, information relating to the mechanism has largely come from other systems. Where appropriate, reference is made to the more specific enzymes or organelles where mechanistic evidence exists.

a) Dehalogenation

One class of reaction frequently encountered in the decomposition process in soils is the removal of a halogen atom from the herbicide molecule. Dehalogenation of an aliphatic acid herbicide is generally the initial detoxification reaction (LEASURE 1964). Dehalogenation of several aromatic herbicides can occur both with the parent molecule or with one or more of the decomposition products. Studies with soils and pure culture solutions show that the number, position and type of halide substitution on acetic, propionic, and butyric acids affect their rate of decomposition (KEARNEY et al. 1965 a). Increasing the number of halo-substituents and the distance from the functional group have been reported to decrease the rate of dehalogenation. Although these observations generally apply, notable exceptions can be cited. In some cases, the halo-compound used in the enrichment cultures may influence the rate at which other compounds are dehalogenated.

Enzymic dehalogenation of the simpler aliphatic acids proceeds by replacement of the halogen by a hydroxy group. Cleavage of the C-F

$$F-CH_2-COO^\ominus \longrightarrow CH_2OH-COO^\ominus$$

$$Br-CH_2CH_2-COO^\ominus \longrightarrow CH_2OH-CH_2-COO^\ominus$$

$$CH_3-\overset{\overset{\textstyle Cl}{|}}{\underset{\underset{\textstyle Cl}{|}}{C}}-COO^\ominus \longrightarrow \left[CH_3-\overset{\overset{\textstyle OH}{|}}{\underset{\underset{\textstyle Cl}{|}}{C}}-COO^\ominus \right] \longrightarrow CH_3\overset{\overset{\textstyle O}{||}}{C}-COO^\ominus$$

$$CCl_3-COO^\ominus \longrightarrow \ ?$$

bond in fluoroacetate or the C-Cl bond in chloroacetate yields glycolate (GOLDMAN 1965, DAVIES and EVANS 1962), the C-Br bond in 3-bromo-propionate yields 3-hydroxypropionate (CASTRO and BARTNICKI 1965), and the C-Cl bonds in 2,2-dichloropropionate (dalapon) yields pyruvate (KEARNEY et al. 1964). The last reaction presumably proceeds by the intermediate formation of the 2-chloro-2-hydroxypropionate which would spontaneously yield the keto acid. Most of these enzymes have a moderately alkaline optimal pH and exhibit the capacity to dehalogenate closely related fluoro-, chloro-, and iodo-substituted acids.

GOLDMAN (1956) has suggested dehalogenation proceeds through a thioether linkage between the enzyme and substrate, and the overall reaction can be considered as a two step process:

$$\text{Enz-S}^- + \text{XCH}_2\text{COO}^- \longrightarrow \text{Enz-S-CH}_2\text{COO}^- + \text{X}^-$$
$$\text{Enz-S-CH}_2\text{COO}^- + \text{OH}^- \longrightarrow \text{HOCH}_2\text{-COO}^- + \text{Enz-S}^-$$

The first reaction would be rate-limiting and depends in part on the nature of the halogen substituent. It is further postulated that a moderately alkaline pH enhances the ionization of the mercapto group on the enzyme rather than the availability of the hydroxyl ion.

Randox, a herbicide, is metabolized enzymically by hydroxylation, then amide cleavage to glycolic acid and diallylamine (HANNAH 1955). Its chlorine atom is equally reactive chemically, being easily replaced by alkoxy, alkylmercapto, amino, etc. groups (JAWORSKI 1956).

The relatively persistent organochlorine insecticide lindane was converted, by dehydrohalogenation, to γ-pentachlorocyclohexene (YULE et al. 1967). The authors attributed breakdown in moist soil to microbial activity. One of the major microbial transformations of p,p'-DDT in soils is to p,p'-DDD (CHACKO et al. 1966). OTT and GUNTHER (1965) have reviewed a number of other systems causing the same conversion. In *Aerobacter aerogenes,* the conversion of DDT to DDD proceeds by reductive dehalogenation and not by dehydrochlorination and subsequent reduction as demonstrated with deuterated DDT (PLIMMER and KEARNEY 1967 and 1968). Retention of the deuterium atom in DDD would exclude the possibility of a two-step reaction involving DDE. Mass spectra of the isolated DDD confirmed the presence of a peak at $m/e = 319$, due to the contribution of the species $C_{14}H_9D^{35}Cl_4$, indicating that the deuterium atom originally present at the 2-position in DDT was retained throughout the reaction.

b) Dealkylation

1. C-R cleavage.—Few herbicides in common use contain alkyl groups directly bonded to the carbon skeleton. Limited evidence suggests that a C-R substituent is resistant to metabolism. When MCPA was degraded by a soil bacterium, the methyl group was retained past the muconic acid stage (GAUNT and EVANS 1961). Similarly, DNC was primarily reduced to amines and their N- and O-conjugates when fed to rabbits, though a small amount was oxidized to 3-amino-5-nitrosalicylic acid (SMITH et al. 1953).

The step-by-step oxidation of a methyl group on the steroid lanosterol was studied in detail by MILLER et al. (1967). Using microsomal enzymes from rat liver, the following pathway was proposed:

$$\overset{A}{RCH_3 \longrightarrow} \overset{B}{RCH_2OH \longrightarrow} \overset{C}{RCHO \longrightarrow} \overset{D}{RCOOH \longrightarrow} RH + CO_2$$

Step A required oxygen and reduced pyridine nucleotide, whereas dehydrogenation and decarboxylation (B-D) were anaerobic and required oxidized pyridine nucleotides. The microsomal dehydrogenases apparently supply reduced pyridine nucleotide for the mixed-function oxidases, since NAD could be added to the aerobic mixture without further addition of exogenous NADPH.

2. N-R cleavage.—N-alkyl groups of varying chain length are encountered in several classes of herbicides. Notable examples include secondary amines such as dialkyl-s-triazines (simazine), tertiary amines such as dimethyl substituted phenylureas (diuron) and dinitrotoluidines (trifluralin), and quarternary ammonium salts such as paraquat.

As pointed out in the section devoted to *Chemical reactions*, the major route of decomposition for the 2-chloro-s-triazines in soils involves the formation of the 2-hydroxy-s-triazines. However, dealkylation of simazine does occur as a secondary reaction catalyzed by several soil microorganisms (Fig. 2), *e.g.*, *Aspergillus fumigatus*

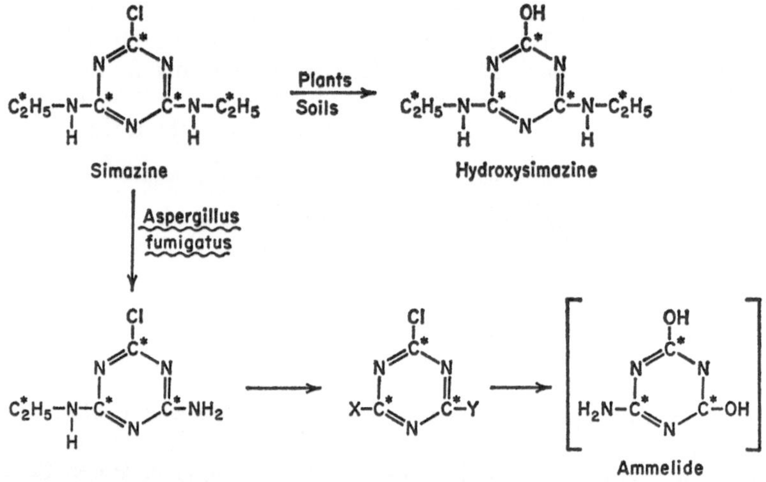

Fig. 2. Proposed pathways of simazine decomposition in soils (from KEARNEY et al. 1965 b)

(KEARNEY et al. 1965 b). Sequential removal of alkyl groups has been demonstrated for the phenylureas and dinitrotoluidines in soils (GEISSBUHLER et al. 1963, PROBST et al. 1967).

In microbial culture solution, paraquat was partially dealkylated

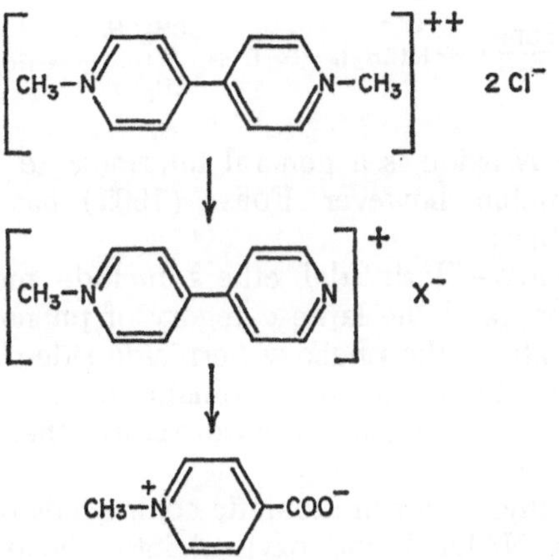

Fig. 3. Proposed pathway of paraquat metabolism by a soil-isolated bacterium (from BOZARTH *et al.* 1966)

and underwent ring cleavage, according to the pathway (Fig. 3) of BOZARTH *et al.* (1966). In soils the degradation rate may be reduced because strong adsorption to soil components greatly diminishes the concentration of paraquat in solution.

Thus far there have been no reports of cell-free preparations from soil microorganisms capable of removing the *N*-alkyl groups from any herbicide. With several *N,N*-dimethyl carbamate insecticides, however, *N*-dealkylation occurred when NADPH and oxygen were provided (HODGSON and CASIDA 1961). The metabolic breakdown of the vinyl phosphate insecticide Bidrin included specifically isolated *N*-hydroxymethyl intermediates:

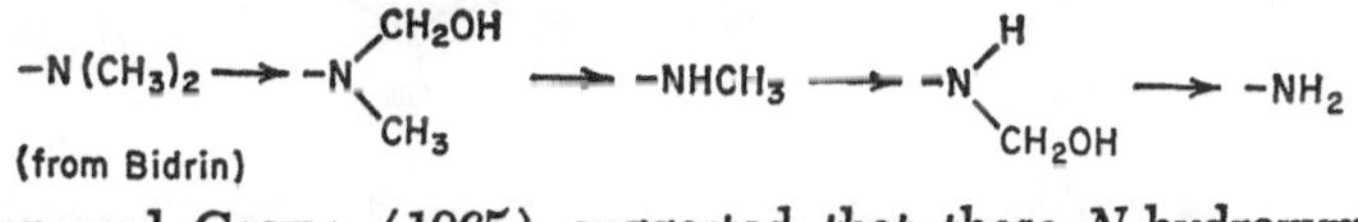

MENZER and CASIDA (1965) suggested that these *N*-hydroxymethyl derivatives of amides may be more stable than those of amines.

The mechanism of *N*-dealkylation may also proceed via the *N*-oxide (FISH *et al.* 1956). The first metabolite of the insecticide schradan is the biologically active schradan oxide, which then isomerizes to the *N*-hydroxymethyl derivative before loss of formaldehyde (CASIDA *et al.* 1954). Using liver microsomes in an aerated NADPH-generating system, PETTIT and ZIEGLER (1963) found demethylation of *N,N*-dimethylaniline produced *N*-methylaniline and formaldehyde. Added *N,N*-dimethylaniline-*N*-oxide was rapidly demethylated in the absence of NADPH and oxygen but required the microsomes. The suggested reaction sequence is:

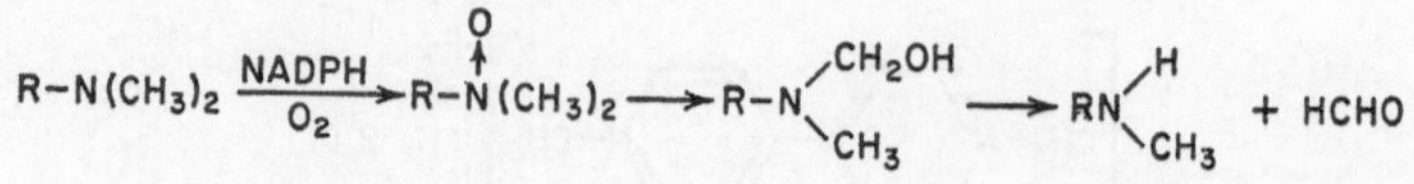

$$R-N(CH_3)_2 \xrightarrow[O_2]{NADPH} R-\overset{O}{\overset{\uparrow}{N}}(CH_3)_2 \longrightarrow R-N\overset{CH_2OH}{\underset{CH_3}{\Big\backslash}} \longrightarrow RN\overset{H}{\underset{CH_3}{\Big\backslash}} + HCHO$$

Whether the N-oxide is a general intermediate in N-dealkylation remains questionable, however. FURST (1965) has summarized evidence on the subject.

3. O-R cleavage.—Herbicidal ethers include methoxylated compounds (dicamba) and the large category of phenoxy acids (MCPA or 2,4-D). The fate of the phenoxy herbicide side-chain has been extensively studied. Typically, the aromatic moiety is released as a phenol in soils, isolated bacterial cultures, and their enzyme systems (KEARNEY et al. 1967).

Methyl and ethyl ethers of aromatic compounds were split, by liver microsomes plus NADPH and oxygen, into phenols and aldehydes (AXELROD 1956). The rate of anisole cleavage depended on the nature and position of other ring substituents. The order of decreasing rate of demethylation was: $-CN > -CHO > -CH_2OH > -NHCOCH_3 > -COOH > -CH_2NH_2 > -CH=CHCH_3 > -NH_2 > -H$. The analogous order for position was: $p > m > o$.

A cell-free system from *Pseudomonas fluorescens* catalyzed the oxidative demethylation and subsequent ring-cleavage of vanillic acid (CARTWRIGHT and SMITH 1967, CARTWRIGHT and BUSWELL 1967). Demethylation required 0.5 mole of oxygen/mole of substrate and as cofactors, reduced glutathione (GSH) and NADPH. The methyl group was removed as formaldehyde and subsequently oxidized to formate and carbon dioxide (Fig. 4).

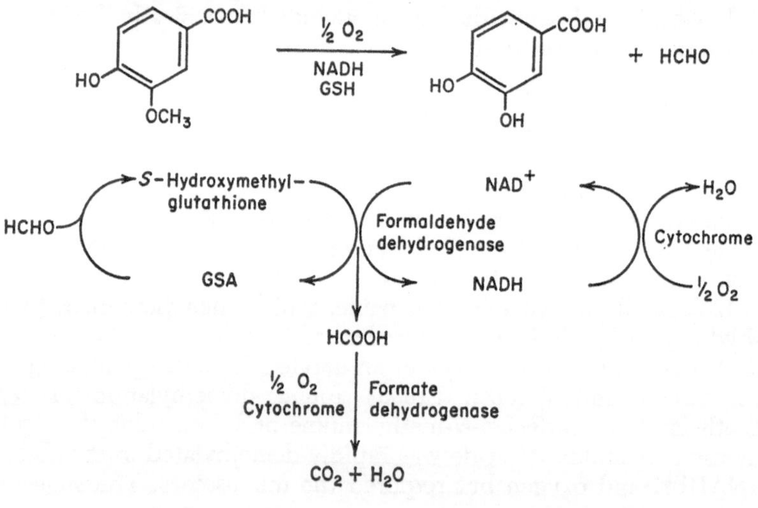

Fig. 4. Demethylation of vanillic acid by *Pseudomonas fluorescens* (from CART-WRIGHT and SMITH 1967, CARTWRIGHT and BUSWELL 1967)

Indirect evidence for the point of ether-oxygen bond cleavage was obtained by RENSON (1964). He found only unlabeled p-hydroxyacetanilide when p-methoxyacetanilide was metabolized by liver microsomes in the presence of $^{18}O_2$ or $H_2^{18}O$; formaldehyde-^{18}O was trapped in the system containing $^{18}O_2$. Later, HELLING et al. (1968) directly confirmed that cleavage occurs between the aliphatic side chain and the ether-oxygen atom. Phenoxy-^{18}O-acetic acid was quantitatively metabolized to phenol-^{18}O by resting cells and cell-free extracts of a soil-isolated *Arthrobacter* sp.

$$\text{(phenyl)-}^{18}O\text{-R} \xrightarrow{\ ^{18}O_2\ } \text{(phenyl)-}^{18}O\text{-H} + \text{RCH}^{18}O$$

c) Amide or ester hydrolysis

Phenylamides of the general structure C_6H_5-NH-CO-R are encountered in a large number of herbicidal compounds. Important members of this class of herbicides include the phenylcarbamates such as CIPC and the acylamides such as propanil. Metabolism of these compounds proceeds by cleavage of the amide or ester linkage to yield aniline, carbon dioxide and alcohol in the case of the phenylcarbamates, and aniline and aliphatic acid in the case of the acylanilides.

Extensive studies have been conducted on the enzymic hydrolysis of CIPC with cell-free preparations obtained from *Pseudomonas striata* isolated from soil enrichment cultures (KEARNEY 1965). A 70-fold purified enzyme, obtained by ammonium sulfate precipitation and column chromatography by gradient elution on DEAE-cellulose, catalyzes hydrolysis of CIPC to 3-chloroaniline, carbon dioxide, and isopropanol. For CIPC the K_m is 8.4×10^{-6} and the optimal pH is 8.5. The purified enzyme has no apparent metallic ion requirement, and is inhibited by diisopropyl fluorophosphate. It is also interesting to note that the CIPC enzyme is strongly inhibited by methylcarbamate insecticides; for carbaryl the K_i is about 4.3×10^{-9}. The enzyme exhibits a broad substrate specificity, since a large number of structurally related phenylcarbamate and acylanilides are hydrolyzed.

Two interesting properties of the enzyme which have been explored in some detail are its substrate specificity and its inhibition by methylcarbamate insecticides. The purified enzyme can hydrolyze a number of alcohol derivatives of the substituted phenylcarbamates. By comparing the relative rates at which various structurally related phenylcarbamates were hydrolyzed *in vitro* we hoped to learn how certain molecular parameters affected their persistence in soils. A survey of 17 compounds revealed that the inductive effects exerted by *meta*-substitution of electron withdrawing groups and the steric effects imposed by increasing the size of the alcohol groups significantly altered the rate of reaction (KEARNEY 1967). A correlation was established between the relative acidities of the phenylcarbamates, as influenced by the inductive effects of *meta*-substituents, and the hydrolytic rate by the microbial enzyme. The more acidic phenylcarbamates were more readily hydrolyzed; e.g., *p*-nitrophenylcarbamates were hydrolyzed more rapidly than *m*-nitrophenylcarbamates. Relative acidities of the substituted phenylcarbamates were determined by titration as acid in *n*-butylamine. This procedure measures the ease with which the proton is removed from the amide nitrogen. Finally, increasing the overall dimensions of the molecule, as tested by substituting a naphthyl moiety for the phenyl group, decreased the rate of reaction. Experiments are underway to determine if these observations can be extended to the general soil persistence of the phenylcarbamates.

The ability of the methylcarbamate insecticides to strongly inhibit the phenylcarbamate hydrolyzing enzyme may offer some insight into the interaction effects which occur when these two classes of compounds are combined in soils (KAUFMAN and KEARNEY 1967). Bioassays of treated soils indicated that the methylcarbamate insecticide carbaryl significantly increased the persistence of CIPC in soils (Fig. 5). Extended persistence of the phenylcarbamate herbicide is desirable

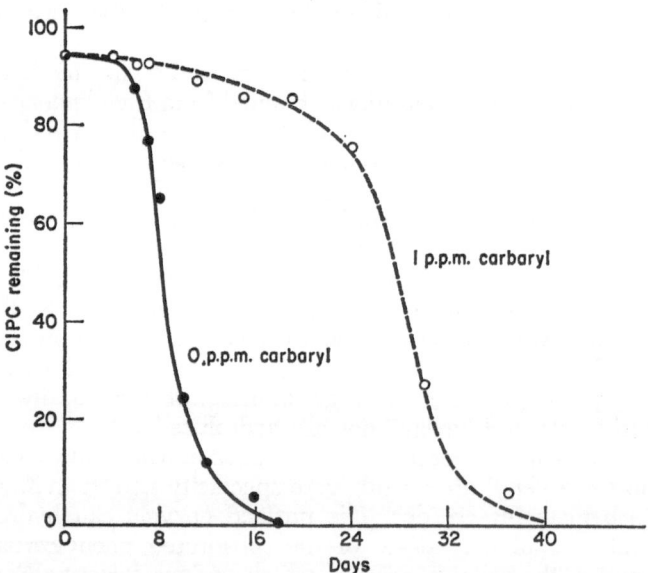

Fig. 5. Effect of carbaryl on microbial decomposition of CIPC (from KAUFMAN and KEARNEY 1967)

in this case, since these compounds are degraded quickly in most soils. Similar results were obtained with combinations of CIPC and other methylcarbamate pesticides, including O-isopropoxyphenyl methylcarbamate, 6-chloro-3,4-xylyl methylcarbamate, 4-dimethylamino-*m*-tolyl methylcarbamate, and 4-dimethylamino-3,5-xylyl methylcarbamate, but not with 2,6-di-*t*-butyl-*p*-tolyl methylcarbamate (terbutol). Soil perfusion studies indicated that the microbial degradation of CIPC was inhibited by carbaryl. Perhaps the methylcarbamate insecticides, in close proximity to the phenylcarbamates in soils, are effectively blocking the metabolism of the herbicides. Whether the inhibition is directly mediated through the phenylcarbamate hydrolyzing enzyme, or through the intact soil microorganism, is not yet clear.

d) Oxidation

Although relatively persistent in soils, certain chlorinated insecticides containing an isolated double-bond are oxidized to an epoxide. Many soil microorganisms are capable of converting, for example, aldrin to dieldrin or heptachlor to heptachlor epoxide (MENZIE 1966).

Thioethers such as prometryne contain a readily-oxidizable sulfur atom. Prometryne is metabolized by soil microbes to the sulfoxide, then to the sulfone (GYSIN 1962). In water, prometryne sulfone hydrolyzes to hydroxypropazine in three days (DELLEY 1964). The systemic insecticides comprising demeton are oxidized to sulfoxides and sulfones in animal, insect, and plant tissues (MARCH *et al.* 1955). MENZIE (1966) cites several references to demeton conversion to the corresponding sulfoxides and sulfones in other systems.

e) Reduction

Pesticides having a nitro substituent frequently undergo reduction to amines. LICHTENSTEIN and SCHULZ (1964) isolated aminoparathion from soil incubated with parathion. The responsible organisms were yeast rather than bacteria, and persistence of parathion was greatly lengthened in dry or autoclaved soil.

Soil aeration status may govern the priority of nitro reduction in the metabolic scheme. With trifluralin, dealkylation was followed by progressive reduction in aerobic soils, whereas this pathway was reversed under anaerobic conditions (PROBST *et al.* 1967). The conversion of aromatic amines to unidentified polar products was rapid and constituted the major route of decomposition. Benefin, an analogue of trifluralin, undergoes similar transformations. The importance of microorganisms in the degradation of these herbicides is uncertain, and, as suggested by PROBST (1968), may affect only intermediate products resulting from prior physiochemical decomposition of the parent compound.

Table I. *Chemical designations of pesticides mentioned in text*

Common name	Chemical name
Aldrin	1,2,3,4,10,10-hexachloro-1,4,4a,5,8,8a-hexahydro-1,4-*endo,exo*-5,8-dimethanonaphthalene
Amitrole	3-amino-1,2,4-triazole
Atrazine	2-chloro-4-ethylamino-6-isopropylamino-*s*-triazine
Benefin	a,a,a-trifluoro-2,6-dinitro-*N*-butyl-*N*-ethyl-*p*-toluidine
Bidrin ®	3-hydroxy-*N,N*-dimethyl-*cis*-crotonamide dimethyl phosphate
Carbaryl (Sevin ®)	1-naphthyl methylcarbamate
CIPC	isopropyl *N*-(3-chlorophenyl) carbamate
2,4-D	2,4-dichlorophenoxyacetic acid
DDD	1,1-dichloro-2,2-bis(*p*-chlorophenyl) ethane
DDT	1,1,1-trichloro-2,2-bis(*p*-chlorophenyl) ethane
Demeton	mixture of *O,O*-diethyl *S* (and *O*)-[2-(ethylthio)-ethyl] phosphorothioates
Diazinon	*O,O*-diethyl *O*-(2-isopropyl-4-methyl-6-pyrimidinyl) phosphorodithioate
Dicamba	2-methoxy-3,6-dichlorobenzoic acid
Dieldrin	1,2,3,4,10,10-hexachloro-6,7-epoxy-1,4,4a,5,6,7,8,8a-octahydro-1,4-*endo,exo*-5,8-dimethanonaphthalene
Disulfoton	*O,O*-diethyl *S*-[2-(ethylthio)ethyl] phosphorodithioate
Diuron	3-(3,4-dichlorophenyl)-1,1-dimethylurea
DNC	4,6-dinitro-2-methylphenol
Dursban ®	*O,O*-diethyl *O*-3,5,6-trichloro-2-pyridyl phosphorothioate
Heptachlor	1,4,5,6,7,8,8-heptachloro-3a,4,7,7a-tetrahydro-4,7-methanoindene
Heptachlor epoxide	1,4,5,6,7,8,8-heptachloro-2,3-epoxy-2,3,3a,7a-tetrahydro-4,7-methanoindene
Imidan	*N*-(mercaptomethyl)phthalimide *S*-(*O,O*-dimethyl phosphorodithioate)
Lindane (γ-BHC)	γ-1,2,3,4,5,6-hexachlorocyclohexane
Linuron	3-(3,4-dichlorophenyl)-1-methoxy-1-methylurea
Malathion	*O,O*-dimethyl dithiophosphate of diethyl mercaptosuccinate
MCPA	4-chloro-2-methylphenoxyacetic acid
Mevinphos	methyl 3-hydroxy-a-crotonate dimethyl phosphate
Paraquat	1,1'-dimethyl-4,4'-bipyridinium salt
Parathion	*O,O*-diethyl *O*-*p*-nitrophenyl phosphorothioate
Phorate	*O,O*-diethyl *S*-(ethylthiomethyl) phosphorodithioate
Picloram	4-amino-3,5,6-trichloropicolinic acid
Prometryne	2-methylmercapto-4,6-bis(isopropylamino)-*s*-triazine
Propanil	3',4'-dichloropropionanilide
Propazine	2-chloro-4,6-bis(isopropylamino)-*s*-triazine
Randox	2-chloro-*N,N*-diallylacetamide
Schradan	octamethylpyrophosphoramide
Simazine	2-chloro-4,6-bis(ethylamino)-*s*-triazine
Trifluralin	a,a,a-trifluoro-2,6-dinitro-*N,N*-dipropyl-*p*-toluidine

Summary

A summary of pertinent reactions associated with pesticide decomposition in soils is presented. Primary emphasis is directed toward reactions of herbicides, although the outline has been expanded to illustrate similar reactions with other pesticides. Recent evidence is presented on the role of the purely chemical reactions taking place on pesticides in soils. In the biochemical section, specific examples of dehalogenation, dealkylation, ester hydrolysis, oxidation, and reduction are considered on a few pesticides. Where pertinent, mechanisms elucidated from other biochemical systems are included to explain analogous reactions in soils.

Résumé *

Réactions des pesticides dans les sols

On présente un résumé de réactions appropriées associées à la décomposition des pesticides dans les sols. On met particulièrement en valeur les réactions des herbicides, mais les données générales ont été reprises en vue de mettre en évidence des réactions analogues chez d'autres pesticides. On a démontré récemment le rôle purement chimique des réactions auxquelles sont soumis les pesticides dans les sols. Dans le chapitre de la biochimie, on donne, pour quelques pesticides, des exemples spécifiques de déshalogénation, désalkylation, hydrolyse d'esters, oxydation et réduction. Dans des cas appropriés, on a inclu des mécanismes élucidés dans d'autres systèmes biochimiques pour expliquer des réactions analogues dans les sols.

Zusammenfassung *

Reaktionen von Pestiziden in Böden

Es wird eine Zusammenfassung von einschlägigen Reaktionen, die mit dem Pestizidabbau in Böden assoziiert sind, gegeben. Die Hauptbetonung ist auf die Reaktionen von Herbiziden gerichtet, obwohl der Ueberblick erweitert worden ist, um ähnliche Reaktionen mit anderen Pestiziden zu illustrieren. Neues Beweismaterial wird über die Rolle von rein chemischen Reaktionen dargeboten, welche an Pestiziden in Böden stattfinden. Im biochemischen Abschnitt werden spezifische

---

* Traduit par S. DORMAL-VAN DEN BRUEL.
* Übersetzt von A. SCHUMANN.

Beispiele von Dehalogenierung, Dealkylierung, Hydrolyse von Estern, Oxidation und Reduktion an einigen Pestiziden betrachtet. Wenn angemessen, sind Mechanismen, welche von anderen biologischen Systemen erläutert worden sind, einbegriffen, um analoge Reaktionen in Böden zu erklären.

References

ARMSTRONG, D. E., G. CHESTERS, and R. F. HARRIS: Atrazine hydrolysis in soil. Soil Sci. Soc. Amer. Proc. 31, 61 (1967).

AXELROD, J.: The enzymic cleavage of aromatic ethers. Biochem. J. 63, 634 (1956).

BOZARTH, G. A., H. H. FUNDERBURK, JR., and E. A. CURL: Studies on the degradation of 1,1'-dimethyl-4,4'-bipyridinium salt by a soil bacterium. Weed Soc. Amer. Abstr., p. 55 (1966).

CARTWRIGHT, N. J., and J. A. BUSWELL: The separation of vanillate O-demethylase from protocatechuate 3,4-oxygenase by ultracentrifugation. Biochem. J. 105, 767 (1967).

——, and A. R. W. SMITH: Bacterial attack on phenolic ethers. An enzyme system demethylating vanillic acid. Biochem. J. 102, 826 (1967).

CASIDA, J. E., T. C. ALLEN, and M. A. STAHMANN: Mammalian conversion of octamethylpyrophosphoramide to a toxic phosphoramide N-oxide. J. Biol. Chem. 210, 607 (1954).

CASTRO, C. E., and E. W. BARTNICKI: Metabolism of 3-bromopropanol by a Pseudomonas sp. Bact. Proc. 65, 10 (1965).

——, and N. O. BELSER: Hydrolysis of cis- and trans-1,3-dichloropropene in wet soil. J. Agr. Food Chem. 14, 69 (1966).

CHACKO, C. I., J. L. LOCKWOOD, and M. ZABIK: Chlorinated hydrocarbon pesticides: Degradation by microbes. Science 154, 893 (1967).

DAVIES, J. I., and W. C. EVANS: The elimination of halide ions from aliphatic halogen-substituted organic acids by an enzyme preparation from Pseudomonas dehalogenans. Biochem. J. 82, 50P (1962).

DELLEY, R.: Unpublished report, Anal. Lab., J. R. Geigy S. A., Basle, Switzerland (1964).

FISH, M. S., N. M. JOHNSON, and E. C. HORNING: t-Amine oxide rearrangements. N,N-Dimethyltryptamine oxide. J. Amer. Chem. Soc. 78, 3668 (1956).

FURST, C. I.: Drug metabolism. In: Annual reports on the progress of chemistry, Vol. 61, p. 465. London: Chem. Soc. (1965).

GAUNT, J. K., and W. C. EVANS: Metabolism of 4-chloro-2-methylphenoxyacetic acid by a soil micro-organism. Biochem. J. 79, 25P (1961).

GEISSBUHLER, J., C. HASELBACH, H. AEBI, and L. EBNER: The fate of N'-(4-chlorophenoxy)-phenyl-N,N-dimethylurea (C-1983) in soils and plants. Weed Research 3, 277 (1963).

GOLDMAN, P.: The enzymatic cleavage of the carbon-fluorine bond in fluoroacetate. J. Biol. Chem. 240, 3434 (1965).

GUNNER, H.: The influence of rhizosphere microflora on the transformation of insecticides by plants. Mass. Agr. Expt. Sta. Annual Rept. Coop. Regional Proj. NE-53 (1967).

GYSIN, H.: Triazine herbicides. Their chemistry, biological properties and mode of action. Chem. & Ind. (London), p. 1393 (1962).

HAMILTON, R. H., and D. E. MORELAND: Simazine: degradation by corn seedlings. Science 135, 373 (1962).

HANCE, R. J.: Decomposition of herbicides in the soil by non-biological chemical processes. J. Sci. Food Agr. 18, 544 (1967).

HANNAH, L. H.: Field studies with a new class of herbicidal chemicals. Proc. NE. Weed Control Conf., p. 15 (1955).

HARRIS, C. I.: Hydroxysimazine in soil. Weed Research 5, 275 (1965).

HELLING, C. S., J.-M. BOLLAG, and J. E. DAWSON: Cleavage of ether-oxygen bond in phenoxyacetic acid by an *Arthrobacter* species. J. Agr. Food Chem., 16, 538 (1968).

HINDIN, E.: Analysis of organic pesticides by gas chromatography. Final Progress Report to PHS Grant No. WP-00215. Washington State Univ. (1963).

HODGSON, E., and J. E. CASIDA: Metabolism of N,N-dialkyl carbamates and related compounds of rat liver. Biochem. Pharmacol. 8, 179 (1961).

JAWORSKI, E. G.: Biochemical action of CDAA, a new herbicide. Science 123, 847 (1956).

KAUFMAN, D. D., and P. C. KEARNEY: Persistence of CIPC in soil treated with methylcarbamate pesticides. Weed Soc. Amer. Abstr., p. 74 (1967).

——, J. R. PLIMMER, P. C. KEARNEY, J. BLAKE, and F. S. GUARDIA: Chemical vs. microbial decomposition of amitrole in soil. Weed Science, 16, 266 (1968).

KEARNEY, P. C.: Purification and properties of an enzyme responsible for hydrolyzing phenylcarbamates. J. Agr. Food Chem. 13, 561 (1965).

—— Metabolism of herbicides in soils. Adv. Chem. Series 60, 250 (1966).

—— Influence of physicochemical properties on biodegradability of phenylcarbamate herbicides. J. Agr. Food Chem. 15, 568 (1967).

——, C. I. HARRIS, D. D. KAUFMAN, and T. J. SHEETS: Behavior and fate of the chlorinated aliphatic acids in soils. Adv. Pest Control Research 7, (1965 a).

——, D. D. KAUFMAN, and M. ALEXANDER: Biochemistry of herbicide metabolism in soils. In: A. D. McLAREN and G. H. PETERSON, Soil biochemistry, p. 318. New York: Dekker (1967).

——, ——, and M. L. BEALL: Enzymatic dehalogenation of 2,2-dichloropropionate. Biochem. Biophys. Research Commun. 14, 29 (1964).

——, ——, and T. J. SHEETS: Metabolites of simazine by *Aspergillus fumigatus*. J. Agr. Food Chem. 13, 369 (1965 b).

KONRAD, J. G., G. CHESTERS, and D. E. ARMSTRONG: Soil degradation of diazinon, a phosphorothioate insecticide. Agron. J. 59, 591 (1967).

KUSAI, K., I SEKUZU, B. HAGIHARA, K. OKUNUKI, S. YAMAUCHI, and M. NAKAI: Crystallization of glucose oxidase from *Penicillium amagasakiense*. Biochem. Biophys. Acta 40, 555 (1960).

LEASURE, J. K.: The halogenated aliphatic acids. J. Agr. Food Chem. 12, 40 (1964).

LICHTENSTEIN, E. P., and K. R. SCHULZ: The effects of moisture and microorganisms on the persistence and metabolism of some organophosphorus insecticides in soils, with special emphasis on parathion. J. Econ. Entomol. 57, 618 (1964).

MARCH, R. B., R. L. METCALF, T. R. FUKUTO, and M. G. MAXON: Metabolism of Systox in the white mouse and American cockroach. J. Econ. Entomol. 48, 355 (1955).

MENN, J. J., J. B. McBAIN, B. J. ADELSON, and G. G. PATCHETT: Degradation of N-(mercaptomethyl)phthalimide-S-(O,O-dimethylphosphorodithioate) (Imidan) in soils. J. Econ. Entomol. 58, 875 (1965).

MENZER, R. E., and J. E. CASIDA: Nature of toxic metabolites found in mammals, insects, and plants from 3-(dimethoxyphosphinyloxy)-N,N-dimethyl-*cis*-crotonamide and its N-methyl analog. J. Agr. Food Chem. 13, 102 (1965).

——, and L. P. DITMAN: Residues in spinach grown in disulfoton- and phorate-treated soil. J. Econ. Entomol. 61, 225 (1968).

MENZIE, C. M.: Metabolism of pesticides. Special Sci. Report No. 96. Washington, D.C. (1966).

MILLER, W. L., M. E. KALAFER, J. L. GAYLOR, and C. V. DELWICHE: Investigation of the component reactions of oxidative sterol demethylation. Study of the aerobic and anaerobic processes. Biochem. **6**, 2673 (1967).

MORTLAND, M. M., and K. V. RAMAN: Catalytic hydrolysis of some organic phosphate pesticides by copper(II). J. Agr. Food Chem. **15**, 163 (1967).

OTT, D. E., and F. A. GUNTHER: DDD as a decomposition product of DDT. Residue Reviews **10**, 70 (1965).

PETTIT, F. H., and D. M. ZIEGLER: The catalytic demethylation of N,N-dimethylaniline-N-oxide by liver microsomes. Biochem. Biophys. Research Commun. **13**, 193 (1963).

PLIMMER, J. R., P. C. KEARNEY, D. D. KAUFMAN, and F. S. GUARDIA: Amitrole decomposition by free radical-generating systems and by soils. J. Agr. Food Chem. **15**, 996 (1967).

——, —— Free radical oxidation of pesticides. Weed Soc. Amer. Abstr., p. 20 (1968).

PROBST, G. W., T. GOLAB, R. J. HERBERG, F. J. HOLZER, S. J. PARKA, C. VAN DER SCHANS, and J. B. TEPE: Fate of trifluralin in soils and plants. J. Agr. Food Chem. **15**, 592 (1967).

——, and J. B. TEPE: Trifluralin. In: P. C. KEARNEY and D. D. KAUFMAN, Degradation of the herbicides, Chapt. 9, in press. New York: Dekker (1968).

RENSON, J.: Studies on the mechanism of microsomal hydroxylation. Fed. Proc. **23**, 325 (1964).

ROTH, W., and E. KNÜSLI: Beitrag zur Kenntnis der Resistenzphänomeme eizelner Pflanzen gegenüber dem phytotoxischen Wirkstoff Simazin. Experientia **17**, 312 (1961).

RUSSELL, J. D., M. CRUZ, J. L. WHITE, G. W. BAILEY, W. R. PAYNE, JR., J. D. POPE, JR., J. I. TEASLEY: Mode of chemical degradation of s-triazines by montmorillonite. Science **160**, 1340 (1968).

SKIPPER, H. D., C. M. GILMOUR, and W. R. FURTICK: Microbial versus chemical degradation of atrazine in soils. Soil Sci. Soc. Amer. Proc. **31**, 653 (1967).

SMITH, J. N., R. H. SMITHIES, and R. T. WILLIAMS: Studies in detoxication. 48. Urinary metabolites of 4:6-dinitro-o-cresol in the rabbit. Biochem. J. **54**, 225 (1953).

STEELINK, C., and G. TOLLIN: Free radicals in soil. In: A. D. McLAREN and G. H. PETERSON, Soil biochemistry, p. 147. New York: Dekker (1967).

TIEDJE, J. M., and M. ALEXANDER: Microbial degradation of organophosphorus insecticides and alkyl phosphates. Abstr., Amer. Soc. Agron. Annual Meetings, Washington, D.C., p. 94 (1967).

YULE, W. N., M. CHIBA, and H. V. MORLEY: Fate of insecticide residues. Decomposition of lindane in soil. J. Agr. Food Chem. **15**, 1000 (1967).

Activation and inactivation
of herbicides by higher plants

by
Shooichi Matsunaka [*]

Contents

I. Introduction

A number of chemicals is metabolized to active or inactive products by higher plants. These metabolic mechanisms sometimes contribute to the selectivity of pesticides.[1] On the other hand, the inactivation mechanism of herbicides will be also used for the purpose of the development of antidotes for herbicides (Matsunaka 1963).

In this paper, after short reviews on activation and inactivation of herbicides by higher plants are presented, two examples used for rice culture in Japan will be described: one is the activation of diphenylether compounds with the help of light irradiation and the other is the inactivation of propanil by rice plants.

II. Activation of herbicides by higher plants

The most famous example of activation of herbicides would be the case of 2,4-DB or of MCPB. The basic studies about this idea has a

---

[*] National Institute of Agricultural Sciences, Konosu, Saitama, Japan.
[1] Pesticides mentioned in text are chemically identified in Table V.

long history from GRACE's (1939) finding that a series of ω-(1-naphthyl) alkanecarboxylic acids with an odd number of methylene groups in the side-chain were effective in promoting the rooting of cuttings, whereas homologues with an even number had negligible activity.

2,4-DB and MCPB themselves have no herbicidal activity, but, when converted to 2,4-D or MCPA by β-oxidation mechanism in higher plant tissues, they are already very active herbicides. So the weeds having high β-oxidation activity such as Canada thistle, annual nettle, or charlock plants would be killed by 2,4-DB or MCPB, but clover plants or celery plants having low β-oxidation activity would be safe (WAIN 1955). By the words of Wain himself (WAIN 1964), such observations strongly supported his idea that a new type of selective weed control based upon the plant's own enzyme make-up was possible.

Bipyridylium compounds, such as paraquat and diquat, had been found to be very useful herbicides. These compounds are active only in the presence of both light and oxygen. One hypothesis on the mode of action of these herbicides seemed to be that the electrons excited by light in the photochemical system in photosynthesis would reduce bipyridylium ions to produce their free radicals which could be further oxidized by molecular oxygen. During the oxidative process, highly reactive peroxide radicals or hydrogen peroxide would be formed, and they might attack intracellular components (MORELAND 1967). MEES (1960) found that monuron, used with diquat, decreased the herbicidal activity of the bipyridylium compound. This comes from the inhibition of photoactivation of diquat by monuron which is very famous inhibitor of the Hill reaction.

Some of the phenylthiourea herbicides which contain a $C = S$ structure in place of a $C = O$ of phenylureas shows low activity in *in vitro* experiments (Hill reaction) but has a high herbicidal activity *in vivo*. For this fact, GOOD (1961) suggested that activation of $C = S$ to $C = O$ would occur similarly to $P = S$ to $P = O$ shown in organophosphorus insecticides.

Although this is not true in the case of higher plants, sesone (2,4-DES), MCPES, 2,4,5-TES, and so on are activated by soil microorganisms. The process is hydrolysis (and oxidation) of these herbicides (AUDUS 1952). No higher plants have such activity, so these herbicides can be used for standing crops by a broadcasting method.

III. Mode of action of diphenylether herbicides and their activation by the help of light

a) Chemical structure and mode of action of diphenylether herbicides

In Japan, diphenylether herbicides (Fig. 1) have been found to be very promising for premergence weed control in transplanted rice

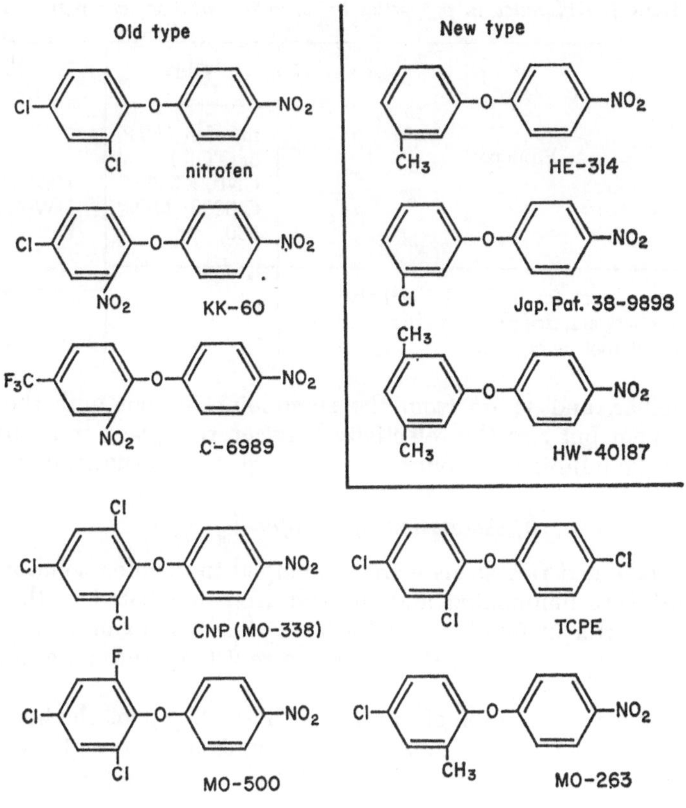

Fig. 1. Classification of diphenylether herbicides

fields and others. Among them, nitrofen and CNP having both wide spectra of weed control and low toxicity to fish, are being used by farmers.

The mode of action of these herbicides would be explained as follows: the young buds of weed seeds would contact the herbicide dissolved in irrigated water or adsorbed on soil particles, and absorb the chemical. The chemical will be converted into a more active form or will yield toxic compound(s) by both light energy and biochemical reaction. These herbicides existing at the root zone do not show any activity (Furuya and Arai 1966). These are the special properties of the first group of diphenylether herbicides.

The relationship between photoactivation and chemical structure was investigated. The test plant was germinating rice seeds. The results showed that one group such as nitrofen or CNP having ortho substituent(s) in one benzene ring required the help of light energy to kill the weeds. This old group could be distinguished from the other new group such as HE-314, HW-40187, 3,5-dichloro-4'-nitrodi-

Table I. *Difference in properties between two diphenylether groups*

Property	Old nitrofen (NIP or TOK), CNP, KK-60, C-6989, MO-500, etc.	++ New HE-314, HW-40187, etc.
Light requirement for herbicidal activity	+	
Selectivity between rice plant and barnyard grass	±	—
Inhibition of root emergence	—	++

phenylether, and so on from the standpoints of not only the light requirement but also the selectivity between rice plant and barnyard grass or the inhibitory activity to root emergence, as shown in Table I.

b) Activation of nitrofen by light

Just-emerged rice seeds were submerged in nitrofen solution, half of them were illuminated and the rest were incubated in the dark. Five days after incubation, the fresh weights of buds and roots were measured. The results are shown in Figure 2 (MATSUNAKA and INADA 1967).

In a low concentration of nitrofen such as one p.p.m., the herbicidal activity was affected very much by the illumination strength, showing a straight decrease in fresh weight of rice plant seedlings by increasing

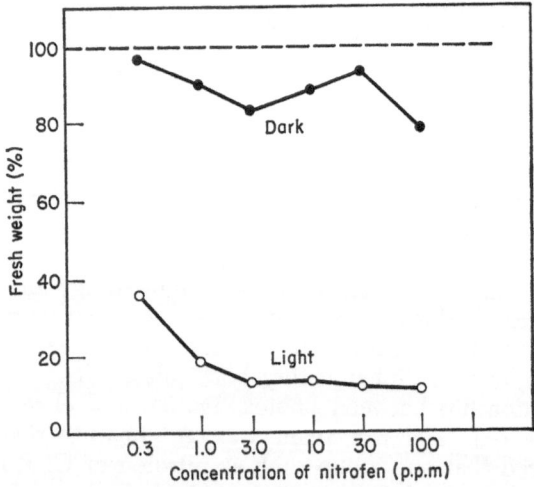

Fig. 2. Effect of light illumination on herbicidal activity of nitrofen (MATSUNAKA and INADA 1967)

illumination. On the other hand, over a concentration of five p.p.m., the effect of illumination was clear even under 1,000 lux.

When nitrofen solution was pre-illuminated without rice plants, the herbicides did not show any activity in the dark. Pre-illumination of nitrofen with riboflavine or fluorescein had also no effect on the herbicidal activity in the dark. Coexistence of manganous ion with nitrofen with rice plants during weak illumination or in the dark showed only additional effect of manganous ion. Further co-existence of simazine with nitrofen did not affect the activity of nitrofen, which showed that the activation of nitrofen by light would be different from the above-mentioned case of bipyridylium compounds, *i.e.*, paraquat and diquat.

By the illumination of germinated rice seeds for one day intervals in the presence of nitrofen (10 p.p.m.), the effect of the illuminated interval on herbicidal activity of nitrofen was investigated. The illumination on the third day seems to be more effective than the first or second days. The area accepting light illumination may be an important factor for the activation of nitrofen. On the fourth day, a part of the seed bud stands out above the surface of the nitrofen solution, then the effect seems to be weaker.

In order to survey the effect of wavelength of illumination, colored fluorescent lamps were used for illumination for the activation of nitrofen. The energy level of the lamps was the same, 0.0166 cal./cm.^{2}/ min. The rice seeds were soaked in nitrofen solution for two days at 30° C. in the dark, transferred into chambers with colored lamps, and illuminated for 50 hours. Red, orange, yellow, green, blue, and white lamps were used. After three days incubation in the dark, the fresh weights of buds and roots were measured. In general, blue light seems to be more effective, but even red light also showed some activity.

As a result of these experiments, it may be concluded that light activation of nitrofen is not a simple conversion to a toxic compound by light, but a photo-biochemical activation after absorption of nitrofen into plant tissue. How nitrofen is changed or yields some toxic compound in weeds by light is now being investigated. The activation mechanism of nitrofen in detail will be presented elsewhere (MATSUN-AKA 1968a).

IV. Inactivation of herbicides by higher plants

As an example of the inactivation of herbicides by higher plants, simazine is the most famous one in corn culture. Corn plants can detoxify simazine, propazine, atrazine, and so on. These s-triazines have a chlorine radical in the 2-position. The triazines having methoxy or methylmercapto radicals in place of chlorine are active against corn plants (McWHORTER and HOLSTON 1961). The mechanism of this selectivity has been explained as follows: 2,4-dihydroxy-7-methoxy-1,4-benzoxazine-3-one contained in corn plants would convert non-enzymatically the chlorine atom of simazine to a hydroxyl radical; the hydroxysimazine produced has no herbicidal activity. On the other hand, a lot of weeds have very low contents of this special compound and are very sensitive to simazine (CASTELFRANCO and BROWN 1962, HAMILTON and MORELAND 1962, McWHORTER and HOLSTON 1961).

It may be also very interesting that this compound has a very similar chemical structure to those of the tolerant factor to European corn borer (*Pyrausta nubilalis*) or to corn stalk-rot fungus (*Diplodia zeae* Lev.) (ANDERSEN 1964).

The second example of inactivation of herbicides in relation to selectivity is the decarboxylation of the side chain of hormonal herbicides. Leaves of the resistant red currant (*Ribes sativum* Syme) oxidized 50 percent of the carboxyl-^{14}C and 20 percent of the methylene-^{14}C from the side chain of 2,4-D in one week; the more sensitive black currant (*Ribes nigrum* L.) degraded only 2 percent of this herbicide in the same period (LUCKWILL and LLOYD-JONES 1960 a). A similar relationship was found in apple varieties of varying susceptibility (LUCKWILL and LLOYD-JONES 1960 b).

Galium aparine can degrade the side chain of MCPA and is very tolerant to this herbicide, but introduction of an α-alkyl group, such as methyl, inhibited the degradation and created an effective herbicide to this weed (LEAFE 1962).

A radioactive metabolite of ^{3}H-pyrazon was detected in the shoots of red beet (*Beta vulgaris* L.) tolerant to this herbicide, but there was no evidence for its presence in the shoots of German millet (*Setaria italica* L.) moderately tolerant to pyrazon and tomato (*Lycopersicon esculentum* Mill) susceptible to pyrazon (STEPHENSON and RIES 1967).

Other examples relate to the formation of conjugated compounds of herbicide, which themselves are inactive, in tolerant species of plants. In the case of 2,4-D, sorghum tolerant to the herbicide formed 2,4-D derivative, while this was not formed in cotton susceptible to 2,4-D (MORGAN and HALL 1963). Soybean which is tolerant to amiben can form N-glucosylamiben [N-(3-carboxy-2,5-dichlorophenyl)-gluco-sylamine](COLBY 1966, SWANSON et al. 1966).

V. Mode of action of propanil

Propanil is one of the most miraculous pesticides, showing high selectivity between rice plants and weeds, especially barnyard grass as shown in Table II (MATSUNAKA 1965).

The mode of action of this herbicide seemed to be the inhibition of photosynthesis as with the s-triazines and phenylureas. In fact, propanil can inhibit the Hill reaction, electron transport in photosynthesis, and carbon dioxide-fixation. The Hill reaction by spinach

Table II. *Selectivity of propanil between rice plants and barnyard grass*
(MATSUNAKA 1965)

Plant	Effective conc. (%) sprayed in	
	50 % inhibition of growth	100 % inhibition of growth
rice plants	2.5	> 6.0
barnyard grass	0.07	0.3

chloroplasts using ferricyanide as an oxidant was clearly inhibited by propanil. Half inhibition occurred at a concentration of $3 \times 10^{-6} M$. The herbicidal activity of paraquat was retarded by the spraying of propanil as shown in Figure 3, which meant that the latter had the

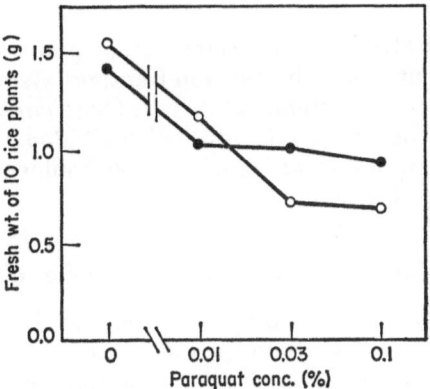

Fig. 3. Retardation of herbicidal activity of paraquat by mixed spraying with propanil (conc. of propanil 0.6 percent): O——O paraquat only, ●——● paraquat plus propanil

same inhibitory action in the photoexcitation of electrons in the chloroplasts as with monuron described above. Carbon dioxide-fixation by both rice plants and barnyard grass (*Echinochloa crusgalli* var. *oryzicola*) was completely inhibited by spraying a practical concentration of propanil, as shown in Figure 4.

However, we can not say that the primary action of propanil on weeds involves only inhibition of photosynthesis because, if so, the herbicidal symptoms should appear slowly. Inhibition of photosynthesis would cause the starvation of carbohydrates, which means death of the plant, and it would take a long time, presumably more than five days. The symptoms from propanil on barnyard grass can be found 24 hours after the spraying. Only this fact shows that propanil has a more drastic action on intracellular components of susceptible weeds.

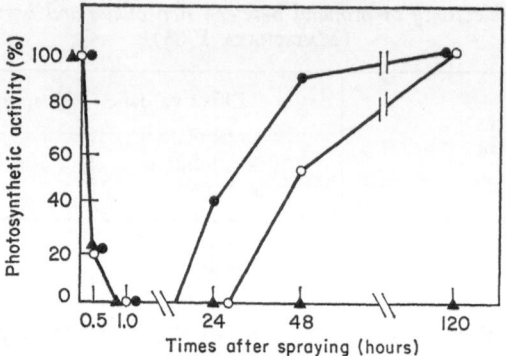

Fig. 4. Inhibition of carbon dioxide-fixation in rice plants and barnyard grass by propanil and the recovery only in rice plants. Propanil concentration in percent: ● rice plant 0.1, ○ rice plant 1.0, ▲ barnyard grass 0.1

ISHIZUKA and MITSUI (1966) showed that propanil can inhibit the respiration of plant roots. The electron transport system between cytochromes b and c in the respiration of yeast (*Saccharomyces cerevisiae*) was also inhibited by propanil, with half inhibition at a concentration of 1.1×10^{-4} M (INOUE *et al.* 1967). These findings seem to be one side of the mode of action of propanil.

VI. Inactivation of propanil by higher plants

If the primary action of propanil is not the inhibition of photosynthesis, there should be found a difference in sensitivity of intracellular components to propanil between rice plant and barnyard grass. From this standpoint, we are now investigating the differences, for instance, in denaturation of intracellular protein, in cellular components, in histological structures, and so on.

On the other hand, in the selectivity of propanil, the inactivation of this anilide compound should not be neglected. As shown by MCRAE *et al.* (1964), ADACHI *et al.* (1966 a), and ISHIZUKA and MITSUI (1966), and more recently by STILL and KUZIRIAN (1967), juice or crude enzyme preparations from rice plants can hydrolyze propanil into 3,4-dichloroaniline and propionic acid. In our experiments, the hydrolyzed product, 3,4-dichloroaniline, showed very low inhibitory activity on the Hill reaction by spinach chloroplasts, where half inhibition occurred at a concentration of $3 \times 10^{-3} M$. This value is much larger than that of $3 \times 10^{-6} M$ in the case of propanil described above.

ADACHI *et al.* (1966 a) sought for the distribution of the inactivation activity in higher plants as shown in Table III. It would be concluded that both rice plant and crabgrass (*Digitaria adscendens*) have high activity while monochoria (*Monochoria vaginalis*) or barnyard grass is very low. ADACHI *et al.* (1966 a) used mature samples,

Table III. *Inactivation of propanil by homogenates of various plants*
(ADACHI *et al.* 1966 a)

Source	Percent hydrolyzed in 24 hours [a]
Monochoria	0.0
Smartweed	5.5
Barnyard grass	7.0
Dayflower	30.0
Crabgrass	58.4
Rice plant	69.6

[a] 100 % = 62.5 μg. of propanil.

but the author found that young crabgrass at age three has the same susceptibility to propanil as with barnyard grass.

On the other hand, ISHIZUKA and MITSUI (1966) had the opinion that even barnyard grass has high hydrolyzing activity for other anilides and that the difference of activity in the case of propanil does not come from the existence or absence of such an anilide-hydrolyzing enzyme, but rather the difference in the specificity of the enzyme on substrates.

Figure 4 also shows the recovery of photosynthetic activity inhibited by propanil in rice plant, but not in barnyard grass. This may become one of the evidences that propanil is degraded *in vivo* by rice plants.

ADACHI *et al.* (1966 b) reported that, in the hydrolysis of propanil into 3,4-dichloroaniline and propionic acid, the optimum pH was at 8.4 (reaction temperature 30° C. and time 24 hours). The homogenate boiled for 30 minutes did not show any activity. STILL and KUZIRIAN (1967) found metabolites other than 3,4-dichloroaniline.

It was a very interesting fact that the inactivation activity was inhibited by the insecticides carbaryl or dipterex, as shown in Table IV.

Table IV. *Effect of pesticides on hydrolysis of propanil by rice plants*
(ADACHI *et al.* 1966 a and b)

Pesticide [a]	Inhibition rate (%)
PMA	14.5
BHC	11.6
Methyl parathion	31.0
Sumithion	38.8
Carbaryl	95.1
Dipterex	100.

[a] Final concentration 10 p.p.m., reaction time 24 hours.

On the other hand, many works showed that severe leaf burn and subsequent yield losses occurred when propanil was mixed with some

insecticides or when they were sprayed just before or after the application of propanil. BOWLING and HUDGINS (1966) reported that insecticides such as aldrin, dieldrin, heptachlor, endosulfan, endrin, or
toxaphene, in combination with propanil, did not increase leaf burn
over that which occurred from propanil alone. Carbaryl, malathion,
phosphamidon, Guthion, or Dylox in combination with propanil increases leaf burn over that which occurred from propanil alone. In
general, they said, increased leaf burn resulted in decreased yields
of rough rice.

These field data could be explained as follows: usually propanil
penetrated into rice plant tissues would inhibit photosynthesis or act
on the intracellular components, but these inhibitions might be recovered in one or two days because the chemical could be inactivated
by the hydrolyzing enzyme. In the presence of some kinds of insecticides, such as carbaryl, the inactivation of propanil by the enzyme is
inhibited (as shown in Table IV) to cause leaf burn.

Table V. *Common or trademark and chemical names of
pesticides mentioned in text*

Aldrin	1,2,3,4,10,10-hexachloro-1,4,4a,5,8,8a-hexahydro-1,4-*endo*, *exo*-5,8-dimethanonaphthalene
Amiben	3-amino-2,5-dichlorobenzoic acid
Atrazine	2-chloro-4-ethylamino-6-isopropylamino-*s*-triazine
BHC	1,2,3,4,5,6-hexachlorocyclohexane
Carbaryl	1-naphthyl *N*-methylcarbamate
CNP (MO-338)	2,4,6-trichloro-4'-nitrodiphenylether
2,4-D	2,4-dichlorophenoxyacetic acid
2,4-DB	4-(2,4-dichlorophenoxy)butyric acid
Dieldrin	1,2,3,4,10,10-hexachloro-6,7-epoxy-1,4,4a,5,6,7,8,8a-octahydro-1,4-*endo*, *exo*-5,8-dimethanonaphthalene
Dipterex	*O,O*-dimethyl-2,2,2-trichloro-1-hydroxyethyl phosphonate
Diquat	1,1'-ethylene-2,2'-bipyridylium dibromide
Dylox	*O,O*-dimethyl-2,2,2-trichloro-1-hydroxyethyl phosphonate
Endosulfan	6,7,8,9,10,10-hexachloro-1,5,5a,6,9,9a-hexahydro-6,9-methano-2,4,3-benzodioxathiepin-3-oxide
Endrin	1,2,3,4,10,10-hexachloro-6,7-epoxy-1,4,4a,5,6,7,8,8a-octahydro-1,4-*endo*, *endo*-5,8-dimethanonaphthalene
Guthion	*O,O*-dimethyl-S-4-oxo-1,2,3-benzotriazine-3(4*H*)-ylmethyl phosphorodithioate
HE-314	3-methyl-4'-nitrodiphenylether
Heptachlor	1,4,5,6,7,8,8-heptachloro-3a,4,7,7a-tetrahydro-4,7-methanoindene
HW-40187	3,5-dimethyl-4'-nitrodiphenylether
Hydroxysimazine	4,6-bis-(ethylamono)-2-hydroxy-*s*-triazine
Malathion	*O,O*-dimethyl-S-(1,2-bis(ethoxycarbonyl)ethyl)-phosphorodithioate
MCPA	4-chloro-2-methylphenoxyacetic acid
MCPB	4-(4-chloro-2-methylphenoxy)butyric acid
MCPES	sodium 4-chloro-2-methylphenoxyethyl sulfate
Methyl parathion	*O,O*-dimethyl-*O-p*-nitrophenyl phosphorothioate
Monuron	3-(*p*-chlorophenyl)-1,1-dimethylurea

Table V. (continued)

Nitrofen (NIP, TOK)	2,4-dichloro-4'-nitrodiphenylether
Paraxon	O,O-diethyl-O-p-nitrophenyl phosphate
Paraquat	1,1'-dimethyl-4,4'-bipyridylium(dichloride)
Parathion	O,O-diethyl-O-p-nitrophenyl phosphorothioate
Phosphamidon	O,O-dimethyl-O-(2-chloro-2-diethylcarbamoyl-1-methylvinyl) phosphate
PMA	phenylmercuric acetate
Propanil	3',4'-dichloropropionanilide
Propazine	2-chloro-4,6-bis(isopropylamino)-s-triazine
Pyrazone	5-amino-4-chloro-2-phenyl-3(2H)-pyridazinone
Sesone	sodium 2,4-dichlorophenoxyethyl sulfate
Simazine	2-chloro-4,6-bis(ethylamino)-s-triazine
Sumioxon	O,O-dimethyl-O-(3-methyl-4-nitrophenyl) phosphate
Sumithion	O,O-dimethyl-O-(3-methyl-4-nitrophenyl) phosphorothioate
2,4,5-TES	sodium 2,4,5-trichlorphenoxyethyl sulfate
Toxaphene	octachlorocamphene (67 to 69 % chlorine)

The author recently found paraoxon or sumioxon showed higher activity both in the inhibition of propanil hydrolysis by rice plant homogenate and in the synergistic herbicidal activity in rice plant with propanil, than either parathion or sumithion. BHC, having no inhibitory activity to acetylcholineesterase, has no joint action with propanil in rice plants (MATSUNAKA 1968b).

As mentioned above, matured crabgrass has high activity for the inactivation of propanil, so the usual dosage of propanil can hardly control matured crabgrass. Fortunately citrus plants are very tolerant to propanil, and a combination of propanil and carbaryl can be effectively used for crabgrass control in citrus orchards (HISADA 1968). This combined herbicide is called WYDAC (trade name) in Japan and was created by the utilization of the above-mentioned mechanisms. In this case carbaryl may be called a synergist to propanil.

As a conclusion, one side of the selectivity of propanil would be illustrated as shown in Figure 5.

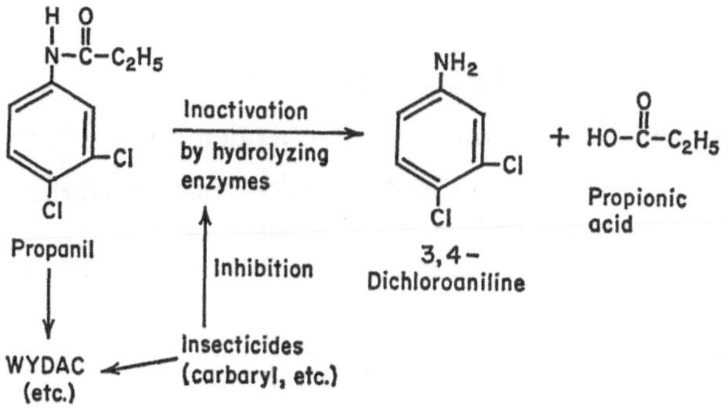

Fig. 5. Inactivation of propanil in tolerant plants

Summary

Examples of activation and inactivation of herbicides by higher plants were reviewed. Diphenylether herbicides can be classified into two groups. One group having ortho-substituent(s) on one phenyl ring requires the help of light to kill weeds. The activation mechanism by light seems to be a photo-biochemical process. The other group is active even in dark.

Propanil (3',4'-dichloropropionanilide) is a miraculous herbicide which shows an excellent selectivity between rice plants and weeds. An inactivation process by rice plants, hydrolysis into 3,4-dichloro-aniline and propionic acid, was inhibited by some kinds of insecticides having inhibitory activity on acetylcholineesterase. Propanil combined with carbaryl is being used practically for crabgrass control in citrus orchards in Japan utilizing the above-described mechanism.

Résumé *

Activation et inactivation des herbicides par les plantes plus hautes

Des exemples d'activation et d'inactivation des herbicides par les plantes plus grandes ont été recherchés.

Les herbicides, type diphényl éther, peuvent être classés en deux groupes. L'un d'eux ayant des substituants en ortho sur le cycle ben-zénique ne détruit les mauvaises herbes qu'à la lumière. Le mécanisme d'activation par la lumière paraît être un processus photo-biochimique. L'autre groupe est actif même dans l'obscurité.

Le propanil (3',4'-dichloropropionanilide) est un herbicide miracu-leux dont la sélectivité est excellente entre les plants de riz et les mauvaises herbes. Un processus d'inactivation par les plants de riz—hydrolyse en 3,4-dichloroaniline et acide propionique—est inhibé par certains insecticides antiacétylcholinestérasiques. Le propanil combiné au carbaryl est ainsi utilisé contre les mauvaises herbes dans les orangeraies au Japon grâce au mécanisme précédent.

Zusammenfassung *

Aktivierung und Inaktivierung von Herbiziden durch höhere Pflanzen

Beispiele von Aktivierung und Inaktivierung von Herbiziden durch höhere Pflanzen weren kritisch besprochen.

Diphenylätherherbizide können in zwei Gruppen eingeteilt werden.

---

* Traduit par R. MESTRES.
* Übersetzt von A. SCHUMANN.

Die eine Gruppe, welche Ortho-Substituenten (ein oder mehrere) an einem Phenylring hat, benötigt die Hilfe von Licht, um Unkraut abzutöten. Der Aktivierungsmechanismus durch Licht scheint ein photobiochemischer Prozess zu sein. Die andere Gruppe ist auch im Dunkeln aktiv.

Propanil [N-(3,4-Dichlorphenyl)-propionamid] ist ein wunderwirkendes Herbizid, welches eine ausgezeichnete Selektivität zwischen Reispflanzen und Unkraut zeigt. Ein Inaktivierungsprozess durch Reispflanzen—Hydrolyse zu 3,4-Dichloranilin und Propionsäure—wurde durch einige Arten von Insektiziden verhindert, welche die Hemmungsaktivität auf Anticholinesterase besitzen. Propanil, verbunden mit Carbaryl, wird in der Praxis zur "Crab"-graskontrolle in Zitrusplantagen in Japan gebraucht, wobei der oben beschriebene Mechanismus ausgenutzt wird.

References

ADACHI, M.: Studies on the selective herbicidal activity of 3,4-dichloropropionanilide. I. Penetration into plants and degradative detoxication.* Nôyakuseisan-gijyutsu (Pesticide and Technique) No. 14, 19 (1966 a).
—— Studies on the selective herbicidal activity of 3,4-dichloropropionanilide. II. Propanil degrading enzyme of rice plant.* Nôyaku-seisan-gijyutsu (Pesticide and Technique) No. 15, 11 (1966 b).
ANDERSEN, R. N.: Differential response of corn inbreds to simazine and atrazine. Weeds 12, 60 (1964).
AUDUS, L. J.: Fate of sodium 2,4-dichlorophenoxy-ethyl-sulphate in the soil. Nature 170, 886 (1952).
BOWLING, C. C., and H. R. HUDGINS: The effect of insecticides on the selectivity of propanil on rice. Weeds 14, 94 (1966).
CASTELFRANCO, P., and M. S. BROWN: Purification and properties of the simazine-resistant factor of Zea Mays. Weeds 10, 131 (1962).
COLBY, S. R.: The mechanism of selectivity of amiben. Weeds 14, 197 (1966).
FURUYA, S., and M. ARAI: Studies on some properties of diphenylether type herbicides.* Zasso-kenkyu [Weed Research (Tokyo)] No. 5, 99 (1966).
GOOD, N. E.: Inhibition of the Hill reaction. Plant Physiol. 36, 788 (1961).
GRACE, N. H.: Physiological activity of a series of naphthyl acids. Can. J. Research 17, 247 (1939).
HAMILTON, R. H., and D. E. MORELAND: Simazine degradation by corn seedlings. Science 135, 373 (1962).
HISADA, T.: Selective contact herbicide for citrus orchard. Proc. Asian-Pacific Weed Control Tech. Interchange Meeting (1968).
INOUE, Y.: Inhibition of respiration of yeast by photosynthesis inhibiting herbicides. Agr. Biol. Chem. 31, 422 (1967).
ISHIZUKA, K., and S. MITSUI: Activation or inactivation mechanisms of biological active compounds in higher plants. II. On anilide degrading enzyme.* Abstr. Ann. Meeting Agr. Chem. Soc. Japan, p. 62 (1966).
LEAF, E. L.: Metabolism and selectivity of plant-growth regulator herbicides. Nature 193, 485 (1962).

---

* In Japanese.

LUCKWILL, L. C., and C. P. LLOYD-JONES: Metabolism of plant growth regulators. I. 2,4-Dichlorophenoxyacetic acid in leaves of red and black currant. Ann. Applied Biol. 48, 613 (1960 a).
—— —— Metabolism of plant growth regulators. II. decarboxylation of 2,4-dichlorophenoxyacetic acid in leaves of apple and strawberry. Ann. Applied Biol. 48, 626 (1960 b).
MATSUNAKA, S.: Development of antidotes for herbicides.* Zasso-kenkyu [Weed Research (Tokyo)] No. 2, 5 (1963).
—— Selective toxicity of herbicides.* In: Shinnôyaku-sôseihô (Methods for development of new pesticides) (R. YAMAMOTO and T. NOGUCHI, eds.) Tokyo: Nankodô (1965).
—— Mode of action of diphenylether herbicides and synergistic action of insecticides on propanil. 155th Nat. Meeting, Amer. Chem. Soc., San Francisco (April 1968a).
—— Propanil hydrolysis: inhibition in rice plants by insecticides. Science 160, 1360 (1968b).
—— and K. INADA: Mode of action of diphenylether herbicides. I. Light for herbicidal activity.* Abstr. Ann. Meeting Agr. Chem. Soc. Japan, p. 47 (1967).
McWHORTER, C. G., and J. T. HOLSTON, JR.: Phytotoxicity of s-triazine herbicides to corn and weeds as related to structural differences. Weeds 9, 592 (1961).
MEES, G. C.: Experiments on the herbicidal action of 1,1'-ethylene-2,2'-dipyridylium dibromide. Ann. Applied Biol. 48, 601 (1960).
MORELAND, D. E.: Mechanism of action of herbicides. Ann. Rev. Plant Physiol. 18, 365 (1967).
MORGAN, P. W., and W. C. HALL: Metabolism of 2,4-D by cotton and grain sorghum. Weeds 11, 130 (1963).
STEPHENSON, G. R., and S. K. RIES: The movement and metabolism of pyrazon in tolerant and susceptible species. Weed Research 7, 51 (1967).
STILL, C. C., and O. KUZIRIAN: Enzyme detoxication of 3', 4'-dichloropropionanilide in rice and barnyard grass, a factor in herbicide selectivity. Nature 216, 799 (1967).
SWANSON, C. R.: Amiben metabolism in plants. I. Isolation and identification of an N-glucosyl complex. Weeds 14, 319 (1966).
WAIN, R. L.: A new approach to selective weed control. Ann. Applied Biol. 42, 151 (1955).
—— The behaviour of herbicides in the plant in relation to selectivity. In: The physiology and biochemistry of herbicides (L. J. AUDUS, ed.). New York-London: Academic Press (1964).

Role of RNA metabolism
in the action of auxin-herbicides

by

J. B. Hanson [*] and F. W. Slife [*]

Contents

I. Introduction

The auxin-herbicides are those herbicidal compounds which in low concentration exhibit the growth promoting properties of the native auxin, indoleacetic acid (IAA). The type compounds here are 2,4-dichlorophenoxyacetic acid (2,4-D) and 2-methyl-4-chlorophenoxy-acetic acid (MCPA). In the early 1940's the selective herbicidal properties of these synthetic growth regulators became known. Their subsequent widespread use for the control of dicotyledonous weeds in graminaceous cereals provided the impetus for development of the herbicide industry. In addition to the several phenoxy-herbicides, substituted benzoic and picolinic acids fall into the auxin-herbicide classification. Discussion here will largely center on 2,4-D, the best studied auxin-herbicide.

In the two decades since the discovery of 2,4-D as a selective herbicide, hundreds of papers have dealt with various aspects of its mode of action and selectivity. Recent summaries are to be found in the textbook of CRAFTS (1961), the review of HILTON et al. (1963), the compilation of 17 review papers in the volume edited by AUDUS (1964), and in the reviews by PENNER and ASHTON (1966) and by MORELAND (1967). With respect to mode of action, the following generalizations can be made:

1. Selectivity is not primarily due to differential absorption, trans-

---

[*] Department of Agronomy, University of Illinois, Urbana.

location or catabolism. Failure to obtain or to maintain adequate internal concentrations of auxin herbicide may occasionally enter as a secondary factor in resistance. 2,4-D is commonly applied to the foliage as sprays or dusts, and though there are variations in absorption and translocation which sometimes correlate with susceptibility these will not consistently explain selectivity. Neither is selectivity established to lie with differential inactivation of the compound, although degradation, binding, and complex formation are known to occur, and a few suggestive correlations have been found. Primarily, the resistant species (largely, but not exclusively, grasses) seem to lack responsive sites or require very high concentrations of herbicide at these sites. Resistance in grasses correlates best with lack of a vascular cambium and pericycle. When injury does occur in grasses it is associated with the intercalary meristems or floral primordia. Herbicidal action thus seems to lie with an active or latent potential for cell division.

2. 2,4-D does not appear to act as a simple toxin, inhibitor, or uncoupler. At high concentrations certain isolated enzymes or enzyme systems can be inhibited, but there is no firm evidence that 2,4-D acts in vivo by directly interfering with intermediary metabolism, respiration or photosynthesis. However, indirect effects are certainly produced since physiological processes such as photosynthesis and ion absorption are inhibited. Such inhibitions seem to be linked to the aberrant growth of susceptible species. A capacity for growth is somehow implicated since young, vigorous, high nitrogen plants with a high growth potential are most susceptible.

3. The auxin-herbicides appear to be acting as auxins. High concentrations of IAA, the native auxin, can produce growth abnormalities very similar to 2,4-D. The failure of IAA as a herbicide probably lies with destruction of excess molecules by indoleacetic acid oxidase. 2,4-D is effective because it is degraded only slowly.

The most positive statements here are those of VAN OVERBEEK (1964): 2,4-D is a persistent synthetic auxin which saturates the cells upsetting the normal balance required for orderly growth and differentiation. The abnormal growth results from hormonal imbalance and it kills like cancer. VAN OVERBEEK (1964) along with HABER (1962) and KEY et al. (1966) suggests that an auxin-cytokinin imbalance is probably implicated. SHANNON et al. (1964) suggest that death occurs because auxin-induced abnormalities in nucleic acid and protein synthesis preclude normal cell development and function.

4. Both inhibition and promotion of growth are involved in the response of a susceptible plant to an auxin-herbicide. Apical meristems of root and shoot are inhibited in cell division and elongation, although cells of the elongation zone often swell laterally. Expansion and differentiation of young leaves is impaired. On the other hand, cells of the stem and taproot show positive responses; first they swell, and

then in the vascular region begin to divide producing root primordia or callus.

These generalizations provide a brief background for the major question to be dealt with here: Why does the plant die?

We propose that the immediate cause of death is physiological disfunction of leaf and root brought about by abnormal growth. In turn, the abnormal growth is believed to be based in an abnormal nucleic acid metabolism.

II. Abnormal RNA metabolism

There is no question but that auxins can alter nucleic acid metabolism. SILBERGER and SKOOG (1953) working with tobacco pith callus made the initial observation of an increase in RNA and DNA with application of IAA. At high concentrations of auxin a depression of nucleic acid synthesis set in. REBSTOCK et al. (1954) found 2,4-D to cause a large increase in the nucleic acid phosphorus of cranberry bean stems, with a concurrent decline in the leaves. WEST et al. (1960) confirmed this observation using cucumber plants. Remobilization of phosphorus and nitrogen from leaves to stem has been observed often in 2,4-D treated plants (PENNER and ASHTON 1966).

The increased RNA in basal stems of 2,4-D treated seedlings is largely ribosomal (WEST et al. 1960, CHRISPEELS and HANSON 1962, KEY et al. 1966). This is equally true of excised, 2,4-D treated tissue (KEY and SHANNON 1964). However, elongation of young stem tissues proves not to depend on ribosomal RNA synthesis, but on the synthesis of a DNA-like RNA, probably messenger RNA (KEY and INGLE 1964). Thus, experiments dealing with low, growth promoting concentrations of 2,4-D may not be fully germane to the herbicidal action of the compound. When herbicidal concentrations are used there is a massive production of ribosomes and other RNA in the affected axis. Excised tissue, which ordinarily degrades ribosomal RNA as it grows, is blocked in its RNA catabolism by high concentrations of 2,4-D (BASLER and NAKAZAWA 1961, KEY 1963, SHANNON et al. 1964). Ribonuclease synthesis, which accompanies cell expansion, is also blocked by high concentrations of 2,4-D (SHANNON et al. 1964).

The most complete description of nucleic acid synthesis as correlated with aberrant growth in 2,4-D treated seedlings is that of KEY et al. (1966). Sprays of 2,4-D on soybean seedlings suppressed synthesis of DNA, RNA and protein, growth, and cell division in the apical region of the hypocotyl up to 48 hours. On the other hand, more basal tissues swelled, began synthesizing nucleic acids and protein, and by 12 hours had initiated cell division. Synthesis of RNA slightly preceded protein synthesis, DNA synthesis and cell division. In 48 hours the ribosomal RNA content of the hypocotyl trebled, with lesser increases in the soluble and large particle fractions. Both in terms of sedimentation constants and base composition the RNA from 2,4-D treated tissue was like that of controls.

This work of KEY et al. (1966) is of exceptional value in that it shows nucleic acid metabolism to be suppressed where growth is suppressed, and to be accelerated where growth is accelerated (albeit abnormally). Other work of KEY and his associates (see above) with low, growth-promoting concentrations of 2,4-D lend strong support to the hypothesis that the correlation is not fortuitous. The native auxin, IAA, also appears to regulate growth through DNA directed RNA and protein synthesis (NOODEN and THIMANN 1963). FAN and MACLACHLAN (1967) report IAA to stimulate RNA synthesis when applied to decapitated pea seedlings.

MALHOTRA (1966) investigated the relationship between species sensitivity to the auxin herbicide Picloram (4-amino,-3,5,6-trichloro-picolinic acid) and the induction of RNA synthesis. He also determined ribonuclease levels. For the five species of graded sensitivity used—barley (very resistant), wheat, corn, cucumber, and soybean (very sensitive)—he found a good positive correlation between induction of RNA synthesis and herbicidal sensitivity. Conversely, there was an inverse correlation between these parameters and bound ribonuclease. Too little is known of the metabolic role of ribonucleases to judge if this latter correlation has physiological significance. It is clear, however, that the capacity to die under the influence of the herbicide is associated in some fashion with a responsive nucleic acid metabolism.

CHRISPEELS and HANSON (1962) drew on the reported affects of auxin in producing swollen nuclei and nucleoli (BAUSOR 1942, DOXEY and RHODES 1949) to propose that 2,4-D acted by renewing nuclear activity, inducing RNA production and cell division. Isolated nuclei do form RNA in response to IAA (ROYCHOUDHURY et al. 1965, MAHESH-WARI et al. 1966).

O'BRIEN et al. (1968) investigated chromatin preparations from control and 2,4-D treated soybean hypocotyls for RNA polymerase activity. The chromatin had an active polymerase but it did not respond to direct addition of 2,4-D. However, two to three hours after spraying seedlings with 2,4-D an increase in polymerase activity could be detected, and by 12 hours the activity was two-to-four fold higher than in control seedlings. Evidently the auxin is not a cofactor for RNA polymerase, but does induce its synthesis.

In summary, evidence provided by a number of investigators suggests that herbicidal levels of the synthetic auxins interfere with the normal nucleic acid metabolism of susceptible species. In apical meristems, nucleic acid metabolism and growth are "frozen." Basal tissues respond with massive nucleic acid synthesis followed by protein synthesis and cell division. The resistant grasses respond only weakly or not at all, a property associated with high levels of ribonuclease. There is still no clue as to the fundamental biochemical act(s) initiated by the auxin herbicides. It seems likely that certain cells are "conditioned" to respond and others are not. Conditioning involves the meristematic state or the potential to revert to the meristematic state.

III. The cause of death

An early witticism was that 2,4-D caused plants to "grow themselves to death," or more exactly, proliferate themselves to death (AUDUS 1959). Field and laboratory observations of 2,4-D treated plants support such opinions: 2,4-D suppresses normal cell division and elongation at growing points but induces swelling, abnormal division, and tissue proliferation in the more mature tissues. Production and development of young leaves are impaired. We suggested that the plant succumbs because the abnormal growth reduces photosynthesis and normal phloem transport, with roots becoming disfunctional in salt and water absorption (HANSON and SLIFE 1961). As noted above, VAN OVERBEEK (1964) agrees.

Oddly enough, despite this early and continuing opinion on 2,4-D action by way of aberrant growth, much of the search has been for a direct toxic action which resistant plants could evade by binding or metabolizing the herbicide. The aberrant growth hypothesis is still far from widely established.

Recently, our laboratory (CARDENAS et al. 1968) undertook to determine if death was actually coordinated with abnormal growth and failure of physiological functions, such as those reported for ion absorption and photosynthesis (see reviews of WORT 1964 a and PENNER and ASHTON 1966). A single mature leaf of three week cocklebur plants was treated with a drop containing 72 μg. of 2,4-D. Over the 10-day period leading to death the plants were analyzed for growth (dry and fresh weight), nitrogen, protein, RNA, DNA, photosynthesis, ion accumulation, and translocation.

Between treatment and two days there was an increase in the weight of the plants, but this was largely in the swelling stem-taproot axis, with very little gain in root and leaf tissue. Gains in leaf tissue were depressed below control levels even in the first day. Between two and seven days the treated plants did not gain in weight (the controls nearly doubled), but the axis did. The growth of the swelling axis proved to be at expense of the leaves which declined in weight. Analyses for nitrogen, protein, RNA, and DNA showed the well-known phenomenon mentioned above of nitrogenous materials declining in the leaves and rising in the stem. The bulk of the extra nucleic acid synthesis, however, had occurred during the first two days.

No new leaves were produced in treated plants. On the tenth day the plants collapsed and withered.

There were thus three phases in the growth toward death: a rapid induction phase (0 to 2 days) when the plant was still capable of net growth; a redistribution phase (2 to 7 days) when the aberrant growth of the axis was at the expense of leaf tissue; a final senescence-collapse phase (7 to 10 days) which could not be described in simple terms of growth or nitrogen metabolism.

It was interesting to find that both ion absorption and photosynthesis were initially accelerated, but the stimulation quickly passed to a strong inhibition by the second day, the point at which the redistribution growth set in. From the beginning the photosynthate and ions were directed to the swelling axis. Roots received very little photosynthate after the first day. Conversely, the leaves received little phosphate or potassium (Rb^{86}-labeled), and did not even retain the photosynthate they manufactured. The stem-taproot axis clearly formed a dominant sink.

Initial stimulation of ion absorption by 2,4-D had been observed earlier by COOKE (1957), while TURNER and BIDWELL (1965) found the initial stimulation of photosynthesis with IAA. If some way could be found to sustain this transient stimulation we might be using 2,4-D for increasing crop yields instead of for killing weeds; WORT (1964 b) has discussed cases of known yield increase.

But to return to the central question: Why does the plant die?

It is our opinion that the plant dies because it fails to be autotrophic. Leaves and roots fail in their physiological function. The failure can be laid to "freezing" of the apical meristems (so that new leaf and root production is curtailed) plus an induced senescence of existing leaves and roots. Induction of senescence follows the demands made by the rapidly proliferating axis. An appropriate analogy here is with the senescence induced by rapidly growing fruits. In a sense the plant has developed a "cancer," as VAN OVERBEEK (1964 a) puts it, and the nutritional demands are unfortunately filled at the expense of the essential vegetative organs. As is usual with tumorous growth an aberrant nucleic acid metabolism is involved. However, we know next to nothing about metabolic events leading to the burst of nucleic acid synthesis, and nothing at all as to why such synthesis should dominate the rest of the plant, signaling premature leaf senescence to nourish the cellular proliferation. The nucleus is affected very early, as is the chromatin-bound RNA polymerase, but this only tells us where to start looking for relevant biochemical events. These may or may not be the same events set into operation by low, growth-promoting concentrations of auxin.

HOLM and ABELES (1968) have very recently added a new dimension to investigations of the biochemistry of 2,4-D action by showing that the stem swelling and nucleic acid synthesis in 2,4-D treated soybeans can be attributed to ethylene production. Ethylene production in response to 2,4-D application was suggested some years ago as a governing factor in the aberrant growth (MORGAN and HALL 1962). However, ethylene in itself is not herbicidal and further work will be needed to determine exactly what relationships exist between auxins, ethylene, and other hormones. As noted by HOLM and ABELES (1968) some factor supplied by the roots may be implicated. If cytokinins are involved in the 2,4-D response, as supposed by HABER (1962), VAN OVERBEEK (1964 a) and KEY et al. (1966) then the cytokinin exudation from bleeding roots (KENDE 1965) suggests one reason why roots may be needed.

Summary

The herbicidal action of the auxin-herbicides such as 2,4-D can best be attributed to the aberrant growth induced in susceptible species. Abnormal nucleic acid metabolism underlies the aberrant growth. The stem root axis forms a dominant metabolic sink, leading to senescence and physiological disfunction of leaves and roots. In consequence, the plant fails to be autotrophic and dies. Although 2,4-D will induce RNA polymerase in treated tissue, there is no evidence for a direct effect of the auxin on polymerase activity. Recent work suggests other hormones, including ethylene, are involved.

Résumé *

Rôle du métabolisme de l'A R N dans l'action des herbicides du type auxinique

L'action herbicide des herbicides du type auxinique tel que le 2,4-D peut le mieux être attribuée à la croissance aberrante induite chez les espèces sensibles. Des modifications dans le métabolisme de l'acide nucleique déterminent la croissance anormale. L'axe racine-tige forme un milieu de transit dominant pour les métabolites, conduisant à la senescence et à des perturbations dans l'activité physiologique des feuilles et des racines. En conséquence, la plante perd son autotrophie et meurt. Bien que le 2,4-D induise la RNA polymerase dans les tissus traités, il ne semble pas y avoir un effet direct de l'auxine sur l'activité de la polymerase. Un travail récent suggère que d'autres hormones, et même l'éthylène sont en cause.

Zusammenfassung **

Die Rolle des RNA-Metabolismus bei der Wirkung von Auxin-Herbiziden

Die herbizide Wirkung von Auxin-Herbiziden wie z. B. 2,4-D kann am besten dem abweichenden Wachstum zugeschrieben werden, das

---

* Traduit par R. MESTRES.
** Übersetzt von A. SCHUMANN.

in empfindlichen Arten veranlasst wird. Dem abweichenden Wachstum liegt ein abnormaler Nukleinsäuremetabolismus zugrunde. Die Hauptwurzelachse bildet einen vorherrschenden Metabolitabzugskanal, dies führt zu Altern und physiologischer Disfunktion der Blätter und Wurzeln. In der Folge versagt die Pflanze, autotroph zu sein und stirbt ab. Obwohl 2,4-D die RNA-Polymerase in behandeltem Gewebe induziert, gibt es keinen Beweis für die direkte Wirkung des Auxins auf die Polymeraseaktivität. Neuere Arbeiten schliessen darauf, dass andere Hormone, einschliesslich Aethylen, darin verwickelt sind.

References

AUDUS, L. J.: Plant growth substances, 2ed. London: Leonard Hill (1959).
—— (Ed.): The physiology and biochemistry of herbicides. New York: Academic Press (1964).
BASLER, E., and K. NAKAZAWA: Effects of 2,4-D on nucleic acids of cotton cotyledon tissue. Bot. Gaz. 122, 228 (1961).
BAUSOR, S. C.: Interrelation of organic materials in the growth substance response. Bot. Gaz. 103, 710 (1942).
CARDENAS, J., F. W. SLIFE, J. B. HANSON, and H. BUTLER: Physiological changes accompanying the death of cocklebur plants treated with 2,4-D. Weed Science 16, 96 (1968).
CHRISPEELS, M. J., and J. B. HANSON: The increase in ribonucleic acid content of cytoplasmic particles of soybean hypocotyl induced by 2,4-dichlorophenoxyacetic acid. Weeds 10, 123 (1962).
COOKE, A. R.: Influence of 2,4-D on the uptake of minerals from the soil. Weeds 5, 25 (1957).
CRAFTS, A. S.: The chemistry and mode of action of herbicides. New York: Interscience (1961).
DOXEY, D., and A. RHODES: The effect of plant growth regulator 4-chloro-2 methylphenoxyacetic acid on mitosis in the onion. Ann. Bot. 13, 105 (1949).
FAN, D. G., and G. A. MACLACHLAN: Massive synthesis of ribonucleic acid and cellulose in the pea epicotyl in response to indoleacetic acid, with and without concurrent cell division. Plant Physiol. 42, 1114 (1967).
HABER, A. H.: Effects of indoleacetic acid on growth without mitosis and on mitotic activity in the absence of growth by expansion. Plant Physiol. 37, 18 (1962).
HANSON, J. B., and F. W. SLIFE: How does 2,4-D kill a plant? Illinois Research 3, 3 (1961).
HILTON, J. L., L. L. JANSEN, and H. M. HULL: Mechanisms of herbicidal action. Ann. Rev. Plant Physiol. 14, 353 (1963).
HOLM, R. E., and F. B. ABELES: The role of ethylene in 2,4-D-induced growth inhibition. Planta 78, 293 (1968).
KENDE, H.: Kinetin-like factors in the root exudate of sunflowers. Proc. Nat. Acad. Sci. U.S. 53, 1302 (1965).
KEY, J. L.: 2,4-D induced changes in ribonucleic acid metabolism in excised corn mesocotyl. Weeds 11, 177 (1963).
——, and J. INGLE: Requirement for the synthesis of DNA-like RNA for growth of excised plant tissue. Proc. Nat. Acad. Sci. U.S. 52, 1382 (1964).
——, and J. C. SHANNON: Enchancement by auxin of ribonucleic acid synthesis in excised soybean hypocotyl tissue. Plant Physiol. 39, 360 (1964).
——, C. Y. LIU, E. M. GIFFORD, JR., and R. DENGLER: Relation of 2,4-D-induced

growth aberrations to changes in nucleic acid metabolism in soybean seedlings. Bot. Gaz. 127, 87 (1966).

MAHESHWARI, S. C., S. GUHA, and S. GUPTA: The effect of IAA on the incorporation of ^{32}P orthophosphate and ^{14}C adenine into plant nuclei in vitro. Biochim. Biophys. Acta 117, 470 (1966).

MALHOTRA, S. S.: Aberrations of the nucleic acid metabolism of plants induced by 4-amino-3,5,6-trichloropicolinic acid. Ph.D. thesis, Univ. of Ill. (1966).

MORELAND, D. E.: Mechanisms of action of herbicides. Ann. Rev. Plant Physiol. 18, 365 (1967).

MORGAN, P. W., and W. C. HALL: Effect of 2,4-dichlorophenoxyacetic acid on the production of ethylene by cotton and grain sorghum. Physiol. Plantarum 15, 420 (1962).

NOODEN, L. D., and K. V. THIMANN: Evidence for a requirement for protein synthesis for auxin-induced cell enlargement. Proc. Nat. Acad. Sci. U.S. 50, 194 (1963).

O'BRIEN, T. J., B. C. JARVIS, J. H. CHERRY, and J. B. HANSON: The effect of 2,4-D on RNA synthesis by soybean hypocotyl chromatin. In: Biochemistry and physiology of plant growth substances. F. WIGHTMAN and G. SETTERFIELDS, eds. Ottawa: Runge Press (in press).

REBSTOCK, T. L., C. L. HAMNER, and H. M. SELL: The influence of 2,4-dichlorophenoxyacetic acid on the phosphorous metabolism of cranberry bean plants. Plant Physiol. 29, 490 (1954).

ROYCHOUDHURY, R., A. DATTA, and S. P. SEN: The mechanism of action of plant growth substances: The role of nuclear RNA in growth substance action. Biochim. Biophys. Acta 107, 346 (1965).

SHANNON, J. C., J. B. HANSON, and C. M. WILSON: Ribonuclease levels in the mesocotyl tissue of Zea mays as a function of 2,4-dichlorophenoxyacetic acid application. Plant Physiol. 39, 804 (1964).

SILBERGER, J., and F. SKOOG: Changes induced by indoleacetic acid in nucleic acid contents and growth of tobacco pith tissue. Science 118, 443 (1953).

TURNER, W. B., and R. G. S. BIDWELL: Rates of photosynthesis in attached and detached bean leaves and the effect of spraying with IAA solution. Plant Physiol. 40, 446 (1965).

VAN OVERBEEK, J.: Survey of mechanisms of herbicide action. In: Physiology and biochemistry of herbicides. L. J. Audus, ed., pp. 387-400. New York: Academic Press (1964).

WEST, S. H., J. B. HANSON, and J. L. KEY: Effect of 2,4-D on nucleic acid and protein content of seedling tissue. Weeds 8, 333 (1060).

WORT, D. J.: Effects of herbicides on plant composition and metabolism. In: Physiology and biochemistry of herbicides. L. J. Audus, ed., pp. 291-334. New York: Academic Press (1964 a).

—— Responses of plants to sublethal concentrations of 2,4-D without and with added minerals. In: Physiology and biochemistry of herbicides. L. J. Audus, ed., pp. 335-342. New York: Academic Press (1964 b).

Mode-of-action of
photosynthesis inhibitor herbicides

by
GUNTER ZWEIG [*]

Contents

I. Introduction

In the commercial development of herbicides during the past 25 years, a relatively few number of classes of chemical compounds have been introduced. Among these are derivatives of halogenated benzoic-, phenoxyalkyl acids, thiocarbamates, aminotriazole, and a series of photosynthesis inhibitors.

The substituted ureas, anilides, uracil, and benznitriles seem to inhibit the primary reaction of photosynthesis—the photolytic decomposition of water—thus stopping the carbon reduction cycle. Although there seems to be some correlation between hydrogen bonding and inhibition of the Hill reaction (oxygen evolution) for the substituted ureas and anilides, no such correlation could be demonstrated for the symmetrical triazines (GOOD 1961). GOOD's hydrogen bonding experiments were carried out under non-physiological conditions by measuring the infrared absorption peak in chloroform solution in the presence of herbicide and acetone (GOOD 1961).

II. Hill reaction inhibitors

A few years ago we observed (ZWEIG *et al.* 1963) that all of the photosynthesis-inhibiting herbicides at a concentration at which they

---

[*] Life Sciences Division, Syracuse University Research Corporation, Syracuse, New York 13210.

totally inhibited oxygen evolution by illuminated *Chlorella*, caused a corresponding increase in fluorescence. The red fluorescence of chlorophyll solutions in ether or acetone by illumination with blue light is a well-known observation. An excellent correlation between oxygen evolution and fluorescence increase was established as illustrated in Table I. The one exception to this general rule seemed to be cyanide ion

Table I. pI_{50} and pF_{50} values of photosynthesis inhibitors (ZWEIG et al. 1963)

Compound	Chlorella p.		Hill reaction- isolated chloroplasts
	$pI_{50}{}^a$	$pF_{50}{}^b$	pI_{50}
Atrazine	6.20	6.52	6.6
Simazine	5.73	6.10	6.4
Monuron	6.15	6.38	6.3
Diuron	6.77	7.92	7.5
Dicryl	6.55	7.07	6.7
KCN	3.6	< 1.87	–

a Oxygen evolution.
b Fluorescence.

which did not stimulate fluorescence and is also not considered a specific inhibitor of the Hill reaction. A possible explanation for the stimulating effect of the herbicides on the fluorescence of *Chlorella* may be by hypothesizing that excess photochemical energy, normally channeled into the carbon-reduction cycle, is now dissipated as radiant (fluorescent) energy.

In a later study (ZWEIG and GREENBERG 1964) utilizing the fluorescence enhancement phenomenon, we measured the diffusibility of photosynthesis-inhibitor-herbicides into and out of *Chlorella* cells. Figure 1 graphically illustrates that free diffusion of the tested herbicides occurred, giving credence to the "weak bonding theory," e.g., hydrogen bonding. The behavior of these herbicides certainly suggested a non-competitive inhibition of the enzyme responsible for the evolution of oxygen from water. No progress can be reported at this time on the identity of this enzyme which we choose to name "photolase."

Additional studies involved the distribution of ^{14}C-labeled photosynthetates by excised bean leaves from plants previously treated with the Hill reaction inhibitor atrazine (ZWEIG and ASHTON 1962). The most interesting finding of these studies is illustrated in Table II, showing that the distribution of ^{14}C-compounds in an atrazine-treated leaf and a corresponding dark control leaf, both exposed to ^{14}C-carbon

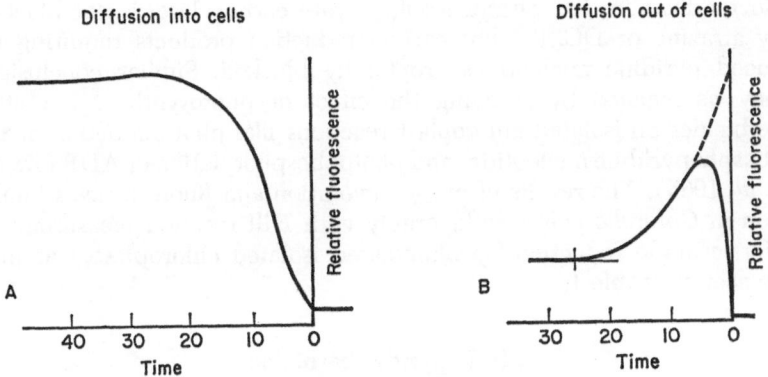

Fig. 1. = movement of $10^{-6}M$ 3-(p-chlorophenyl)-1,1-dimethylurea (CMU) into Chlorella cells as measured by fluorescence increase, B = movement of $10^{-6}M$ CMU out of *Chlorella* cells as measured by fluorescence decline. Turner instrument, time in seconds

Table II. *Light and dark $^{14}CO_2$-fixation of excised bean leaves* (ZWEIG and ÁSHTON 1962)

| Compound | Total radioactivity of alcohol-soluble compounds (%) | | | |
| | Light | | Dark | |
	Control	Atrazine [a] treated	Control	Atrazine [b] treated
Malic acid	3.7	10.6	8.8	5.2
Serine	8.8	9.5	4.0	10.0
Glycine	10.0	0.6	0	0
Aspartic acid	1.6	59.6	66.6	56.0
Glutamic acid	0.8	9.5	10.5	16.8
Asparagine	– [c]	3.8	1.5	1.6
Sucrose	70.8	0	0.7	0.6

[a] Plants treated for 72 hours.
[b] Plants treated for 48 hours.
[c] Not resolved at high concentrations of sucrose.

dioxide, gave similar patterns, *e.g.*, low sucrose and high aspartic and glutamic acids. Similar results have been recently found (CHO and ZWEIG 1968) on *Chlorella* with DCMU (3,4-dichlorophenyl *N,N*-dimethylurea). The explanation for these results is that dark carbon-dioxide fixation, *i.e.*, phosphoenolpyruvate carboxylase, is not blocked by atrazine or DCMU, but carbon reduction products requiring reduced pyridine nucleotides are totally blocked. Similar conclusions may be reached by studying the effect of photosynthesis-inhibitor-herbicides on isolated chloroplast reactions like photoreduction or triphosphopyridine nucleotides and photophosphorylation of ADP (ZWEIG *et al.* 1965). The results of oxygen evolution and fluorescence stimulation in *Chlorella* compare favorably with Hill reaction measurements (ferricyanide reduction by illuminated isolated chloroplasts) as may be seen in Table I.

III. Dipyridyl herbicides

Quite a different story is the mode-of-action of dipyridyl herbicides such as diquat and paraquat. MEES (1960) demonstrated with leaf discs that diquat was effective only in light and presence of oxygen, and that a Hill reaction inhibitor like DCMU stopped the activity of diquat. MEES (1960) and DAVENPORT (1963), postulated that the herbicidal action of dipyridyl was due to the formation of free radicals during photosynthesis.

Recently it could be demonstrated that a stable free radical of diquat formed under strictly anaerobic conditions by the action of illuminated chloroplasts (ZWEIG *et al.* 1965) (cf. Figure 2); DCMU inhibited the photoreduction of diquat. These observations lent support to MEES' (1960) theory that a toxic free radical could explain the phytotoxicity of diquat. However, further experiments showed that, although diquat did not inhibit the Hill reaction, it did indeed stop the photoreduction of triphosphopyridine nucleotide and partially the photophosphorylation of ADP to yield ATP. Thus, there could be an alternate explanation for the phytotoxicity of dipyridyl compounds —deprivation of ATP and $NADPH_2$, thus inhibiting the carbon-reduction cycle.

To verify this point, ARRIAGA-DIAZ *et al.* (1966) studied the effect of diquat on the distribution of ^{14}C-compounds by ^{14}C-carbon dioxide-fixation of *Chlorella, Elodea,* and excised tobacco leaves. The most significant results were from *Chlorella* and showed that total ^{14}C-carbon dioxide-fixation was partially suppressed, compared with DCMU which inhibited 99 percent of this fixation. The most significant decrease in carbon-labeling occurred in alanine and to a lesser extent in glycine and serine, but sucrose synthesis seemed to proceed without much interference. The decrease in the label of amino acids formed by reductive amination may be explained by the observation of the inhibition of $NADPH_2$ formation by diquat (ZWEIG *et al.* 1965). *Chlorella* may not be the most suitable test organism for herbicides, as was shown recently (ZWEIG *et al.* 1968). During the time period

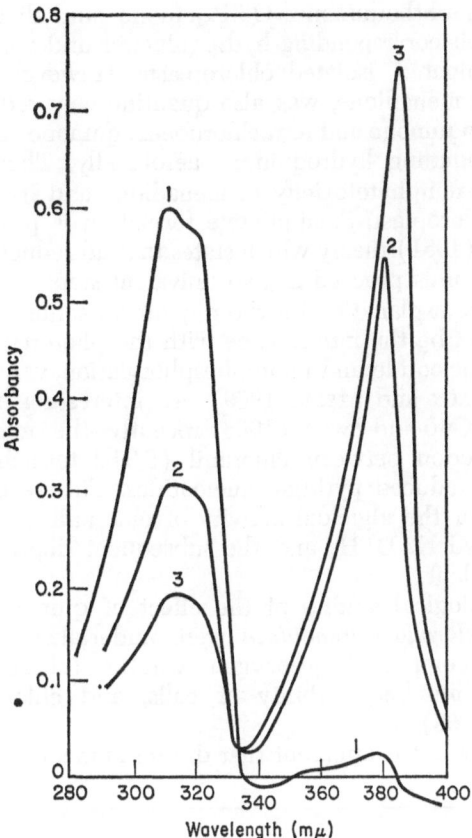

Fig. 2. Photoreduction of diquat by isolated chloroplasts. Curve 1, diquat, 3.3 x $10^{-5}M$, before illumination; curve 2, five-minute illumination; curve 3, 10-minute illumination

when algicides like dichlone and chloranil totally bleached *Chlorella pyrenoidosa* (24 to 48 hours), diquat showed little toxicity.

IV. Quinone herbicides

An investigation was begun on the mode-of-action of quinone algicides dichlone (2,3-dichloro-1,4-naphthoquinone) and O6K-quinone (3-chloro-2-amino-1,4-naphthoquinone), since it appeared logical to investigate the free-radical theory proposed for the mode-of-action of the dipyridyls. It was well-known that quinones were capable of being converted to semiquinones polarographically. CHO *et al.* (1966) could demonstrate that many quinones which show algicidal activity like naphthoquinone, O6K-quinone, and dichlone could be reduced to their corresponding hydroquinones under anaerobic conditions by illuminated, isolated chloroplasts. However, the nonphytotoxic quinone, menadione, was also quantitatively reduced to hydroquinone. Benzoquinone and tetrachlorobenzoquinone could be reduced to the corresponding hydroquinone aerobically. Thus, it is difficult to explain the nonphytotoxicity of menadione and its photoreduction by isolated chloroplast. Semiquinone formation is possible according to MICHAELIS' (1951) theory which states that all reductions of bivalent organic compounds proceed in two univalent steps.

An alternate explanation for the phytotoxic action of O6K-quinone herbicide might be the interference with the photoreduction of phosphopyridine nucleotide and photophosphorylation, very similar to that of diquat (BLACK and MYERS 1966). An interesting observation has been made by CHO and ZWEIG (1968) recently—the stoichiometric nonenzymatic reaction between chloranil (2,3,5,6-tetrachloro-1,4-benzoquinone) and reduced pyridine nucleotides. This reaction suggested a hypothesis for the algicidal activity of chloranil, *i.e.*, the oxidation of photoreduced $NADPH_2$ and the subsequent deprival of reducing power to the plant.

The physiological studies of the effect of quinones, diquat, and DCMU on *Chlorella pyrenoidosa* were undertaken to measure the effect of these compounds on oxygen evolution (short- and long-term contact), cell number, viability of cells, and chlorophyll content (ZWEIG *et al.* 1968).

From Figure 3 it is apparent that diuron immediately stops oxygen

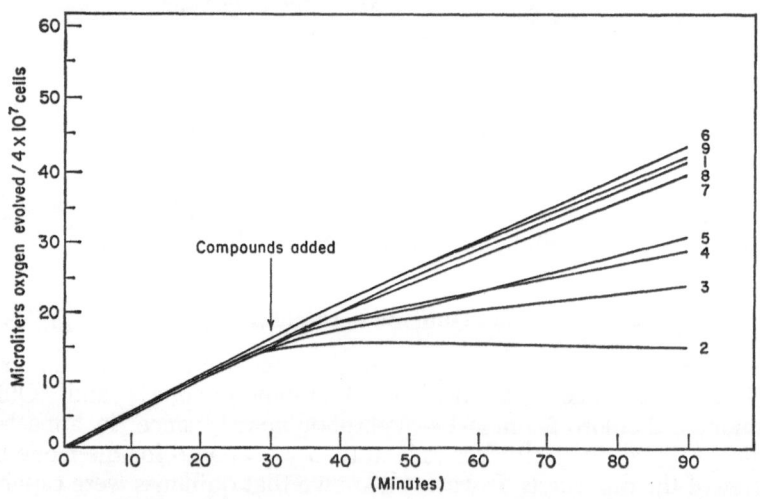

Fig. 3. Short-term effect of test compounds ($3 \times 10^{-5}M$) on oxygen evolution of *Chlorella*: *1* = control, *2* = diuron, *3* = dichlone, *4* = O6K-quinone, *5* = 1,4-naphthoquinone, *6* = 1,4-benzoquinone, *7* = menadione, *8* = diquat, and *9* = chloranil

evolution by illuminated *Chlorella,* and that dichlone, O6K-quinone, and 1,4-naphthoquinone exert a partial inhibitory effect. The other quinones studied, benzoquinone, menadione, chloranil, as well as the dipyridyl diquat have little or no inhibitory effect during 90 minutes illumination. The effect of diuron is completely reversible by a simple wash of the cells (ZWEIG and GREENBERG 1964, ZWEIG *et al.* 1968). From these results it is clear that diuron has an immediate, reversible effect on the oxygen-evolving step in green algae. It cannot be ruled out that some of the quinones might also inhibit a reaction close to the primary photochemical step, but this may not be their primary mode-of-action of their phytotoxicity.

The long-term effect (48 hours) of diuron and benzoquinone is algistatic as seen in Figure 4. The three parameters studied, oxygen-

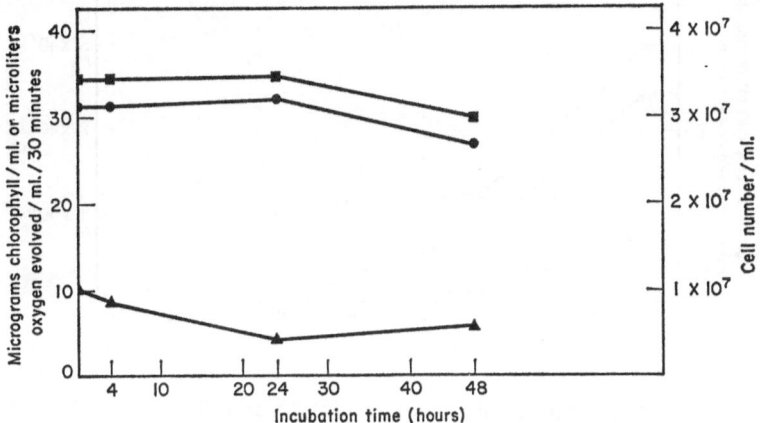

Fig. 4. Long-term effect of diuron ($3 \times 10^{-5} M$) on cell number (■), chlorophyll concentration (●), and oxygen evolution (▲) of *Chlorella*

evolution, cell number, and chlorophyll concentration remained constant during this period. Viability studies on algae, treated with diuron for 65 hours gave 90 percent viable cells, confirming other observations that during this time period only a small number of cells had been killed (Table III). This is in contrast to dichlone which effected total death of a similar *Chlorella* culture within the same time period. This is not to suggest that diuron is not a herbicide, since the culture from the experiment shown in Table III did become completely bleached and was killed after seven days treatment with diuron.

Dichlone seems to be the most active algicide as is shown in Figure 5 depicting the dramatic decrease in oxygen evolution, number of cells (measured as total number including non-viable with Coulter counter) and chlorophyll concentration already within 24 hours of treatment. The culture became completely bleached within 48 hours.

Table III. *Viability studies of* Chlorella *treated with diuron and dichlone*

Treatment	Percent of viability of 100 cells after		
	0 hr.	65 hr.	90 hr.
Control	90	90	78
Diuron [a]	90	90	39
Dichlone [a]	90	0	0

[a] $3 \times 10^{-5} M$.

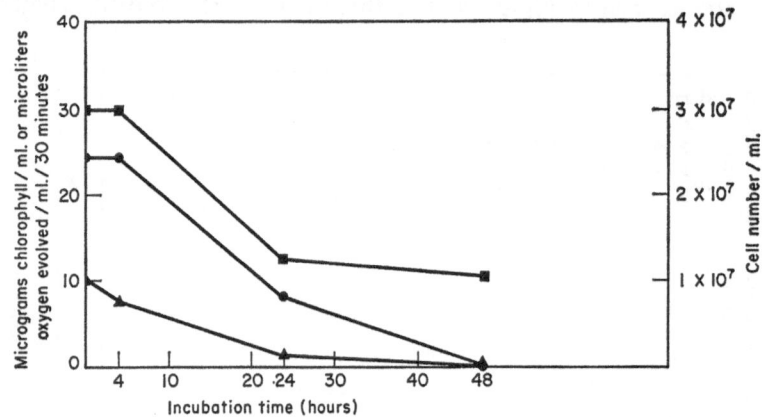

Fig. 5. Long-term effect of dichlone ($3 \times 10^{-5} M$) on cell number (■), chlorophyll concentration (●), and oxygen evolution (▲) of *Chlorella*

O6K-quinone, naphthoquinone, and chloranil caused a similar irreversible decrease in oxygen-evolution even within four hours after treatment. It appeared that the decrease in oxygen evolution always preceded chlorophyll destruction suggesting that the two events are not necessarily related (as for example, the bleaching by aminotriazole).

Thus by simple and relatively rapid experiments (two days) one can determine the effect of potential herbicides on oxygen evolution, chlorophyll concentration, and cell population. From the results, one can determine if a potential herbicide falls into a class of Hill reaction inhibitors, algistats, or algicides. Naturally, no general conclusions can be made *a priori* on the effect of these compounds on higher plants. For example, diquat is a very effective herbicide but not a good algicide. By studying the effect on algae, one overcomes difficulties like root uptake and translocation. Also the aquatic environment provides a constant concentration of the test compounds if no photodecomposition occurs, as in the case of benzoquinone when red light must be used.

Acknowledgment

The studies reported here from my laboratory have been supported by several grants from the U.S.P.H.S., EF 00357, EF 00896, and ES 00223 whose aid is gratefully appreciated. My former and present associates whose contributions I want to acknowledge in chronological order are: I. TAMAS, E. GREENBERG, J. HITT, and D. H. CHO. I also want to acknowledge the financial assistance by the NSF, for a travel grant to Japan to present this paper.

Summary

In summary, it may be stated that excised leaves, isolated chloroplasts, and algae provide excellent tools for studying the possible mode-of-action of herbicides affecting the primary photosynthetic act or subsequent electron transport. The distribution of ^{14}C-labeled compounds following ^{14}C-carbon dioxide-fixation by *Chlorella* has not been consistent, and the use of a synchronized cell culture or isolated chloroplasts prepared by the method of JENSEN and BASSHAM (1966) is recommended. The effect of herbicides on intact plants must follow these preliminary investigations. It is possible that the initial observations from the greenhouse or field might suggest the first clue to the primary mode-of-action of herbicides.

Résumé *

Mode d'action des herbicides inhibiteurs de croissance

En résumé, il peut être établi que les feuilles coupées, les chloroplastes isolés et les algues fournissent d'excellents instruments pour l'étude du mode d'action possible des herbicides affectant l'action primaire de la photosynthèse ou le transport ultérieur des électrons. La distribution des composés marqués en C^{14} faisant suite à la fixation du ^{14}CO$_2$ par Chlorella n'a pas été concluante, mais l'usage d'une culture cellulaire synchronisée ou de chloroplastes isolés, préparés par la méthode de JENSEN et BASSHAM (1966) est recommandée. L'effet des herbicides sur les plantes intactes doit suivre ces recherches préliminaires. Il est possible que les observations initiales provenant de

---

* Traduit par S. DORMAL-VAN DEN BRUEL.

serres ou de champs puissent permettre de découvrir le mode d'action
primaire des herbicides.

Zusammenfassung *

Wirkungsweise von Photosynthese-hemmenden Herbiziden

Zusammenfassend kann gesagt werden, dass herausgeschnittene
Blätter, isolierte Chloroplasten und Algen ausgezeichnete Werkzeuge
zum Studium der möglichen Wirkungsweise von Herbiziden sind,
welche den primären Photosyntheseakt oder folgenden Elektronen-
transport beeinflussen. Die Verteilung von C^{14}-markierten Verbindungen, welche der $C^{14}O_2$-Fixierung durch Chlorella folgt, ist nicht über-
einstimmend, und der Gebrauch einer zeitlich abgestimmten Zellkultur
oder isolierten Chloroplasten, die mit der Methode von JENSEN und
BASSHAM (1966) bereitet worden sind, wird empfohlen. Die Wirkung
von Herbiziden auf intakte Pflanzen muss diesen vorläufigen Untersuchungen folgen. Es ist möglich, dass die Anfangsbeobachtungen im
Gewächshaus oder im Feld den ersten Anhaltspunkt für die primäre
Wirkungsweise von Herbiziden andeuten.

References

ARRIAGA-DIAZ, C.: Some effects of diquat on respiration and amino acid metabolism in *Chlorella* and *Elodea*. M.S. Thesis, Univ. of Calif., Davis (1966).

BLACK, C. C., and L. MYERS: Some biochemical aspects of the mechanisms of
herbicidal activity. Weeds 14, 331 (1966).

CHO, D. H., L. PARKS, and G. ZWEIG: Photoreduction of quinones by isolated
spinach chloroplasts. Biochim. Biophys. Acta 126, 200 (1966).

CHO, D. H. and G. ZWEIG: Unpublished data (1968).

DAVENPORT, H. E.: Mechanism of cyclic phosphorylation by illuminated chloroplasts. Proc. Roy Soc. (London), Ser. B 157, 332 (1963).

GOOD, N. E.: Inhibitors of the Hill reaction. Plant Physiol. 36, 788 (1961).

JENSEN, R. G., and J. A. BASSHAM: Photosynthesis of isolated chloroplasts. Proc.
Nat. Acad. Sci. (U.S.). 56, 1095 (1966).

MEES, G. C.: Experiments on the herbicidal action of 1,1'-ethylene-2,2-dipyridy-
lium dibromide. Ann. Applied Biol. 48, 601 (1960).

MICHAELIS, L.: Theory of oxidation-reduction. In: Enzymes, J. B. Sumner and
K. Myrback, eds., Vol. 2, pp 1-54, New York: Academic Press (1951).

ZWEIG, G., and F. M. ASHTON: The effect of 2-chloro-4-ethylamino-6-isopropyl-
amino-*s*-triazine (Atrazine) on distribution of ^{14}C-compounds following CO_2-
fixation in excised kidney bean leaves. J. Exp. Bot. 13, 5 (1962).

ZWEIG, G., and E. GREENBERG: Diffusion studies with photosynthesis inhibitors
on *Chlorella*. Biochim. Biophys. Acta 79, 226 (1964).

ZWEIG, G., J. HITT, and R. MCMAHON: Effect of certain quinones, diquat, and
diuron on *Chlorella pyrenoidosa* Chick. (Emerson strain), Weed Science 16,
69 (1968); also presented at Ann. Meeting Weed Soc., Washington, D.C.
(1967).

* Übersetzt von A. SCHUMANN.

Zweig, G., N. Shavit and M. Avron: Diquat (1,1'-ethylene-2,'2-dipyridylium dibromide) in photoreactions of isolated chloroplasts. Biochim. Biophys. Acta **109**, 332 (1965).

Zweig, G.: Effect of photosynthesis inhibitors (herbicides) on CO_2-fixation by *Chlorella pyrenoidosa*. Ann. Meeting Weed Soc., New Orleans, (1968).

Zweig, G., I. Tamas, and E. Greenberg: The effect of photosynthesis inhibitors on oxygen evolution and fluorescence of illuminated *Chlorella*. Biochim. Biophys. Acta **66**, 196 (1963).

The Strategy of Finding Fungicides

by
JAMES G. HORSFALL [*] and R. J. LUKENS [*]

Contents

I. Introduction

An old English proverb says that, "You get what you ask for." The fungicides we have are those we have asked for. Unless we change the form of asking occasionally, we shall run out of candidates. This is strategy.

Up to now the chief form of asking has been to poison the fungus —to poison its food or to kill its spores. The spore-killing technique on glass or on the leaf has produced bordeaux mixture, chloranil, zineb, ferbam, captan, and a host of others. The poisoned food technique has produced creosote, pentachloronitrobenzene, diphenyl, nitrophenol, and, of course, many others. A sophisticated modern variant of the poisoned food technique is to use a soil column as an analogue of chromatography. This has selected out many effective soil fungicides.

II. Rate of introducing fungicides

We wondered if the asking methods had run their course, if they were producing as many new fungicides as before. Accordingly, we compiled a list of fungicides that have found use in the field with

---

[*] The Connecticut Agricultural Experiment Station, New Haven, Connecticut.

their approximate dates of adoption beginning with mercuric chloride in 1705 and ending with dichloran in 1960. The data in Figure 1

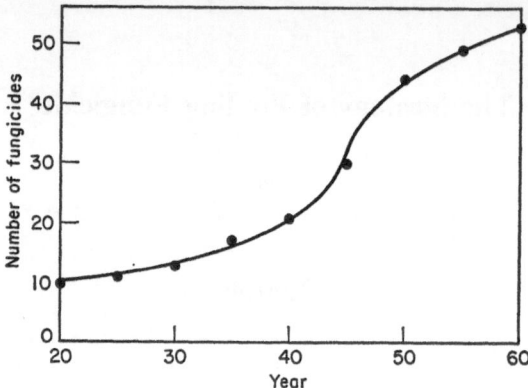

Fig. 1. Rate of producing useful fungicides

suggest that the curve in fact is flattening and that, perhaps, the old asking methods have run their course. If so, new methods of asking are needed. In the modern idiom, a "quantum jump" in method is needed. The poisoned food and spore-killing methods were such jumps. Now we need new ones.

We have tried three possibilities—lethal synthesis, increasing host resistance, and antisporulation. In this paper we deal lightly with the first two, and give more emphasis to the third.

III. Lethal synthesis

Very often a fungus kills itself by synthesizing a lethal compound from a non-lethal one. We suspect that this phenomenon is much more common than suspected. It is often overlooked because if a compound kills, we call it toxic without taking the considerable pains to discover if lethal synthesis is involved.

Our first brush with lethal synthesis was in 1942 when we discovered (Horsfall 1956) that 2-methyl-1,4-diacetoxynaphthalene is as fungitoxic as 2-methyl-1,4-naphthoquinone. It seemed unlikely that the diester would be a toxic compound. The work of Byrde and Woodcock (1953) opened the road to show that this is a case of lethal synthesis. They hold that the fungus deesterifies the ester and forms the diol which is toxic. On the basis of the work of Rich and Horsfall (1954), we believe that the fungus proceeds further and oxidizes the compound to the more toxic quinone form to which its toxicity corresponds.

In 1949 we accidentally discovered what turns out to be our second case of lethal synthesis (RICH and HORSFALL 1949). The fungicide is 2,4-dinitro-(1-methyl)heptylphenyl crotonate. This compound has come to be the well known antimildew fungicide, Karathane. This compound is astonishing because dinitrophenols are weed killers of note. Karathane should be intensely phytotoxic but it is not. The answer seems to lie in differential lethal synthesis. The fungus can hydrolyze the crotonic ester, create the free nitrophenol, and poison itself. The host cannot. This elementary fact of chemistry makes commercial exploitation possible. The compound kills the fungus by lethal synthesis and leaves the host alone.

One of our test fungi, *Alternaria solani*, is like the green host. It does not hydrolyze the ester linkage. It does not synthesize its own death potion, and cannot be controlled by Karathane in the field.

We have tested a few other Karathane esters on *Alternaria solani*. The inactive ester, as we have said, is the crotonic ester, R-OOC-CH = CH-CH$_3$, which has in the chain a double bond conjugated with the acid group. The end carbon can be dropped off as in the methacrylate ester [-OOC-C(CH$_3$) = CH$_2$] and the activity is still nil. If, however, a carbon be added to the end of the chain to give five carbons and the double bond pushed outward to the end carbon, we have the 4-pentenoate ester [-OOC-CH$_2$-CH$_2$-CH = CH$_2$]. This kills *A. solani*. Presumably enzymes of *Alternaria* can deesterify the non-conjugated ester, form the dinitrophenol, and thus *Alternaria* kills itself.

HAMILTON and SZKOLNIK (1958) have shown that to convert actidione into its semicarbazone is to make it safer for the host, but not less fungitoxic. Presumably the fungus can unblock the carbonyl group, the host cannot. This, too, is differential lethal synthesis.

We have searched for other useful cases of differential lethal synthesis without much success, however. One could hope to find such compounds in other esters such as phosphate, borate, and benzoate, or among other classes of compounds in which the active toxophore is blocked chemically, but which can be unblocked by the fungus not by the host.

IV. Increasing host resistance

For many years we have tried to introduce compounds into plants such as tomatoes to render them resistant to fungal invasion. Two of the most dramatic of the compounds we have found are 2-carboxymethylthiobenzothiazole and maleic hydrazide. The former increases the resistance of tomatoes to *Fusarium lycopersici* (DIMOND *et al.* 1952) but the latter makes the plant fatally susceptible to the same fungus. We know that *Fusarium* attacks plants low in sugar. Maleic hydrazide is known to reduce sugar movement through stems to roots and we suspect that the benzothiazole compound increases sugar movement downward.

In any case there does seem to be a great possibility of discovering useful new fungicides by this method of asking. We warn beginners, however, that the road is rocky. This method aims to starve the fungus, but it must function without the comfortable aid of phagocytosis which complements so much the activity of drugs in curing human diseases.

We suspect that the search for effective systemic chemotherapeutants for plant diseases would have been much more successful by now than it has been, had plants been possessed of a phagocytosis system. Our colleagues in this laboratory, DIMOND et al. (1952 et seq.) were able to reduce fusarial wilt of tomato to a very low level with therapeutants but as soon as medication was stopped, the disease set in again and killed the plants. If the tomato host had been possessed of white blood cells in its sap, these would have killed off the stragglers and the plant would have recovered. Penicillin would not produce as many dramatic cures, perhaps none at all, without the help of the phagocytes in the blood.

V. Antisporulation

The year of this conference (1967) marks the twentieth year of our search for useful antisporulants (RICH and HORSFALL 1948). We have found numerous effective antisporulants in the laboratory and we have tested a dozen or so in the field. Only hexachloroisopropanol (HORSFALL and RICH 1960) seems to have shown any promise in the field so far. It has been no zineb, however.

To control disease by inhibiting pathogen reproduction is admittedly a hazardous effort. One dare not permit very many spores to be produced. As DIMOND et al. (1941) first showed and RICHARDSON and MUNNECKE (1964) have confirmed, disease is proportional to the logarithm of the number of propagules. This means that one can inhibit the production of a vast number of spores at the source of inoculum without reducing disease very much at the infection court. A logarithmic input of effort will be required.

As long as we are on this philosophical course, we could say that the likelihood of devising a new method of finding fungicides will also require a logarithmic input of effort. The easier methods of finding new fungicides will have been found first.

Despite the difficulties, we still are interested in devising a method of finding effective and useful antisporulants. In the end, however, they will probably have to be combined with protectants in the farmers' spray tank.

In seeking antisporulants we use *Alternaria solani* and LUKENS' (1960) method of producing the spores. Chopped and washed mycelial fragments are distributed over moistened filter paper and incubated in petri dishes for 30 hours in the light to form conidiophores.

This paper with its conidiophores attached is then placed on top of another paper previously soaked in the test compound and dried. The combination is moistened, put in the dark to form spores, and examined in the morning.

An antisporulant is one that inhibits spore formation without killing the fungus.

a) Trichloromethyl compounds

While conducting a search for other useful compounds, we have explored the mechanism by which 1,1,1,3,3,3-hexachloroisopropanol (I) inhibits the differentiation of spores. The compound has a $-CCl_3$ group adjacent to an $-OH$ group.

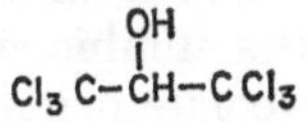

$$\underset{\text{I}}{Cl_3 C-\underset{\overset{|}{OH}}{CH}-CCl_3}$$

A search turned up a number of other compounds having similar structures and similar activity. One of these is trichloromethylbenzyl alcohol (II).

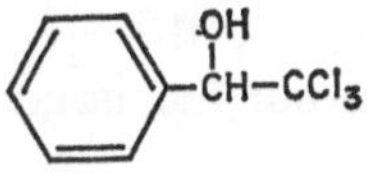

II

Another is α-trichloromethyl-2-pyridine ethanol (III). Both of these were effective antisporulants although somewhat less active than the isopropanol analogue.

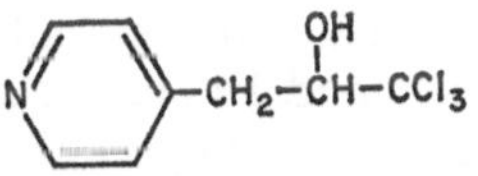

III

Compound III reminded us of another compound that Dr. ZELITCH (1963) of our laboratory had shown to inhibit glycolic acid oxidase and thus close stomates on green plants. His compound is α-hydroxy-2-pyridinemethane sulfonic acid (IV).

According to Dr. Zelitch, the sulfonic acid group inhibits the activity

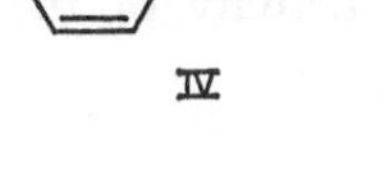

IV

of glycolic acid oxidase because it acts as an antimetabolite for the carboxyl moiety of glycolic acid (V).

$$
\begin{array}{c}
\text{OH} \\
| \\
\text{HC—COOH} \\
\text{V}
\end{array}
$$

Was it possible that the $-CCl_3$ in our pyridine compound (III) acts similarly as an antimetabolite for the carboxyl moiety of glycolic acid. Some of our colleagues thought that it probably would not because the three chlorine atoms occupy so much more space than the $-COOH$, that the molecule would be held too far away from the surface of the enzyme. We could find out by testing. Dr. Zelitch tested our compound. It does inhibit stomatal movement. We then found (a) that it inhibits glycolate oxidase in a partially purified cell-free preparation from *Alternaria* mycelia, (b) that glycolic acid will strongly reverse the antisporulant effect of the compound on *Alternaria*, and (c) that the benzyl alcohol and isopropanol analogues act in the same way.

Therefore, we conclude that compounds with a $-CCl_3$ group *alpha* to a hydroxyl group can act as antimetabolites for glycolic acid which is probably heavily involved in the sporulation of *A. solani*. Presumably the $-CCl_3$ group is not too large to replace the carboxyl in the reaction.

b) Phenoxyacetic acids

This raises the immediate question whether one can inhibit sporulation and glycolate oxidation by tampering with the hydroxyl moiety of glycolic acid as well. 2,4-D and its analogues are, of course, such compounds in which an ether linkage replaces the hydroxyl group in glycolic acid (see LUKENS and HORSFALL 1968).

We tested the following six analogues of phenoxyacetic acid which are numbered to correspond with the points in the two curves that follow: (1) 2,4-dichloro-, (2) 2,4,6-trichloro-, (3) 2,3,5,6-tetrachloro-, (4) pentachloro-, (5) 4-chloro-2-cyclohexyl-, and (6) 4-nitro-.

That they inhibit glycolic acid oxidase is shown by three typical examples in Figure 2. The data were obtained by mixing partially purified cell-free enzymes with the test compound and assaying for glycolate oxidation.

The question is whether antiglycolate activity is related to antisporulant activity. The data illustrated in Figure 3 show that it is. The fit is reasonably good. The graph does indicate, however, that some of the antisporulant activity of the trichloro-analogue (com-

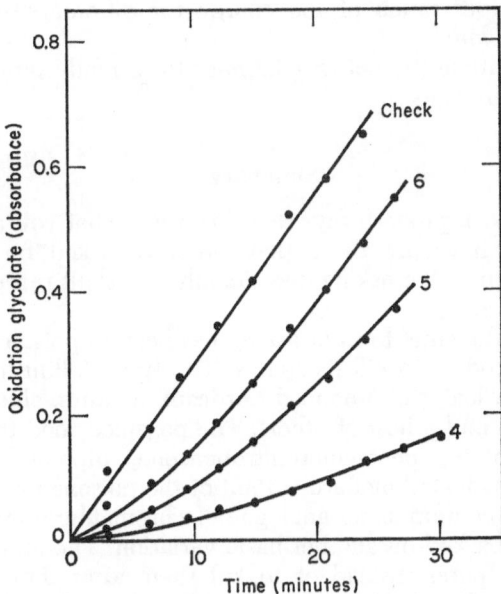

Fig. 2. Inhibition of glycolic acid oxidase by phenoxyacetic acids. No. 4 is pentachlorophenoxyacetic acid, no. 5 is 4-chloro-2-cyclohexylphenoxy-acetic acid, and no. 6 is 4-nitrophenoxyacetic acid

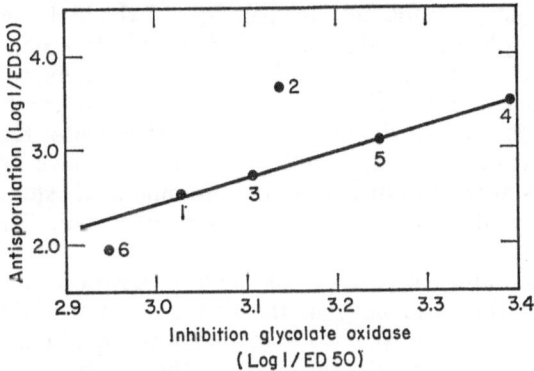

Fig. 3. Relating antiglycolate activity to antisporulant activity

pound 2) lies outside of attacking glycolate oxidase. The 2,4,6-trichlorophenyl-moiety increases antisporulation activity of phenyl-hydrazines and phenol as well. These compounds hit other targets in the sporulating process. Error of measuring the weak antisporulant activity of compound 6 can account for the slight digression of that compound from the curve. Figure 3 helps us conclude that glycolate oxidation provides much of the energy for differentiation of spores of *Alternaria solani*.

In the meantime the search continues for a really satisfactory antisporulant for the field.

Summary

An old English proverb says that, "You get what you ask for." The fungicides we have are those that we have asked for. Unless we change the form of the asking occasionally, we shall run out of candidates. This is strategy.

Up to now the chief form of asking has been to poison the fungus—to poison its food or to kill its spores. The spore killing technique on glass or on the leaf has produced bordeaux mixture, chloranil, zineb, ferbam, captan and a host of others. The poisoned food technique has produced creosote, pentachloronitrobenzene, diphenyl, and nitrophenol. A sophisticated modern variant of the poisoned food technique is to use a soil column as an analogue of chromatography.

We are embarked on another basic variation. The aim is to inhibit production of spores instead of to kill them after they are formed. At present our data indicate that the glycolic acid pathway is intimately involved in producing the energy for sporulation. We find that we can inhibit both sporulation and oxidation of glycolic acid with $-SO_3H$ and $-CCl_3$ analogues of the carboxyl moiety of glycolic acid and with phenoxy compounds as analogues of the hydroxyl moiety.

The principle of lethal synthesis offers a fairly new method of asking. Molecules can be designed that will be converted to a poison by the fungus, not by the host. Karathane illustrates the principle even though it was not discovered through that route. Karathane is a highly successful field fungicide for powdery mildews because powdery mildews appear to hydrolyze the crotonic acid ester and liberate the fungitoxic dinitrophenol but the host does not. The fungus is killed, the host is not.

These techniques of asking all neglect, however, the possibility of influencing the physiology of the host so that it is better able to throw the invader out. We work in that field, too. For example, 2-carboxymethylthiobenzothiazole increases the resistance of the tomato to *Fusarium lycopersici*, but it is not fungitoxic. On the contrary, maleic hydrazide makes the same plant fatally susceptible to the same fungus. Our present hypothesis is: (a) *Fusarium* attacks roots of plants low in sugar, (b) MH reduces sugar movement to the root, and (c) we suspect that the benzothiazole compound increases sugar movement to the root. This method of asking should give us many more useful compounds.

Résumé *

La stratégie de recherche des fongicides

Nous disposons des fongicides que nous avons cherchés. A moins de changer un jour la forme de notre demande, nous n'aurons plus de nouveaux produits.

Jusqu'ici les principes envisagés ont été: d'empoisonner la moisissure—d'empoisonner son aliment ou de détruire ses spores. La technique de destruction des spores a donné la bouillie bordelaise, le chloranil, le zineb, le ferbam, le captane et nombre d'autres produits. La méthode d'empoisonnement des aliments a produit la creosote, le pentachloronitrobenzène, le biphényle et le nitrophénol. Une variante moderne de cette dernière méthode consiste à utiliser une colonne de terre en guise de chromatographie.

Nous travaillons sur un principe différent. Le but est d'inhiber la production de spores au lieu de les détruire après leur formation. Actuellement nos résultats montrent que le cycle de l'acide glycolique est intimément lié à la production de l'énergie de sporulation. Nous trouvons que l'on peut inhiber à la fois la sporulation et l'oxydation de l'acide glycolique avec les analogues $-SO_3H$ et CCl_3 de la partie carboxylique de l'acide glycolique et avec les composés phénoxy comme analogues de la partie hydroxyle.

Le principe de la syntèse léthale offre une toute nouvelle méthode de recherche. Il est possible de concevoir des molécules qui seront converties en poison par le champignon et non par l'hôte. Le dinocap illustre ce principe bien qu'il nait pas été découvert par cette voie. Le dinocap est un excellent fongicide sur le terrain pour les oïdiums parce que les oïdiums semblenthydrolyser l'ester de l'acide crotonique et libérer le dinitrophénol fongicide tandis que l'hôte ne le fait pas. Le champignon est tué et l'hôte non.

Ces techniques négligent toutes cependant, la possibilité d'influencer la physiologie de l'hôte afin qu'il soit plus apte à rejeter le parasite. Nous travaillons également ce sujet. Par exemple, le 2-carboxyméthylthiobenzothiazole augmente la résistance de la tomate au *Fusarium lycopersici*, mais n'est pas fongicide. Au contraire, l'hydrazide maléique rend la même plante fatalement sensible au même champignon. Notre hypothèse actuelle est que: (a) le *fusarium* attaque les racines des plantes pauvres en sucre, (b) l'hydrazide maléique réduit la migration des sucres dans la racine, (c) nous pensons que le benzothiazole augmente le transfert des sucres vers la racine. Cette méthode de recherche devrait nous donner beaucoup d'autres composés utiles.

---

* Traduit par R. Mestres.

Zusammenfassung *

Die Taktik, Fungizide zu finden

Ein altes englisches Sprichwort sagt: "Du bekommst, wonach Du fragst!" Die Fungizide, die wir haben, sind die, nach denen wir gefragt haben. Ausser wenn wir die Form der Fragestellung gelegentlich ändern, werden wir bald ohne Kandidaten sein. Dies ist Taktik.

Bis heute ist die Hauptform der Fragestellung die gewesen, den Pilz zu vergiften—seine Nahrung zu vergiften oder seine Sporen zu töten. Die Sporenabtötungstechnik auf Glas oder auf dem Blatt hat die Bordeaux-Mischung, Chloranil, Zineb, Ferbam, Captan und ein Heer von anderen hervorgebracht. Die Futtervergiftungstechnik hat Creosote, Pentachlornitrobenzol, Diphenyl und Nitrophenol produziert. Eine intellektuelle, moderne Variante der Futtervergiftungstechnik ist der Gebrauch einer Bodensäule als ein Analogon zur Chromatographie.

Wir haben eine andere grundlegende Variation begonnen. Unsere Absicht ist es, die Produktion der Sporen zu hemmen anstatt sie nach ihrer Bildung abzutöten. Zurzeit deuten unsere Daten daraufhin, dass der Glykolsäureweg wesentlich in die Energieproduktion zur Sporenbildung verwickelt ist. Wir finden, dass wir sowohl die Sporenbildung als auch die Oxidation der Glykolsäure mit $-SO_3H-$ und $-CCl_3-$ Analogen (am Carboxylteil) der Glykolsäure und mit Phenoxyverbindungen als Analoge des Hydroxylteiles hemmen können.

Das Prinzip der lethalen Synthese bietet eine ziemlich neue Methode der Fragestellung. Moleküle können entworfen werden, die durch den Pilz, nicht durch den Wirt, in einen Giftstoff umgewandelt werden. Karathan veranschaulicht dies Prinzip, obwohl es nicht über diesen Weg entdeckt worden ist. Karathan ist ein sehr erfolgreiches Feldfungizid für stäubende Meltaus, da stäubende Meltaus offenbar Crotonsäureester hydrolysieren und das fungitoxische Dinitrophenol in Freiheit setzen, während das der Wirt nicht tut. Der Pilz wird abgetötet, der Wirt nicht.

Diese Techniken der Fragestellung übersehen jedoch alle die Möglichkeit, die Physiologie des Wirtes so zu beeinflussen, dass er besser in der Lage ist, den Eindringling hinauszuwerfen. Wir arbeiten auch auf diesem Gebiet. Zum Beispiel lässt 2-Carboxymethyl-thio-benzthiazol die Resistenz der Tomate gegenüber *Fusarium lycopersici* zunehmen, aber es ist nicht fungitoxisch. Im Gegensatz dazu macht Maleinhydrazid dieselbe Pflanze tötlich anfällig für denselben Fungus. Unsere gegenwärtige Hypothese ist: a) *Fusarium* greift Wurzeln von Pflanzen, die niedrig im Zuckergehalt sind, an, b) MH reduziert den Zuckertransport zu den Wurzeln, c) wir nehmen an, dass die Benz-

---

* Übersetzt von A. SCHUMANN.

thiazolverbindung den Zuckertransport zu den Wurzeln erhöht. Diese Methode der Fragestellung sollte uns viele weitere nützliche Verbindungen geben.

References

BYRDE, R. J. W., and D. WOODCOCK: Fungicidal activity and chemical constitution. II. Compound related to 2:3-dichloro-1:4-naphthaquinone. Ann. Applied Biol. 40, 675 (1953).

DIMOND, A. E., J. G. HORSFALL, J. W. HEUBERGER, and E. M. STODDARD: Role of the dosage-response curve in the evaluation of fungicides. Bull. 451 Conn. Agr. Expt. Sta. (1941).

DIMOND, A. E., DAVID DAVIS, R. A. CHAPMAN, and E. M. STODDARD: Plant chemotherapy as evaluated by the *Fusarium* wilt assay on tomatoes. Bull. 557 Conn. Agr. Expt. Sta. (1952).

HAMILTON, J. M., and M. SZKOLNIK: Control of *Coccomyces hiemalis* by systemic movement of cycloheximide semicarbazone in sour cherry following root or leaf absorption. Phytopathol. 48, 262 (1958).

HORSFALL, J. G.: Principles of fungicidal action. Waltham, Mass.: Chronica Botanica (1956).

——, and S. RICH: Antisporulant action of hexachloro-2-propanol. Phytopathol. 50, 640 (1960).

LUKENS, R. J.: Conidial production from filter paper cultures of *Helminthosporium vagans* and *Alternaria solani*. Phytopathol. 50, 867 (1960).

——, and J. G. HORSFALL: Glycolate oxidase, a target for anti-sporulants. Phytopathol. In press (1968).

RICH, SAUL, and J. G. HORSFALL: Metal reagents as antisporulants. Phytopathol. 38, 22 (1948).

—— —— Fungicidal activity of dinitrocaprylphenyl crotonate. Phytopathol. 39, 19 (1949).

—— —— Relation of polyphenol oxidases to fungitoxicity. Proc. Nat. Acad. Sci. 40, 139 (1954).

RICHARDSON, L. T., and D. E. MUNNECKE: Effective fungicide dosage in relation to inoculum concentration in soil. Can. J. Bot. 42, 301 (1964).

ZELITCH, I.: The control and mechanisms of stomatal movement. Bull. 664, Conn. Agr. Expt. Stat., p. 18 (1963).

Mode of action of
agricultural antibiotics developed in Japan

by

TOMOMASA MISATO [*]

Contents

I. Introduction

Much work has been done on possible agricultural antibiotics during the past 20 years. In western countries, however, only a few of these antibiotics have been developed for practical use. These are streptomycin, terramycin, cycloheximide, and griseofulvin. Streptomycin, the first antibiotic introduced in agriculture, was first used in the United States for the control of pear fire blight. This antibiotic and a mixture of streptomycin and terramycin have been used for the control of bacterial plant diseases, while cycloheximide and griseofulvin have been used for fungal plant disease control.

In Japan, on the other hand, not only these four antibiotics have been used against diseases of fruit trees and vegetables, but also

---

[*] The Institute of Physical and Chemical Research, Tokyo, Japan.

93

several other antibiotics have been used for the control of plant diseases. Thus, blasticidin S and kasugamycin have been in practical use for the rice blast control and cellocidin and chloramphenicol have been practically used for the control of rice bacterial leaf blight. Furthermore, polyoxin has been put to practical use this year for the control of rice sheath blight and some fruit tree diseases. This review deals with mode of action of blasticidin S, kasugamycin, cellocidin, and polyoxin, all of which have been discovered and developed in the author's country.

II. Blasticidin S

Blasticidin S (TAKEUCHI et al. 1958) is a weakly basic compound produced by Streptomyces griseochromogenes which was isolated from a soil of Saigasaki, Wakayama (FUKUNAGA et al. 1955). This antibiotic inhibits various species of bacteria and fungi at five to 100 p.p.m. It gives excellent control of rice blast when sprayed at 10 to 20 p.p.m. on rice plants. Toxicity of blasticidin S is rather high, its oral LD50 to rats being 39.5 mg./kg. Accidental exposure to its dust particles causes conjunctivitis of eyes. No accident has been reported, however, with emulsions and wettable powder suspensions.

Blasticidin S free base has a molecular formula of $C_{17}H_{26}O_5N_8$. The chemical structure was established by OTAKE et al. (1965) as shown in Figure 1. It consists of a new nucleoside designated as

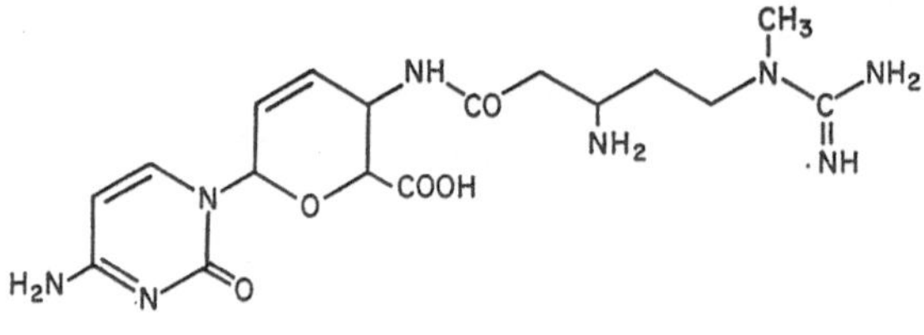

Fig. 1. Chemical structure of blasticidin S

cytosinene ($C_{10}H_{12}O_4N_4$) and a new amino acid, blastidic acid ($C_7H_{16}O_2N_4$).

a) Curative effects on rice blast

Blasticidin S exhibits a higher curative effect than phenylmercury acetate on rice blast. A smaller number of disease spots were found on rice plant leaves which were sprayed, one or two days after inoculation with P. oryzae spores, with a five p.p.m. solution of blasticidin S than those on leaves treated with 34 p.p.m. of phenylmercury acetate (Table I). An opposite relationship was obtained when these compounds were sprayed four days before inoculation, indicating that the protective effect of blasticidin S is somewhat inferior to that of the mercuric compound (MISATO et al. 1050).

Table I. *Effect of blasticidin S against leaf blast*

Fungicides	Conc. act. ingredient (mcg./ml.)	No. disease spots/pot [a]	
		Protective effect (spray 4 days before inoculation)	Therapeutic effect (spray 1 day after inoculation)
	2	96	19.3
	5	64	3.7
Blasticidin S	10	100	1.5
(wettable powder)	20	76	0.2
	50	42	0
	100	11	0
	200	10	0
Phenylmercuric acetate	10 [b]	48	10.0
(emulsifiable concentrate)	20 [b]	28	7.8
Untreated check	—	86	76.3

[a] Three pots/treatment, 10 seedling/pot, four leaves/seedling.
[b] As Hg.

b) Effect on protein synthesis of Piricularia oryzae

Blasticidin S inhibits oxygen consumption of *P. oryzae*. Oxygen uptake was inhibited to a greater extent in mycelia than in spores when glucose, pyruvate, succinate, or glutamate was used as the substrate (MISATO *et al.* 1961 a). The extent of inhibition by blasticidin S was almost constant throughout a concentration range between 0.1 and 100 p.p.m. On the other hand, this antibiotic did not inhibit succinic acid oxidation by mitochondria of *P. oryzae* even at 500 p.p.m. Blasticidin S had no effect on glycolysis, succinic acid dehydrogenation, oxidative phosphorylation, or incorporation of ^{32}P into nucleic acids (Table II). The effects of blasticidin S on oxygen consumption were, therefore, considered as secondary.

MISATO *et al.* (1961 b) noticed that blasticidin S caused a marked decrease in the rate of glutamic acid-^{14}C incorporation into the pro-

Table II. *Effect of blasticidin S on the metabolism of Piricularia oryzae*

Growth: agar dilution streak method, minimum inhibitory conc., five p.p.m.
     Shaking liquid culture, minimum inhibitory conc., 0.1 p.p.m.
Respiration: spore, mycelium, partially inhibited at 1 to 100 p.p.m.
Electron transport system: no inhibition at 100 p.p.m.
Succinic dehydrogenase: no inhibition at 100 p.p.m.
Oxidative phosphorylation: no inhibition at 100 p.p.m.
Incorporation of ^{32}P into the nucleic acid: no inhibition at two p.p.m.
Incorporation of ^{14}C-amino acids into the protein fraction: 100 percent inhibition
    at one p.p.m.

tein fraction of the mycelium. This inhibitory effect of blasticidin S was observed almost at the same level as the minimum growth-inhibitory concentration (one p.p.m.). This fact suggests that the primary effect of blasticidin S is the inhibition of protein synthesis. HUANG et al. (1964 a) found that the incorporation of amino acids-^{14}C was almost completely prevented by the antibiotic at one p.p.m. Amino acid activation and the transfer of activated amino acids to soluble RNA were inhibited to smaller extents at this concentration. The probable site of action of blasticidin S is the final step in protein synthesis which occurs on the ribosomal fraction (Fig. 2).

c) Resistant strains

NAKAMURA and SAKURAI (see HUANG et al. 1964 b) reported that P. oryzae developed resistance to blasticidin S when a normal blasti-cidin S-sensitive strain was grown successively in media containing

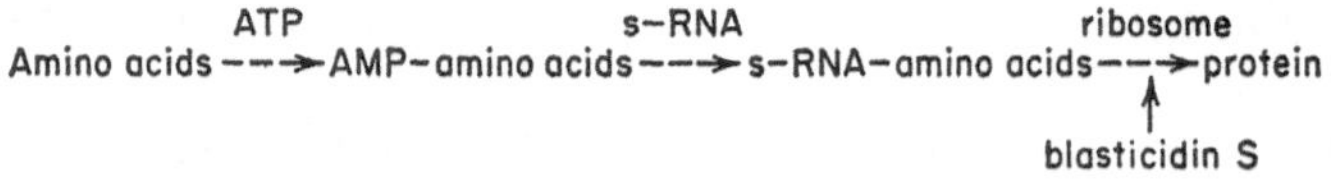

$$\underset{\text{Amino acids}}{} \overset{\text{ATP}}{\dashrightarrow} \underset{\text{AMP-amino acids}}{} \overset{\text{s-RNA}}{\dashrightarrow} \underset{\text{s-RNA-amino acids}}{} \overset{\text{ribosome}}{\underset{\underset{\text{blasticidin S}}{\uparrow}}{\dashrightarrow}} \text{protein}$$

Fig. 2. The action site of blasticidin S on protein synthesis

this antibiotic of increasing concentrations. The development of resis-tance to the antibiotic was associated with a loss in pathogenicity to the rice blast. The tolerance of the derived strain to blasticidin S seems to be based on its lower permeability than that of the normal sensitive parent strain (HUANG et al. 1964 b). Amino acid incorpora-tion into ribosomal protein by intact cells of the tolerant strain was not inhibited by this antibiotic even at 1,000 p.p.m., whereas protein synthesis in cell-free extracts of this strain was inhibited by the same antibiotic at one p.p.m. On the other hand, blasticidin S equally in-hibited the protein synthesis in the intact cell of P. oryzae and the cell-free extract.

d) Action on plants

Blasticidin S free base is toxic to plants at high concentrations. Yellowish (chlorotic) spots are produced on rice plant leaves within several days after spray application with blasticidin S above 40 p.p.m. Chlorotic spots are also observed on leaves of apple, pear, and peach trees and of cucumber, soybean, tomato, and other plants when they are sprayed with blasticidin S at or above 10 p.p.m. Of various deriva-tives tested, benzylaminobenzenesulfonate was found by ASAKAWA et al. (1963) to be least phytotoxic to the rice plant. HASHIMOTO et al. (1963) found that the relative content of nonprotein nitrogen, which consists mainly of amino acids and RNA, increased after treatment with blasticidin S. The increase in the tryptophane content was especially marked. The chlorophyll content in rice plants also increased when these rice plants were treated with blasticidin S at low concentrations. However, a marked decrease was observed in its content in leaf blades where chlorotic spots had been produced.

e) Effect on plant virus

In a preliminary report, KITANI et al. (1963) stated that the rate of infection of rice plants with stripe virus, which is transmitted by a leafhopper, Delphacodes striatella Fallen, was reduced by blasticidin S. HIRAI and SHIMOMURA (1965) observed that blasticidin S had a high inhibitory effect at a low concentration on the synthesis of tobacco mosaic virus (TMV). The antibiotic at 0.05 p.p.m. inhibited TMV multiplication in tobacco leaf disks by about 50 percent but caused no injury to leaves. At the same concentration, the antibiotic almost completely inhibited local lesions caused by TMV either on Nicotiana glutinosa or pinto bean. The same authors suggested that the synthesis of enzymes associated with the polymerization of TMV-RNA might be inhibited by blasticidin S.

f) Antagonistic action of detoxin to blasticidin S

Detoxin, which was found in the broth of Streptomyces caespitosum by YONEHARA et al. (1967), reduced phytotoxicity and irritant effect on eyes of blasticidin S without reducing the fungicidal activity of the antibiotic on rice blast. Detoxin also showed antagonistic action against antibacterial activity of blasticidin S on Bacillus cereus.

Experiments were carried out on the effect of detoxin on the incorporation of ^{14}C-blasticidin S into P. oryzae and B. cereus. The results suggest that detoxin specifically inhibits the penetration of blasticidin S into B. cereus but does not affect the incorporation of this antibiotic into P. oryzae.

III. Kasugamycin

Kasugamycin (UMEZAWA et al. 1965), a water-soluble basic antibiotic, is obtained from the culture broth of Streptomyces kasugaensis isolated from a soil of kasuga Shrine in Nara. Kasugamycin selectively inhibits P. oryzae and some bacteria including Pseudomonas species. This antibiotic controls rice blast disease when sprayed at about 20 p.p.m. It can be safely used without injuring rice plants and other crops. It has very low toxicity to mammals (oral LD_{50} to the mouse is > two g./kg.) and to fish (TLM to the carp is > one mg./ml.). The chemical structure was established by SUHARA et al. (1966) (Fig. 3.).

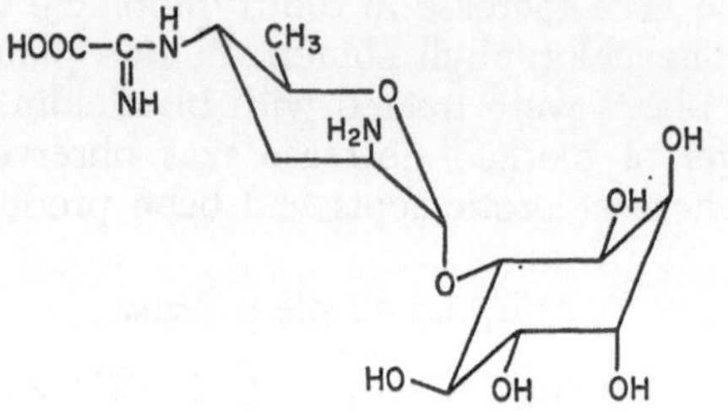

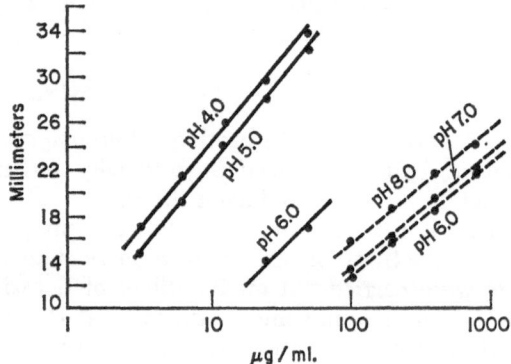

Fig. 3. Chemical structure of kasugamycin

a) Preventive effect on rice blast

Kasugamycin inhibits the growth of *P. oryzae* in acidic media (pH 5.0) but hardly inhibits it in neutral media (pH 7.0) (HAMADA *et al.* 1965). On the contrary, kasugamycin showed stronger inhibition against *Pseuedomonas* at pH 7.0 than at pH 5.0 or 6.0 (Fig. 4).

Fig. 4. Inhibition curve of kasugamycin hydrochloride: ——— *P. oryzae* (rice plant juice medium), ... *Ps. tabaci* (glucose peptone medium)

Kasugamycin has a preventive effect on rice blast. Its lowest effective concentration is almost similar to that of blasticidin S (ISHIYAMA *et al.* 1965). Unlike blasticidin S, however, kasugamycin has no toxicity to rice plants. Kasugamycin can be used safely to prevent rice blast at concentrations 20 times as high as the lowest effective concentration. Thus, when rice plants are infected heavily, kasugamycin can be used at high concentrations giving high effectiveness but causing no plant injury.

b) Inhibition of polypeptide synthesis

Kasugamycin markedly inhibited polyuridylate-stimulated poly-phenylalanine synthesis in a cell-free system derived from *Escherichia coli* or from *Psuedomonas flouorescens* (TANAKA *et al.* 1965). This antibiotic inhibits the incorporation of leucine into protein by the bacterial cell-free extract system and the incorporation of amino acids into protein by the growing cells. However, the inhibitory effects on the incorporation of these amino acids were lower than those on the polyphenylalanine synthesis. On the other hand, the bacterial RNA or DNA synthesis was not significantly affected.

These results indicate that kasugamycin interferes with the bacterial protein synthesis, without affecting the synthesis of nucleic acids. The inhibitory activity of kasugamycin on polyphenylalanine synthesis by the rat liver system was much lower than that on the bacterial system. The selective toxicity, therefore, seems to be attributable to the differential sensitivity of the mammalian and bacterial ribosomes to this antibiotic.

c) Resistant strains

Kasugamycin has minimum growth-inhibitory concentrations of 0.78 and 0.39 p.p.m. to Noken P-2 and C-1 strains, respectively, of *P. oryzae*, when assessed by the agar dilution method. OMORI (1966), however, found several colonies growing on potato-dextrose-agar media containing 100 p.p.m. of kasugamycin, when 10^{-6} to 10^{-7} spores of these strains were seeded to 10 ml. of the media. New strains obtained from these colonies were able to grow on similar media containing 100 p.p.m. of this antibiotic. Both susceptible strains and resistant strains showed similar morphological characteristics. These resistant strains were phytopathogenic to the rice plant, though the activities were somewhat lower than those of susceptible strains (OMORI 1966).

IV. Cellocidin

Cellocidin is an antibiotic produced by *Streptomyces chibaensis* isolated from a soil of Chiba Prefecture (SUZUKI *et al.* 1958). It showed an excellent preventive effect against rice bacterial leaf blight when sprayed on rice plant leaves at 100 to 200 p.p.m. Cellocidin has a very simple chemical structure of acetylenedicarboxamide containing only four carbon atoms. Technical cellocidin for commercial formulations is now synthesized chemically.

$$C - CONH_2$$
$$|||$$
$$C - CONH_2$$
Cellocidin

Cellocidin exhibits high mammalian toxicity when injected intravenously, its LD_{50} to the mouse being 11 mg./kg. In oral administration and cutaneous application, however, this antibiotic shows very low mammalian toxicity, LD_{50} to the mouse being 89 to 125 and 667 mg./kg., respectively. Toxicity of cellocidin to fish is lower than that of DDT. The antibiotic sometimes causes plant injuries when solutions above 200 p.p.m. are sprayed. Lower concentrations are, therefore, preferred in the spray application of cellocidin.

Effects of cellocidin on the metabolism of *Xanthomonas oryzae* were studied by OKIMOTO and MISATO (1963 a and b) (Table III).

Table III. *Effect of cellocidin on Xanthomonas oryzae*

Growth: agar dilution streak method, M. I. C. 10 p.p.m.
        shaking liquid culture, M. I. C. four p.p.m.
Respiration: endogenous respiration, about 50 percent inhibition at 100 p.p.m.
        exogenous respiration, strong inhibition at 10 to 100 p.p.m.
        (except succinate)
Glucose utilization: EMP glycolytic system, 100 percent inhibition at 100 p.p.m.
                                 60 percent inhibition at   10 p.p.m.
        pentose phosphate pathway, no inhibition at 100 p.p.m.
Dehydrogenase: DPN requiring dehydrogenase, strong inhibition at 100 p.p.m.
        succinic dehydrogenase, urease, no inhibition at 100 p.p.m.
Electron transport system: DPNH oxidase $\left.\vphantom{\begin{matrix}a\\b\end{matrix}}\right\}$ no inhibition at 100 p.p.m.
                            cytochrome oxidase
Nucleic acid metabolism: $\left.\vphantom{\begin{matrix}a\\b\end{matrix}}\right\}$ no inhibition at
Incorporation of ^{14}C-amino acids into the protein fraction: four p.p.m.
Transaminase: glutamate – oxaloacetate transaminase $\left.\vphantom{\begin{matrix}a\\b\end{matrix}}\right\}$ no inhibition at 10 p.p.m.
                 glutamate – pyruvate transaminase
α-Ketoglutarate $\rightarrow$ succinyl CoA $\rightarrow$ succinate: 100 percent inhibition at one p.p.m.

The growth of *X. oryzae* in shaken liquid media was completely inhibited by cellocidin at four p.p.m. when cellocidin was added six hours after innoculation, whereas the minimum growth-inhibitory concentration of this antibiotic against this bacterium was 10 p.p.m. as assessed by the agar dilution streak method. Cellocidin at 10 p.p.m. inhibited endogenous respiration of *X. oryzae* resting cells by about 50 percent. The inhibition was still not complete at 100 p.p.m. This antibiotic inhibited exogenous respiration of *X. oryzae* in the presence of glucose, gluconate, pyruvate, or acetate mostly at 10 p.p.m. and completely at 100 p.p.m. When succinate was used as the substrate, cellocidin only slightly inhibited the exogenous respiration even at 100 p.p.m.

The pathway in glucose utilization by *X. oryzae* and the effects of cellocidin on the pathway were studied by measuring the amounts of lactic acid and pyruvate formed by resting cells or by cell-free extracts of the bacterium in the presence of various metabolic intermediates of the EMP glycolytic system or the 6-PG splitting system. The results suggest that *X. oryzae* has two metabolic pathways, the

EMP pathway and the 6-PG splitting system, in glucose utilization. Cellocidin inhibited the EMP pathway but not the 6-PG splitting system at 10 to 100 p.p.m.

Oxidation of various intermediates in the Krebs' cycle by cell-free extracts of X. *oryzae* and effects of cellocidin on the oxidation were examined by the triphenyltetrazolium chloride reduction method. Cellocidin selectively inhibited NAD-requiring dehydrogenases, such as α-ketoglutarate dehydrogenase, glutamic dehydrogenase, and malic dehydrogenase, at the minimum growth inhibitory concentration of 10 p.p.m., but did not inhibit dehydrogenases which did not require NAD, such as succinic dehydrogenase and urease. The antibacterial effect of cellocidin on X. *oryzae* was antagonized by cysteine or glutathione. Cellocidin was, therefore, considered to interact with SH groups. Oxygen uptake by cell-free extracts of X. *oryzae* was inhibited by cellocidin at 100 p.p.m. when substrates, such as α-ketoglutarate, isocitrate, etc., which require NAD as a coenzyme for oxidation, were used, but was not inhibited when succinate was used as the substrate.

Cellocidin inhibited the oxidation of α-ketoglutarate by cell-free extracts of X. *oryzae* in the presence of NAD and malonate. The inhibition was observed at one p.p.m. of cellocidin, which was lower than the minimum growth-inhibitory concentration. Cellocidin was, therefore, considered to act as inhibitor on the α-ketoglutarate → succinate system.

Cellocidin did not inhibit electron transport systems, such as $NADH_2$ oxidase and NAD reductase, of cell-free preparation of X. *oryzae* at 100 p.p.m. This antibiotic inhibited the incorporation of ^{14}C-amino acids into protein fraction of X. *oryzae* at 10 p.p.m. neither.

V. Polyoxin

Polyoxin is an antibiotic produced by *Streptomyces cacaoi* var. *asoensis* which was isolated from a soil in the Aso area of Kumamoto Prefecture (SUZUKI *et al.* 1965). It is a mixture of several closely related compounds (ISONO *et al.* 1966 and 1967). Nine components, polyoxin A through I, have been isolated. Alkaline hydrolysis of polyoxin A afforded four degradation products liberating each one mole of ammonia and carbon dioxide; that is, 5-hydroxymethyluracil, polyoxin C, polyoximic acid, and polyoxamic acid (Fig. 5). Similar hydrolyses of polyoxins B, D, E, F, G, H and I clarified their molecular constituents, which are summarized in Table IV. In contrast to polyoxins A, B, C, G, and I having 5-hydroxymethyluracil as their chromophore, polyoxins D, E, and F have uracil-5-carboxylic acid and polyoxin H has thymine. Polyoxins A, F, H, and I have polyoximic acid,

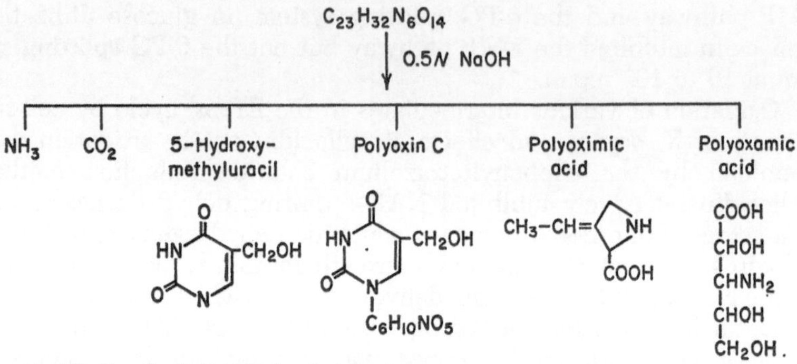

$$C_{23}H_{32}N_6O_{14}$$

$$\downarrow \text{0.5}N \text{ NaOH}$$

| NH₃ | CO₂ | 5-Hydroxy-methyluracil | Polyoxin C | Polyoximic acid | Polyoxamic acid |

Fig. 5. Alkaline hydrolysis of polyoxin A

whereas others did not. Polyoxins A, B, D, F, and H had a trihydroxy amino acid, polyoxamic acid, while polyoxins E and G did not; in-

Table IV. *Molecular Constituents of Polyoxins*

Polyoxin	Chromophore	Polyoximic acid	Polyoxamic acid	Desoxy-polyoxamic acid
A	5-Hydroxymethyluracil	+	+	–
B	5-Hydroxymethyluracil	–	+	–
C	5-Hydroxymethyluracil	–	–	–
D	Uracil-5-carboxylic acid	–	+	–
E	Uracil-5-carboxylic acid	–	–	+
F	Uracil-5-carboxylic acid	+	+	–
G	5-Hydroxymethyluracil	–	–	+
H	Thymine	+	+	–
I	5-Hydroxymethyluracil	+	–	–

stead, the latter two gave an amino acid which was designated as desoxypolyoxamic acid.

Polyoxin components are not toxic to mammals or fish. Oral administration with polyoxin at 15 g./kg. to mice did not cause any adverse effect. Polyoxin does not affect human eyes or skin either. Spray application with 200 p.p.m. of polyoxin had no injurious effects on many kinds of plants tested. Particularly, no injury was observed on rice plant leaves when sprayed with this antibiotic at 800 p.p.m.

These components show marked differences in their activities toward each of various fungi (Table V). Thus, polyoxin D is most active to the rice sheath blight fungus, while polyoxin B and G are most active to the pear black spot fungus and to the apple cork spot fungus. Spray application of polyoxin B at 50 to 100 p.p.m. gave excellent control of rice sheath blight, pear black spot, and apple cork spot in field trials. Polyoxin has been put on the market quite recently.

Table V. Antifungal Activities of Polyoxins

Test-organism	M.I.C. (μg./ml.)								
	A	B	C	D	E	F	G	H	I
Piricularia oryzae	3.12	6.25	>100	3.12	12.5	25	6.25	3.12	>100
Cochliobolus miyabeanus	3.12	3.12	>100	6.25	12.5	6.25	3.12	25	>100
Pellicularia sasakii	12.5	1.56	>100	<1.56	1.56	50	1.56	50	>100
Alternaria kikuchiana	50	12.5	>100	50	50	>100	6.25	12.5	>100
Glomerella cingulata	>100	>100	>100	>100	>100	>100	>100	>100	>100
Physalospora laricina	25	3.12	>100	100	50	>100	6.25	12.5	>100
Cladosporium fulvum	3.12	1.56	>100	100	25	25	3.12	6.25	>100
Fusarium oxysporum	>100	>100	>100	>100	>100	>100	>100	>100	>100

Polyoxin components, except inactive polyoxin C and I, have very specific high activities to some phytopathogenic fungi. These components were, on the other hand, found to be inactive to all kinds of bacteria tested (SASAKI et al. 1966).

Polyoxin B has high inhibitory effects on the mycelial growth and sporulation of sensitive fungi, but has only slight inhibitory effects on the spore germination of these fungi. Although polyoxin-treated spores of sensitive fungi do germinate, their germ-tubes do not enlongate but swell to take global forms having diameters two to three times larger than those of the original spores. These deformed spores have no pathogenicity. Similar phenomena of swelling are also observed with fungal mycelia. Studies are now in progress on the effect of polyoxin on the metabolism of several sensitive fungi.

Summary

This review deals with mode of action of the four agricultural antibiotics, blasticidin S, kasugamycin, cellocidin, and polyoxin, all of which have been discovered and developed for practical use in Japan. Blasticidin S or kasugamycin exhibits an excellent curative effect on rice blast and the inhibition of protein synthesis in *Piricularia oryzae* is considered as the primary effect of the antibiotic. Cellocidin shows a preventive effect against rice bacterial leaf blight and is considered to act as inhibitor on the α-ketoglutarate $\rightarrow$ succinate system in the Krebs' cycle in *Xanthomonas oryzae*. Polyoxin has very specific high activities to the rice sheath blight, the pear black spot, or the apple cork spot and interferes with the biosynthesis of cell walls in their sensitive fungi.

Résumé *

Mode d'action des antibiotiques à utilisation agricole produits au Japon

Cette mise au point concerne le mode d'action des quatre antibiotiques à utilisation agricole: blasticidine S, kasugamycine, cellocidine et polyoxine, tous découverts et rendus utilisables au Japon. La blasticidine S ou la kasugamycine ont un effet curatif excellent sur la brunissure du riz: l'inhibition de la synthèse protei que dans *Piricularia oryzae* est considérée comme l'effet primordial de ces antibiotiques. La cellocidine agit préventivement contre la nécrose bactérienne de la feuille de riz et il est admis qu'elle inhibe le système α-cétoglutarate $\rightarrow$ succinate dans le cycle de Krebs chez *Xanthomonas oryzae*. La polyoxine présente de grandes activités très spécifiques de la flétrissure de la gaine du riz, de la tavelure de la poire ou de la maladie de la tache de liège de la pomme: elle interfère avec la biosynthèse des parois cellulaires des champignons sensibles.

---

* Traduit par R. MESTRES.

Zusammenfassung *

Die Wirkungsweise von in Japan entwickelten landwirtschaftlichen Antibiotika

Dieser Ueberblick behandelt die Wirkungsweise der vier landwirtschaftlichen Antibiotika Blasticidin S, Kasugamycin, Cellocidin und Polyoxin, welche alle in Japan entdeckt und für den praktischen Gebrauch entwickelt wurden. Blasticidin S oder Kasugamycin zeigen eine ausgezeichnete heilende Wirkung bei Reisbrand, und die Hemmung der Proteinsynthese in *Piricularia oryzae* wird als die Hauptwirkung des Antibiotikums angesehen. Cellocidin zeigt eine Schutzwirkung gegen bakteriziden Blattmeltau an Reis, und man glaubt, dass es als Hemmer des α-Ketoglutarat → Succinatsystems im Krebszyklus des *Xanthomonas oryzae* wirkt. Polyoxin hat eine sehr spezifische hohe Aktivität gegen den Reisstengelmeltau, die Birnen-Schwarzfleckenkrankheit oder den "apple cork spot" und hindert die Biosynthese der Zellwände in ihren empfindlichen Pilzen.

References

ASAKAWA, M., T. MISATO and K. FUKUNAGA: Studies on the prevention of the phytotoxicity of blasticidin S. Pesticide and Technique (Tokyo) 8, 24 (1963).

FUKUNAGA, K., T. MISATO and M. ASAKAWA: Blasticidin, a new antiphytopathogenic fungal substance. Bull. Agr. Chem. Soc. Japan 19, 181 (1955).

HAMADA, M., T. HASHIMOTO, T. TAKAHASHI, S. YOKOYAMA, M. MIYAKE, T. TAKEUCHI, Y. OKAMI and H. UMEZAWA: Antimicrobial activity of kasugamycin. J. Antibiotics (Tokyo), Ser. A. 18, 104 (1965).

HASHIMOTO, K., M. KATAGIRI and T. MISATO: Studies on the phytotoxic action of an antiblastic antibiotic, blasticidin S. I. The effect of blasticidin S on the content of free amino acids, proteins and nucleic acids in rice leaf and yeast. J. Agr. Chem. Soc. Japan 37, 245 (1963).

HIRAI, T., and T. SHIMOMURA: Blasticiden S, an effective antibiotic against plant virus multiplication. Phytopathology 55, 291 (1965).

HUANG, K. T., T. MISATO and H. ASUYAMA: Effect of blasticidin S on protein synthesis of Piricularia oryzae. J. Antibiotics (Tokyo), Ser. A 17, 65 (1964a).

——, Selective toxicity of blasticidin S to Piricularia oryzae and Pellicularia sasakii. J. Antibiotics (Tokyo), Ser. A 17, 71 (1964b).

ISHIYAMA, T., I. HARA, M. MATSUOKA, K. SAITO, S. SHIMADA, R. IZAWA, T. HASHIMOTO, M. HAMADA, Y. OKAMI, T. TAKEUCHI and H. UMEZAWA: Studies on the preventive effect of kasugamycin on rice blast. J. Antibiotics (Tokyo), Ser A 18, 115 (1965).

ISONO, K. and S. SUZUKI: Studies on polyoxins, antifungal antibiotics, Part II. Degradative study of Polyoxin A. Agr. Biol. Chem., 30, 813 (1966).

ISONO, K., J. NAGATSU, K. KOBINATA, K. SASAKI and S. SUZUKI: Studies on polyoxins, antifungal antibiotics, Part II. Isolation and characterization of polyoxins C, D, E, F, G, H and I. Agr. Biol. Chem., 31, 190 (1967).

KITANI, S. and A. KISO: Studies on the chemical prevention against rice stripe virus disease. Ann. Phytopath. Soc. Japan 28, 293 (1963) (Abstr.).

---

* Übersetzt von A. SCHUMANN.

MISATO, T., I. ISHII, M. ASAKAWA, Y. OKIMOTO and K. FUKUNAGA: Antibiotics as protectant fungicides against rice blast. II. The therapeutic action of blasticidin S, Ann. Phytopath. Soc. Japan 24, 302 (1959).

——, Antibiotics as protectant fungicides against rice blast. III. Effect of blasticidin S on respiration of Piricularia oryzae. Ann. Phytopath. Soc. Japan. 26, 19 (1961a).

——, Antibiotics as protectant fungicides against rice blast. IV. Effect of blasticidin S on the metabolism of Piricularia oryzae. Ann. Phytopath. Soc. Japan. 26, 25 (1961b).

OKIMOTO, Y. and T. MISATO: Antibiotics as protectant bactericide against bacterial leaf blight of rice plant. (2) Effect of cellocidin on growth, respiration, and glycolysis of Xanthomonas oryzae. Ann. Phytopath. Soc. Japan. 28, 209 (1963).

——, Antibiotics as protectant bactericide against bacterial leaf blight of rice plant. (3) Effect of cellocidin on TCA cycle, electron transport system, and metabolism of protein in Xanthomonas oryzae. Ann. Phytopath. Soc. Japan. 28, 250 (1963).

OMORI, K.: Kasugamycin-resistant strains of Piricularia oryzae. Nōgyōgijutsu 21, 479 (1966).

OTAKE, N., S. TAKEUCHI, T. ENDO and H. YONEHARA: The structure of blasticidin S. Tetrahedron Letters No 19, 1411 (1965).

SASAKI, S., T. AKASHIBA, J. EGUCHI, N. OHTA, T. TSUCHIYAMA and S. SUZUKI: Studies on Polyoxins. (1) Preventive effect on Sheath Blight. Ann. Phytopathol, Soc. Japan 32, 99 (1966) (in Japanese).

SUHARA, Y., K. MAEDA, H. UMEZAWA and M. OHNO: Chemical Studies on kasugamycin. V. The structure of kasugamycin. Tetrahedron Letters 12, 1239 (1966).

SUZUKI S., G. NAKAMURA, K. OKUMA and Y. TOMIYAMA: Cellocidin, a new antibiotic. J. Antibiotics (Tokyo), Ser. A 11, 81 (1958).

SUZUKI, S., K. ISONO, J. NAGATSU, T. MIZUTANI, Y. KAWASHIMA and T. MIZUNO: A new antibiotic, Polyoxin A. J. Antibiotics (Tokyo), Ser. A 18, 131 (1965).

TAKEUCHI, S., K. HIRAYAMA, K. UEDA, H. SAKAI and H. YONEHARA: Blasticidin S, a new antibiotic. J. Antibiotics (Tokyo), Ser. A 11, 1 (1958).

TANAKA, N., T. NISHIMURA, H. YAMAGUCHI, C. YAMAMOTO, Y. YOSHIDA, K. SASHIKATA and H. UMEZAWA: Mechanism of action of kasugamycin. J. Antibiotics (Tokyo) Ser. A 18, 139 (1965).

UMEZAWA, H., Y. OKAMI, T. HASHIMOTO, Y. SUHARA, M. HAMADA and T. TAKEUCHI: A new antibiotic, kasugamycin. J. Antibiotics (Tokyo) Ser. A 18, 101 (1965).

YONEHARA, H., H. SETO, S. AIZAWA, T. HIDAKA, A. SHIMAZU and N. OTAKE: The detoxin complex, selective antagonists of blasticidin S. J. Antibiotics (Tokyo) Ser. A 21, 369 (1968).

The fungitoxic mechanisms in
quinoline compounds and their chelates

by
GEORGE L. McNEW * and HERMAN GERSHON *

Contents

I. Introduction

It must be remembered that any change in the chemical substituents of a molecule has many ancillary effects other than chemical
reactivity which may have a bearing upon its pesticidal effectiveness.
The replacement of a hydrogen by a halogen may make the compound
into an alkylating agent, but it does much more. There are inevitable
shifts in electron density that change the reactivity of adjacent groups
or even the charge distribution on the entire molecule so its movement in solution and affinity for other molecules and surfaces will
be affected. There may be obvious shifts induced in hydrophilic and
lipophilic properties induced by certain well recognized types of substituent groups. Even the size of the molecule and its steric configuration may be affected so the molecule will have different physical
dimensions.

Because of such far-reaching secondary changes in the chemical
and physical attributes of the molecule, much of the reasoning on the
relationship of chemical structure to mechanism of action should be
viewed suspiciously. Ideas presented as proof that there has been a

---

* Boyce Thompson Institute, Yonkers, New York.

change in chemical reactivity induced by the use of particular substituent groups may frequently be more logically explained by changes in the physical behavior of the molecule in approaching, becoming attached to, and permeating cell walls and cell membranes.

With this broad viewpoint in mind, we would like to reconsider some of the data available on the chelation potential of substituted quinoline compounds as a fungitoxic mechanism. The series of events that have occurred over the past 25 years reveal both the potential and the weaknesses of studying a series of compounds for fungitoxicity in order to arrive at a concept as to the mechanism of their action. Some data obtained by one of us (H.G.) will point the way toward which studies must be oriented in determining distribution patterns of the molecules in establishing the validity of new hypotheses on the basic relation of structure to activity.

ZENTMYER (1943) advanced the hypothesis that 8-quinolinol (8-hydroxyquinoline, oxine) may be a fungicide because of its capacity to chelate heavy metal ions. If the available metals are removed from the living cell, its essential enzymes would become non-functional and the cell would be either destroyed or immobilized by failure of its basic physiological functions. Strong support for this simple, straightforward hypothesis was soon forthcoming in the research on bactericidal activity of hydroxyquinoline compounds by ALBERT and MAGRATH (1947), ALBERT et al. (1947 and 1953) and RUBBO et al. (1950). They were able to demonstrate that only 8-quinolinol of a series of seven hydroxyquinoline isomers was capable of chelating metallic ions by virtue of the relationship of the hydroxyl group to the heterocyclic nitrogen and was also active against the test bacteria. A typical reaction involving the use of copper ions as a representative metal would be as follows:

8-Quinolinol   2:1 Complex (active)

The fact that there are many organic compounds with potential chelating ability that are not fungicides need not detract from this hypothesis. For example, it is conceivable that a strong chelator of essential metals such as EDTA (ethylenediamine tetraacetic acid) or citric acid may not penetrate the cell in sufficient quantity to immobilize its enzymes or conceivably could be degradated before it could interfere with vital processes.

The first very valid argument against the hypothesis came when POWELL (1946) and MASON (1948) obtained evidence that the 2:1 complex of 8-quinolinol with copper [copper(II) 8-quinolinolate or copper oxine] was a better fungicide than 8-quinolinol. Since the organic moieties' chelation potential had been fully satisfied, the molecule should not operate further as a chelating agent unless it dissociated in the cell environment. Proof was advanced by ALBERT et al. (1953) and GREATHOUSE et al. (1954) that 8-quinolinol was inactive in media free of metallic ions where microorganisms could grow in its presence. This theory was fairly questioned by HORSFALL (1956) who pointed out that in the absence of essential bivalent metals the organisms could not have sustained growth that could have been inhibited by the chelation of such metals. It was shown that an excess of copper ions, as well as excess 8-quinolinol in the presence of copper (II) 8-quinolinolate, would prevent its fungitoxity (ALBERT et al. 1953, GREATHOUSE et al. 1954).

To explain these several facts, the following rationalizations of the dissociation theory advanced by ALBERT et al. (1953) have been offered. The copper(II) 8-quinolinolate is more fungitoxic than either copper or 8-quinolinol because (1) the organic moiety imparts a lipid solubility to the copper that accelerates cell permeation and (2) once inside the cell, the 2:1 complex dissociates into the 1:1 complex of 8-quinolinol and copper and frees an equivalent of 8-quinolinol. The 1:1 complex could become the active toxicant by combining with and blocking metal binding sites in enzymes.

1:2 Complex   1:1 Half complex

In the presence of an excess of copper ions, the equilibrium would be shifted to the right and the major portion of the available material would be in a highly hydrophilic condition so that it would not penetrate the lipid membrane; whereas, excess 8-quinolinol which could also pentrate the cell would shift the equilibrium to the left, thereby diminishing the release of the toxic 1:1 complex.

Although significant progress has been made in understanding the mechanism of antimicrobial action of 8-quinolinol type compounds, most of the explanatory material is in the form of hypotheses. Definite proof that the antimicrobial action takes place within the cell is lacking. Proof that the 1:1 copper(II) 8-quinolinol complex which results from dissociation of the 1:2 complex is the active toxicant has not been established with certainty. Does the antifungal activity of 8-quinolinol depend on the formation of a copper(II) complex? Whether chelation is the sole mode of antimicrobial action is still subject to question. Finally, what role is played by substituents on the antimicrobial activity of these compounds?

A very crucial point for clearing up some of the remaining reservations would be to prepare 8-quinolinol complexes which would not release 8-quinolinol in the first dissociation and would have different stability constants. The opportunity to make such experiments was made possible by the announcement of GERSHON et al. (1962) of a series of antimicrobial mixed chelates of copper, substituted 8-quinolinols, and substituted arylhydroxy acids such as 4-bromo-3-hydroxy-2-naphthoic acid and 3,5-diiodosalicylic acid. These mixed chelates are fungitoxic but upon dissociation they release the two organic acids which are relatively non-toxic according to the following scheme:

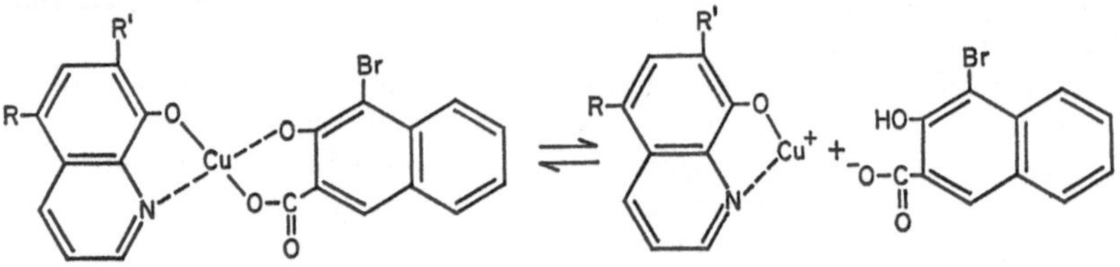

1:1:1 Complex   1:1 Complex   Bromohydroxy
                   of copper (II)   naphthoic acid
                   8-quinolinol   (inactive)

II. Materials and methods

A series of 8-quinolinol compounds was synthesized with different electron-attracting and electron-donating groups in the 5- and 7-positions so as to change the electron densities around the ligand ring. This should modify the dissociation constants of any ligands and complexes formed. These substituted hydroxyquinoline compounds were then used to prepare their respective copper(II) bis chelates and a series of mixed chelates (1:1:1) with copper and either 4-bromo-3-hydroxy-2-naphthoic acid or 3,5-diiodosalicylic acid. For studies on possible non-chelating mechanisms of fungitoxicity, a number of 8-methoxyquinoline derivatives were prepared with substituents in the 5- and 5,7-positions.

These compounds were dissolved in dimethyl sulfoxide, added to Sabouraud Dextrose Broth (Difco) in graded levels from one to 100 p.p.m. A final concentration of one percent of dimethyl sulfoxide was used after preliminary tests had shown that the solvent at this concentration did not materially affect the results. In order to observe any possible effects of the macromolecules of the medium on the activity of the test chemicals, selected compounds in the several series were further tested in a simple medium by employing Czapek Dox Broth (Fisher). This was adjusted to pH 5.7 with 10 percent hydrochloric acid. Results were in close agreement to those obtained in Sabouraud Dextrose Broth.

Spores of five test organisms grown on Sabouraud Dextrose Agar (Difco) were transferred by pipette to inoculate flasks containing the test chemicals. These fungi were *Aspergillus niger* van Tieghem, *Aspergillus oryzae* (Ahlburg) Cohn, *Trichoderma viride* Persoon ex Fries, *Myrothecium verrucaria* Ditmar ex Fries, and *Trichophyton mentagrophytes* (Robin). After incubation for six days on a rotary shaker at 28° C. the flasks were examined for evidence of mycelial growth. Those showing no growth were diluted one to 100 with sterile medium and incubated for two weeks to ascertain whether the chemical had been lethal or merely fungistatic in its effects. All data were recalculated as micromoles per liter.

III. Results and discussion

a) Differential reaction of the test fungi

The five species of fungi did not react in the same fashion. The two species of *Aspergillus* were sensitive to several test compounds but as

Table I. *Relative toxicities of a series of substituted 8-quinolinols, their copper (II) bis chelates (2:1), and mixed copper(II) chelates with aryl hydroxy acids (1:1:1) to Aspergillus niger in Sabouraud Dextrose Broth*

Substituents on 8-quinolinol ring	Threshold conc. (μmoles/l.) for toxic effect [a]							
	Fungistatic effect				Fungicidal effect			
	Non-chelated	Bis chelated	Mixed chelated [b]		Non chelated	Bis chelated	Mixed chelated [b]	
			Naph.	Salicyl.			Naph.	Salicyl.
None	76	8.5	8.8	8.4	NA	NA	NA	NA
5-Fluoro	74	10	10	9.8	NA	NA	NA	NA
5-Chloro	17	NA	24	25	380	NA	NA	NA
5-Bromo	8.9	NA	58	62	21	NA	NA	NA
5-Iodo	11	NA	NA	NA	NA	NA	NA	NA
5-Nitro	32	NA	NA	NA	NA	NA	NA	NA
5-Nitroso	NA	—	—	—	—	—	—	—
5-Amino(2HCl)	NA	—	—	—	—	—	—	—
5,7-Dichloro	9.4	NA	—	—	NA	—	—	—
5,7-Dibromo	9.9	NA	—	—	NA	—	—	—
5,7-Diiodo	88	NA	—	—	NA	—	—	—
5,7-Dinitro	NA	NA	—	—	—	—	—	—
5,7-Diamino(HCl)	NA	NA	—	—	—	—	—	—

[a] NA = not active below 100 p.p.m.

[b] Naph. = 4-bromo-3-hydroxy-2-naphthoic acid; Salicyl. = 3,5-diiodosalicylic acid.

shown by the data on A. *niger* in Table I; the active compounds were primarily fungistatic but rarely fungicidal. Although some members were inhibitory at dosages as low as < 10 μmoles/l., the inoculum recovered and grew well after exposure to concentrations of 100 p.p.m. when the concentration was reduced to 10 μmoles or less. The data on A. *oryzae* presented in Table II agree in nearly all essentials to

Table II. *Relative toxicities of a series of substituted 8-quinolinols, their 2:1 copper(II) bis chelates, and their mixed (1:1:1) chelates with copper(II) and aryl hydroxy acids for Aspergillus oryzae growing in Sabouraud Dextrose Broth*

Substituents on 8-quinolinol nucleus	Threshold conc. (μmoles/l.) for toxic effect [a]							
	Fungistatic effect				Fungicidal effect			
	Non-chelated	Bis chelated	Mixed chelated [b]		Non chelated	Bis chelated	Mixed chelated [b]	
			Naph.	Salicyl.			Naph.	Salicyl.
None	120	8.5	11	13	NA	8.5	34	34
5-Fluoro	200	10	12	13	NA	13	110	NA
5-Chloro	39	NA	51	51	NA	NA	NA	NA
5-Bromo	36	NA	92	110	NA	NA	NA	NA
5-Iodo	15	NA	NA	NA	NA	NA	NA	NA
5-Nitro	16	NA	NA	NA	NA	NA	NA	—
5-Nitroso	NA	—	—	—	—	—	—	—
5-Amino(2HCl)	NA	—	—	—	—	—	—	—
5,7-Dichloro	23	NA	—	—	NA	NA	—	—
5,7-Dibromo	20	NA	—	—	NA	NA	—	—
5,7-Diiodo	NA	NA	—	—	NA	NA	—	—
5,7-Dinitro	NA	NA	—	—	NA	NA	—	—
5,7-Diamino(HCl)	NA	NA	—	—	NA	NA	—	—

[a] NA = not active below 100 p.p.m.

[b] Naph. = 4-bromo-3-hydroxy-2-naphthoic acid; Salicyl. = 3,5-diiodosalicylic acid.

those for A. *niger* and differ only in the magnitude of dosage response. The other three species, on the other hand, had a greater propensity

Table III. *Relative toxicities of a series of substituted 8-quinolinols, their copper (II) bis chelates (2:1), and their mixed chelates (1:1:1) with copper(II) and two substituted aryl hydroxy acids for Trichoderma viride in Sabouraud Dextrose Broth*

Substituents on 8-quino-linol ring	Threshold conc. (μmoles/1.) for toxic effect [a]							
	Fungistatic effect				Fungicidal effect			
	Non-che-lated	Bis che-lated	Mixed chelated [b]		Non che-lated	Bis che-lated	Mixed chelated [b]	
			Naph.	Sálicyl.			Naph.	Salicyl.
None	21	8.5	6.3	6.7	620	NA	44	37
5-Fluoro	25	7.8	8.2	6.5	520	NA	NA	NA
5-Chloro	11	7.2	16	17	130	NA	NA	NA
5-Bromo	18	NA	40	41	85	—	NA	NA
5-Iodo	18	NA	NA	NA	66	—	—	—
5-Nitro	16	NA	NA	NA	53	—	—	—
5-Nitroso	NA	—	—	—	—	—	—	—
5-Amino(2HCl)	NA	—	—	—	—	—	—	—
5,7-Dichloro	14	NA	—	—	56	NA	—	—
5,7-Dibromo	17	NA	—	—	30	NA	—	—
5,7-Diiodo	23	NA	—	—	NA	NA	—	—
5,7-Dinitro	NA	NA	—	—	NA	NA	—	—
5,7-Diamino(HCl)	NA	NA	—	—	NA	NA	—	—

[a] NA = not active below 100 p.p.m.

[b] Naph. = 4-bromo-3-hydroxy-2-naphthoic acid; Salicyl. = 3,5-diiodosalicylic acid.

to succumb to the chemical at inhibitory or somewhat higher dosages, as shown by the data in Tables III, IV, and V. Their growth also was inhibited at slightly lower dosages of the toxic chemicals than were the two species of *Aspergillus*.

The substituents on the unchelated 8-quinolinol had a very decided effect on fungitoxicity for all five species so the following generalizations would seem to be valid. In the unchelated compounds, a fluorine in the 5-position had no appreciable effect on either fungistatic or fungicidal activity for any of the species, as compared with 8-

Table IV. *Relative toxicities of a series of substituted 8-quinolinols, their copper (II) bis chelates, and their mixed (1:1:1) chelates with copper(II) and an aryl hydroxy acid for Myrothecium verrucaria in Sabouraud Dextrose Broth*

Substituents on 8-quino-linol ring of ligand	Threshold conc. (μmoles/l.) for toxic effect [a]							
	Fungistatic effect				Fungicidal effect			
	Non-che-lated	Bis che-lated	Mixed chelated [b]		Non che-lated	Bis che-lated	Mixed chelated [b]	
			Naph.	Salicyl.			Naph.	Salicyl.
None	28	8.8	13	17	34	8.5	13	17
5-Fluoro	25	7.8	14	18	37	1.0	14	18
5-Chloro	5.6	7.2	26	27	5.6	79	80	71
5-Bromo	4.5	NA	47	47	8.9	NA	54	120
5-Iodo	11	NA	NA	NA	11	NA	NA	NA
5-Nitro	11	NA	NA	NA	16	NA	NA	NA
5-Nitroso	NA	—	—	—	—	—	—	—
5-Amino(2HCl)	NA	—	—	—	—	—	—	—
5,7-Dichloro	4.5	10	—	—	4.5	132	—	—
5,7-Dibromo	9.9	NA	—	—	9.9	NA	—	—
5,7-Diiodo	20	NA	—	—	28	NA	—	—
5,7-Dinitro	NA	—	—	—	—	—	—	—
5,7-Diamino(HCl)	NA	—	—	—	—	—	—	—

[a] NA = not active below 100 p.p.m.
[b] Naph. = 4-bromo-3-hydroxy-2-naphthoic acid; Salicyl. = 3,5-diiodosalicylic acid.

hydroxyquinoline; chlorine, bromine, iodine, and the nitro group in this position enhanced activity very appreciably for all species. This was especially conspicuous for the species of *Aspergillus*. Either a nitroso or amino group in this position markedly diminished fungitoxicity of the 8-hydroxyquinoline. The 5,7-dichloro and dibromo analogs also were highly active. The 5,7-diiodo analog was as active as the unsubstituted compound for all fungi except *A. oryzae*. The 5,7-dinitro and diamino compounds were essentially inactive.

Without exception all five species were injured more severely by the 2:1 copper(II) complexes containing the unsubstituted and 5-

Table V. *Relative toxicities of a series of substituted 8-quinolinols, their copper (II) bis chelates, and their mixed (1:1:1) chelates with copper(II) and an aryl hydroxy acid for Trichophyton mentagrophytes in Sabouraud Dextrose Broth*

Substituents on 8-quinolinol ring of ligand	Threshold conc. (μmoles/l.) for toxic effect[a]							
	Fungistatic effect				Fungicidal effect			
	Non-che-lated	Bis che-lated	Mixed chelated[b]		Non che-lated	Bis che-lated	Mixed chelated[b]	
			Naph.	Salicyl.			Naph.	Salicyl.
None	21	5.7	8.5	8.4	21	5.7	8.5	12
5-Fluoro	25	5.7	10	6.5	25	7.8	10	8.1
5-Chloro	11	12	5.9	6.3	13	17	12	19
5-Bromo	8.9	NA	31	30	8.9	NA	92	45
5-Iodo	15	NA	NA	NA	15	NA	NA	NA
5-Nitro	5.3	NA	NA	NA	5.3	NA	NA	NA
5-Nitroso	NA	—	—	—	—	—	—	—
5-Amino(2HCl)	NA	—	—	—	—	—	—	—
5,7-Dichloro	9.4	NA	—	—	9.4	NA	—	—
5,7-Dibromo	9.9	NA	—	—	17	NA	—	—
5,7-Diiodo	28	NA	—	—	110	NA	—	—
5,7-Dinitro	NA	NA	—	—	NA	NA	—	—
5,7-Diamino(HCl)	NA	NA	—	—	NA	NA	—	—

[a] NA = not active below 100 p.p.m.
[b] Naph. = 4-bromo-3-hydroxy-2-naphthoic acid; Salicyl. = 3,5-diiodosalicylic acid.

fluoro substituted ligands than by the corresponding unchelated quinoline compound. As a matter of fact, these two representatives were the most active members of the substituted series of chelates being the only active members against the two species of *Aspergillus* and being matched only by the 5-chloro derivative against the other three species of fungi. None of the other 2:1 chelates with 5- or 5,7-substituents were active at concentrations up to 100 p.p.m. The three fungistatic compounds were also fungicidal against *M. verrucaria* and *T. mentagrophytes*.

b) Second mechanism of action

These data leave very little doubt that we are dealing with multiple mechanisms of action in this series of substituted hydroxyquinolines. The halogens exert a tremendous influence on the nature of the toxic action and this seems to operate by some rule apart from the chelation potential of the several compounds or by action of the chelated products. It would be premature to ascribe this to an alkylating reaction or to some other specific mechanism before all details have been investigated on the stability constants of the molecules and their rates of dissociation of the chelates in representative environments. Methods are being employed currently to determine these physical constants and to evaluate the effects of variable electron density on biological effects.

The activity of the mixed chelates of substituted 8-quinolinol, copper and substituted aryl hydroxy acids was fundamentally the same as for the copper(II) complex of 8-quinolinol regardless of whether 4-bromo-3-hydroxy-2-naphthoic acid or 3,5-diiodosalicylic acid was part of the complex. Against all five species of fungi, the mixed chelates with unsubstituted, 5-fluoro, 5-chloro, and 5-bromo-8-quinolinols were fungistatic and usually were fungicidal except for A. niger. The inhibitory dosage levels were surprisingly close if not identical to those of the copper(II) bis chelate of the respective substituted 8-quinolinols.

Tests made on the five species of fungi concurrently with the evaluation of the mixed chelates revealed that neither the 4-bromo-3-hydroxy-2-naphthoic acid, the 3,5-diiodosalicylic acid, nor their chelates of copper(II) (2:1) were fungitoxic. There was no inhibition by dosages up to 100 p.p.m. This evidence leaves little room for doubt that if the mixed chelates (1:1:1) were to dissociate so as to release either of the free acids or their 1:1 copper chelates, these moieties would not contribute to the fungitoxic mechanism.

One may safely conclude from these data that the 1:1:1 mixed chelates and the 2:1 bis 8-quinolinols have the same mechanism of action and the same dosage response requirements. Since the mixed chelates could not have released 8-quinolinol upon dissociation of the copper chelate, it follows that free 8-quinolinol probably is not essential for the action of copper 8-hydroxyquinoline. The only dissociation product common to the bis and mixed chelate is the 1:1 complex of copper and 8-quinolinol. We feel this is very strong supporting evidence to the hypothesis of Albert et al. (1953), as nearly conclusive as data can be without demonstrating the actual presence of the toxophore moiety inside the cell.

Additional evidence of a second mechanism of fungitoxicity in the substituted 8-quinolinols was obtained by eliminating the chelating potential by substitution of an 8-methoxy for the hydroxyl group. In keeping with the observation of ALBERT et al. (1953) on bacteria, the 8-methoxyquinoline was inactive. However, the 5- and 5,7-substituted 8-methoxyquinolines were fungistatic at 120 to 2,500 μmoles/l. This represents about one percent of the total activity of the corresponding substituted 8-quinolinols. It is believed that this may explain, in part, the changes in activity of the 8-quinolinols due to the presence of substituents.

One of the very striking features of the data in Tables I to V is the consistent phenomenon that a molecule is toxic at very low dosages or not at all up to a dosage of 100 p.p.m.

c) Physical dimensions of molecules

The most logical hypothesis is that there is some physical attribute of the molecule which absolutely prevents it from reaching its site of action. Attention was focused, therefore, upon two attributes of the series of molecules under consideration—their physical dimensions and the distribution of charges on the molecule.

The copper complexes of the 8-quinolinols are known to be planar in structure. Furthermore, the 2:1 complexes must possess an equal charge distribution ($\delta-$) at the opposite ends of the long axis of the molecule. An entirely different distribution of charges exist in the 1:1:1 mixed chelates where there are different groups at opposite ends of the axis. The substituent groups in the 5-position may be expected to increase the length of the molecule progressively as we advance through the series H, F, Cl, Br, I, and NO_2.

If the fungous cell should have perforations of the wall or membranes through which these molecules must pass on their way to the cytoplasm, such pores or channels could prevent passage of the molecules to the point where they first encounter a lipid barrier. It follows that any such opening is very likely to have alternate negative and positive charges at its periphery so it could either attract or repel molecules carrying a charge ($\delta-$) in proportion to the magnitude of $\delta-$. Under these conditions, a molecule carrying equal polar charges would approach the barrier broadside so its long axis would be crucially important.

We postulate, therefore, that there may well be a series of such pores, holes, or channels of a more or less uniform size which can serve more or less as a molecular sieve. A survey of the literature fails to reveal any electron microscopy of these or other species of fungi that depicts such uniform perforations. However, the type of structure to fit this hypothesis is to be found in the outer envelope or capsule of the bacterium *Lampropedia hyalina* according to the report of CHAPMAN et al. (1963). In the outer of two layers there is a honeycomb network of hexagonally distributed pores 75 Å in diameter spaced 145 Å from center to center.

Research is in progress to determine whether there are any analogous structures in the fungous wall which usually is far different from a bacterial envelope. This research is being advanced by two techniques. A careful electron microscopic study is being made of the cell wall membrane. In addition, representative tritiated 8-quinolinol compounds have been synthesized and are being used to make radioautographic studies of ultrathin sections of the treated spores and mycelium to see if non-active members accumulate at specific points in the wall and membrane and to what extent active representatives penetrate the cytoplasm and aggregate on cytoplasmic bodies.

In order to gain more perspective on what to look for calculations were made on the size of molecules involved when chelated 8-quinolinol contained different substituents in the 5-position. Calculations kindly provided by Dr. Y. Okaya of *International Business Machine Corp.* of Yorktown, Heights, N. Y. are presented in Table VI alongside

Table VI. *Relationship of length of long axis to fungitoxicities in a series of substituted 8-quinolinols serving as the ligand in 1:2 copper complexes and 1:1:1 mixed complexes with aryl hydroxy acids*

Substituent in 5-position of 8-quinolinol	1:2 Complex		1:1:1 Complex (naph.) [a]		1:1:1 Complex (salicyl.) [b]	
	Length (Å)	Tox-icity [c] (μmoles/1.)	Length (Å)	Tox-icity [c] (μmoles/1.)	Length (Å)	Tox-icity [c] (μmoles/1.)
H	14.2	5.7-8.5	16.9	6.3-13	16.1	6.7-17
F	15.0	5.7-14	17.4	10-14	16.6	6.5-18
Cl	16.6	7.2-NA [d]	18.2	5.9-51	17.4	6.3-51
Br	17.4	NA	18.6	31-92	17.8	30-110
I	18.0	NA	18.9	NA	18.1	NA
NO$_2$	18.3	NA	18.9	NA	18.1	NA

[a] 1:1:1 Complex (naph.) is chelate of a substituted 8-quinolinol, copper and 4-bromo-3-hydroxy-2-naphthoic acid.

[b] 1:1:1 Complex (salicyl.) is chelate of a substituted 8-quinolinol copper and 3,5-diiodo salicylic acid.

[c] Toxicity is range of minimum inhibitory (fungistatic) doses for five species of fungi.

[d] NA = not active.

the threshold concentration for fungistasis. If the terms of the above hypothesis are to be met, there should be an effective pore size of approximately 15 to 17 Å but less than 17.4 Å. The activity of the mixed chelates having a longer axis can be explained as probably due to their approaching the spore surface at an angle rather than parrellel to the membrane and/or wall, since the charge ($\delta -$) is somewhat less than equally distributed. The greater the angle assumed the longer a molecule could be before it would be held back by the sieve-like action.

IV. Conclusions

Judging from the response of five species of fungi in shake culture, it has been possible to: (1) verify that the 2:1 copper complex of 8-quinolinol is more fungitoxic than the free compound; (2) demonstrate that fungitoxicity of 8-quinolinol is not materially affected by substituting fluorine in the 5-position but substitution of Cl, Br, I, or a nitro group in this position enhances toxicity decisively as does 5,7-disubstitution of either chlorine or bromine but not iodine and nitro, and the substitution of either nitroso or amino groups in the 5- or 5,7-positions squelches toxicity; (3) there appear to be at least two mechanisms of action involved in these substituted 8-quinolinols, one dependent upon chelation potential which can be suppressed by replacing the hydroxyl group with a methoxyl group in the 8-position and the other associated with the substituent groups, and the substituent groups exert an entirely different influence on the copper chelates of 8-quinolinol than on the free ligands; and (4) fungitoxicity of copper chelates was not materially affected by replacing one 8-quinolinol group with a non-toxic aryl hydroxy acid. In case such 1:1:1 complexes dissociate in the fungous cell they would release either the aryl hydroxy acid or its 1:1 copper chelate, neither of which has appreciable fungitoxicity at 100 p.p.m. It is concluded therefore that the toxicity of copper quinolinolate must reside in the 1:1 copper complex formed by its dissociation.

These data provide additional evidence to support a theory that copper quinolinolate does operate by releasing the 1:1 complex as proposed by ALBERT et al. (1953). The present interpretation of this hypothesis is as follows. The 2:1 copper chelate of 8-quinolinol is more active than either copper or the free quinolinol because of its lipid solubility properties and ability to carry into the cell the preformed copper 1:1 complex which will be released as soon as dissociation occurs. The 8-quinolinol moiety released probably has a very minor toxic effect, if any. Substituent groups could be placed on the 5- or 5,7-positions of 8-quinolinol to accentuate its activity after it is released but if such substituent groups increase the size of the chelate, it will destroy activity completely, possibly by blocking the uptake of the molecule by some mechanical means that depends upon steric and electrostatic properties of the molecule.

Research is being continued on this series of compounds to determine their stability constants, their distribution and sites of activity in the cell, and the relation of electrostatic and steric properties to fungitoxicity.

Summary

The conflicting hypotheses advanced to explain the mechanism of fungitoxicity of 8-hydroxyquinoline and its 2:1 copper(II) chelate have been further clarified by synthesizing a series of substituted 8-hydroxyquinolines, their copper(II) chelates, and mixed 1:1:1 chelates with copper(II), and a relatively poor antifungal moiety such as 4-bromo-3-hydroxy-2-naphthoic acid or 3,5-diiodosalicylic acid. The release of free 8-hydroxyquinoline from copper(II) 8-hydroxyquinolinolate is not essential to fungitoxicity, but the 1:1 copper(II) 8-hydroxyquinolinolate from the preformed chelate is the toxicant. It was shown that certain substituent groups placed in the 5- or 5,7-positions of 8-hydroxyquinoline suppressed the fungitoxicity of the 2:1 chelates, and their inactivity was explained by postulating the existence of a mechanical barrier of approximately 15-17 Å pore size that supplements lipophylichydrophylic balance in regulating spore permeation.

Résumé *

Les mécanismes fongitoxiques des composés de la quinoléine et de leurs chelates

Les hypothèses contradictoires avancées pour expliquer mécanisme de la fongitoxicité de la 8-hydroxyquinoléine et de son chelate 2:1 de cuivre (II) ont été éclaircies ultérieurement par la synthèse d'une série de 8-hydroxyquinoléines substituées, de chelates de cuivre (II) et de chelates mixtes 1:1:1 de cuivre (II), et d'un antifongique relativement faible tel que l'acide 4-bromo-3-hydroxy-2-naphtoïque ou l'acide 3,5-diiodosalicylique. La libération de 8-hydroxyquinoléine libre à partir de 8-hydroxyquinoléinolate de cuivre (II) n'est pas essentielle pour la fongitoxicité, mais le 8-hydroxyquinoléinolate 1:1 de cuivre (II) formé à partir du chelate est l'élément toxique. Il a été démontré que certains substituants placés en position 5- ou 5,7- dans la 8-hydroxyquinoléine supprimaient la fongitoxicité des chelates 2:1, et leur inactivation a été expliquée en postulant l'existence d'une barrière mécanique d'une dimension de pore d'environ 15 à 17 Å, qui complète l'équilibre lipo-hydrophilique en réglant l'imprégnation de la spore.

---

* Traduit par S. DORMAL-VAN DEN BRUEL.

Zusammenfassung *

Die fungitoxischen Mechanismen in Chinolinverbindungen und ihren Chelaten

Die widerstreitenden Hypothesen, welche vorgebracht worden sind, um den Mechanismus der Fungitoxizität von 8-Hydroxychinolin und seines 2:1 Kupfer(II)-Chelates zu erklären, sind weiter aufgeklärt worden durch die Synthese einer Serie von substituierten 8-Hydroxy-chinolinen, ihrer Kupfer(II)-Chelate, und gemischten 1:1:1 Chelaten mit Kupfer(II) und eines relativ geringen Antipilzteiles wie 4-Brom-3-hydroxy-2-naphthalinsäure oder 3,5-Dijodsalicylsäure. Die Abgabe von freiem 8-Hydroxychinolin von Kupfer(II)-8-hydroxychinolat ist nicht wesentlich für die Fungitoxozität, sondern das 1:1 Kupfer(II)-8-hydroxychinolat aus dem vorher gebildeten Chelat ist die toxische Substanz. Es wurde gezeigt, dass gewisse Substituentengruppen, welche in den 5- oder 5,7-Stellungen von 8-Hydroxychinolin stehen, die Fungitoxizität der 2:1 Chelate underdrückten, und ihre Inaktivität wurde erklärt durch die Voraussetzung der Existenz einer mechanischen Barriere von ungefähr 15 bis 17 Å Porenweite, welche das lipophilisch-hydrophilische Gleichgewicht bei der Regulierung der Sporenpermeation ergänzt.

Acknowledgment

The authors are deeply appreciative of support from U. S. Public Health Service Grant No. AI-05808.

References

ALBERT, A., and D. MAGRATH: The choice of a chelating agent for inactivating trace metals. 2. Derivatives of oxine (8-hydroxyquinoline). Biochem. J. 41, 534 (1947).

——, S. D. RUBBO, R. J. GOLDACRE, and B. G. BALFOUR: The influence of chemical constitution on antibacterial activity. Part III: A study of 8-hydroxy-quinoline (oxine) and related compounds. Brit. J. Exptl. Pathol. 28, 69 (1947).

——, M. I. GIBSON, and S. D. RUBBO: The influence of chemical constitution on antibacterial activity. Part VI. The bactericidal action of 8-hydroxyquinoline (oxine). Brit. J. Exptl. Pathol. 34, 119 (1953).

CHAPMAN, J. R., R. G. E. MURRAY, and M. R. J. SALTON: The surface anatomy of Lampropedia hyalina. Proc. Roy. Soc. (London), Ser. B. 158, 498 (1963).

---

* Übersetzt von A. SCHUMANN.

Gershon, H., R. Parmegiani, and W. J. Nickerson: Antimicrobial activity of metal chelates of salts of 8-quinolinols with aromatic hydroxycarboxylic acids. Applied Microbiol. 10, 556 (1962).

Greathouse, G. A., S. S. Block, E. G. Kovach, D. E. Barnes, C. W. Byron, G. G. Long, D. Gerber, and J. McClenny: Research on chemical compounds for the inhibition of growth of fungi. 2nd Ann. Rept., 1953-1954. Contract DA-44-099-Eng-1258, Proj. No. 8-91-02-001 Eng. Res. and Dev. Lab. Virginia (1954).

Horsfall, J. G.: Principles of fungicidal action. Waltham, Mass.: Chronica Botanica Co. (1956).

Mason, C.: A study of the fungicidal action of 8-quinolinol and some of its derivatives. Phytopathol. 38, 740 (1948).

Powell, D.: Copper 8-quinolinolate, a promising fungicide. Phytopathol. 36, 572 (1946).

Rubbo, S. D., A. Albert, and M. I. Gibson: The influence of chemical constitution on antibacterial activity. Part V. The antibacterial action of 8-hydroxyquinoline (oxine). Brit. J. Exptl. Pathol. 31, 425 (1950).

Zentmyer, G. A.: Inhibition of metal catalysis as a fungistatic mechanism. Science 100, 294 (1944).

On the fungicidal action of phenylmercuric compounds

by
TETSUJI ISHIYAMA *

Contents

I. Introduction

In Japan, phenylmercuric compounds such as phenylmercuric acetate (PMA), iodide (PMI), fixtan (PMF), N-phenylmercuric-p-toluensulfonanilide (PMTS), etc. have been widely used as sprays (dose: solution of about 20 p.p.m. as metallic mercury, 1,000 L./ha or dusts of 0.2 percent as metallic mercury, 30 to 40 kg./ha) for the control of rice blast caused by *Pyricularia oryzae* since 1953. These compounds are highly effective not only to rice blast but also to stem rot caused by *Helminthosporium sigmoideum* and *H. sig. var. irregu-*

---

* Hokko Chemical Industry Co., Atsugi, Kanagawa, Japan.

lare. PMI is fairly effective to brown spot caused by *Cochliobolus miyabeanus.* These phenylmercuric compounds as sprays, however, have been replaced by non-mercuric compounds, according to the recommendation of the Japanese Government, though they have not yet proved to have chronic toxicity to mammal or to be toxic by residual mercury in rice grains applied by the present routine method. These phenylmercuric compounds have superior characteristics as blast-control chemicals. Here, I will describe the relationship between the effectiveness to rice blast of mercuric compounds and their chemical structure, and also the fungicidal action of phenylmercuric compounds studied from phytopathological standpoint by us, the researchers in our research laboratories.

II. The effectiveness to rice blast of mercuric compounds in relation to their chemical structure

According to our tests, the effect of the R radical of mercuric compounds R-Hg-X (58 compounds were tested) on the effectiveness to rice blast was higher than that of the X radical, and the order of the effectiveness of R radical groups were: phenyl > *o*-, *p*-tolyl > ethyl, methoxyethyl, dimethylphenyl > naphthyl. When compared among the compounds, however, the order of their effectiveness was not always the same as the above-mentioned order of groups, being interacted by X radical. The range of the effect of the X radical in individual R radical groups was narrow, compared with the wide range of the effect of the R radical in individual X radical groups.

In phenylmercuric (PM) compounds, the order of the effectiveness to rice blast in the field was as follows: PMI > PMA, PM-bromide, PM-chloride, PMF, PM-oleate, PMTS > PM-acetylide, PM-dinaphthyl-methane-disulfonate (PMD). The specific characteristics of the effectiveness of PM compounds such as PMA, PMI, etc. are the effectiveness to rice blast both on sprayed and unsprayed parts (leaves and ears) of a rice plant, specific durability to rainfall, and persistent inhibition of sporulation.

Fungicidal actions to which their effectiveness is attributed are divided as below:
   (1) Prevention of infection and inhibition of mycelial growth in tissues of sprayed parts (leaves and ears) of a rice plant.
      (a) Prevention of infection induced by inhibition of conidial germination, formation of appressoria, and penetration.
      (b) Inhibition of growth of mycelia in tissues.
   (2) Prevention of infection or inhibition of mycelial growth in tissues of unsprayed parts (leaves and ears) of a rice plant.
      (a) Same as (1)(a) above. (According to our tests, this characteristic was not recognized in PM compounds at practical routine doses.)

(b) Inhibition of mycelial growth in tissues.

(3) Resistance to rainfall, due to sorption by the rice plant.

(4) Inhibition of sporulation.

The fungicidal actions (1-a, b, 2-b, 3, 4) of PMI were all strong; they were all weak in PM-acetylide and PMD. In the intermediately effective compounds such as PMA, PMB, PMC, PMF, PMO, and PMTS, however, the strength of the individual action was not the same, and varied according to the compounds. In PMO, for instance, (1) was weak but (4) was strong, whereas in PMTS, (1) was very strong but (4) was a little weak.

III. Analysis of fungicidal actions of phenylmercuric compounds and others

The effectiveness of a fungicide against a disease of a plant is the composed results of its fungicidal actions expressed under the interaction of fungicide, host plant, and causal fungus. Accordingly, the activity of a fungicide against a disease is not to be evaluated only by the direct action of the fungicide to the causal fungus, but is to be evaluated under the interaction of three factors.

IV. Prevention of infection induced by inhibition of conidial germination, formation of appressoria, and penetration

Phenylmercuric compounds, for instance PMA, showed strong inhibitory action to conidial germination of blast fungus, tested in a hanging drop or slide glass (germination was 0 percent in the solution of 3.2 p.p.m. as Hg), with high positive correlation to the preventive effectiveness against rice blast. On the contrary, the antibiotic Kasugamycin, pentachlorobenzylalcohol, alkyl rhodans ($C_{16}H_{33}SCN$), etc. showed no inhibitory action to conidial germination, even at their practical routine concentrations commonly used in the field, although they are highly effective on the prevention of rice blast. These phenomena show that the inhibitory action to the conidial germination in the hanging drop or other tests without the host plant *in vitro* does not always correlate with the preventive effect against the disease. Therefore the preventive activity against blast fungus should be tested at the sites of conidial germination, formation of appressoria, and penetration (at least, at the site of penetration in relation to the preventive effectiveness against blast). The tests on rice leaves or sheaths showed that PMA showed slight or nearly no effect on the conidial germination even at the solution inhibitory action concentration of five p.p.m. as Hg, but showed strong inhibitory action on the formation of appressoria and especially on the penetration through epidermis, when conidia were added 30 to 60 minutes after the fungicide application on the leaves or sheaths. The same results were recognized in the tests on onion epidermis killed with alcohol, too. Alkylrhodan and pentachlorobenzyl alcohol showed high inhibitory action only on penetration at a concentration of 500 p.p.m. or much less. The inhibitory action of PMA to the formation of appressoria and penetration was not lost even by one hour washing by tap water, after 30 to 60 minutes keeping in the undried condition after the treatment with PMA. Remarkable decrease of free PMA on the leaves kept undried for 30 to 60 minutes after the treatment with PMA solutions was recognized by chemical analysis. The chemical analysis of Hg in the solution of PMA dipped with rice leaves at the 15 minute intervals showed rapid sorption (mainly adsorption at the beginning of dipping, and later, after saturation of adsorption, increase of sorption including perhaps absorption) of PMA. The velocity and amount of PMA sorption by rice leaves varied according to the varieties, with high correlation with the phytotoxicity of PMA. Most of the Indica varieties of high sensitivity to PMA toxicity showed higher velocity and amount of sorption than Japonica varieties resistant to the toxicity. The above-mentioned results may show that the mycelia which contact with the leaf surface absorb some part of PMA adsorbed by leaf surfaces and inhibit their own growth, resulting in the inhibition of formation of appressoria and penetration through epidermis, or may show, in some instances, chemical change to preventive structures in the cuticula layer by PM compounds such as PMA or PMI. At any rate, further investigation should be required for the real causes of these phenomena.

V. The effect of adsorbed PMA and PMI against the infection of brown spot fungus

On the other hand, all the three steps of infection (germination, formation of appressoria, and penetration) of the brown spot fungus, *Cochliobolus miyabeanus* were inhibited only by free PMA and were not inhibited by the PMA adsorbed on leaves. Thus the fungicidal activity of applied PMA on leaves, after being adsorbed by the leaf surfaces, varies according to the species of fungus in effectiveness. PMI of low solubility (four μg./cc. at 15° C.) on rice leaves, sheaths, or onion epidermis inhibited the conidial germination of both *Pyricularia oryzae* and *Cochliobolus miyabeanus* with the suspension of 10 p.p.m. as Hg, whereas PMA of high solubility (1,800 μg./cc. at 15° C.) inhibited very slightly the germination, tested after 30 to 60 minutes undried after PMI or PMA application. One to two days after PMI application, PMI also lost its inhibitory action against germination, formation of appressoria, and penetration of the brown spot fungus on rice leaves and sheaths. Low soluble PMI will, perhaps, remain freely unadsorbed, to some extent, for longer periods than soluble PMA and will inhibit effectively the germination of conidia of these fungi during the early period. At a later period, however, PMI will lose its inhibitory action against germination, formation of appressoria, and penetration of only brown spot fungus on rice leaves and sheaths, perhaps due to sorption. Even at the later period, blast fungus was strongly prohibited in its penetration, but not in conidial germination. It is an interesting fact that the inhibitory action of adsorbed PM compounds against fungal penetration varies according to the fungi. When emulsified PMI was applied on rice leaves, conidial germination, formation of appressoria, and penetration of brown spot fungus inoculated 60 minutes after its application were not all inhibited, whereas conidial germination of blast fungus was not inhibited but its penetration was inhibited as in the case of PMA.

VI. Specific volatilization of low volatile ethylmercuric phosphate on rice leaves

Ethylmercuric phosphate (EMP), chloride (EMC), and methoxyethylmercuric chloride (MMC) are not so effective in blast control. The lesser effectiveness of EMC and MMC is attributed, to some extent, to their high volatility. The volatility of EMP (vapor pressure 0.025 mm. of Hg at 100° C.) is much lower than EMC (vapor pressure 0.81 mm. Hg at 100° C.), and a little lower than PMA (v. p. 0.09 mm. of Hg at 100° C.). But this is also less effective to blast. We conducted some tests to make clear the causes of its lesser effectiveness.

According to the results, EMP is easily changed to volatile EMC in the presence of chloride ion on rice leaf surfaces, excreted from the tissues. It will be a causes of low preventive effectiveness of EMP; nevertheless, the fungicidal activity (inhibition of conidial germination) is high and persistent owing to its low volatility as tested on a glass plate. Thus, EMP is also a specific chemical whose evaluation of fungicidal activity is to be observed under the interaction of fungicide, host plant, and fungus. The fungicidal action of the vapor of EMP, EMC, MMC, and PM compounds such as PMA was recognized against conidial germination in Petri dishes or in a bell jar. In the open air, however, the action was not recognized at all. Therefore, the vapor action of these compounds has no effect on the prevention of blast in the field.

VII. Prevention of infection induced by inhibition of conidial germination, formation of appressoria, and penetration on unsprayed parts of a rice plant

The inhibitory action of PM compound such as PMA to the three steps of infection was not recognized on unsprayed parts of a rice plant or on new parts (leaves and ears) emerged after its application, by its translocation in the tissues, from sprayed parts.

VIII. Inhibition of mycelial growth in the tissues of sprayed and unsprayed parts of a rice plant

There are a number of reports which proved the protective effect of PM compounds such as PMA, PMI, etc. against rice blast occurrence not only on sprayed parts (leaves and ears) but also on unsprayed parts or on new parts emerged after the spray, when applied to a rice plant at the routine dose, since the report of Okamoto and Saito (1953). According to our tests, PMA applied on either the upper or back sides of the leaf, zero to two days after the completion of fungal penetration (kept one day in a humid chamber after conidial inoculation on the upper side of leaf) inhibited remarkably the emergence of lesions. The blast occurrence on unsprayed leaves and ears was protected by the application of PMA on other parts of a rice plant with the routine dose, only one day before inoculation. Blast occurrence even on new leaves and ears emerged after the spray was also protected by the previous spray of PMA to a plant.

PMA easily permeates into the tissues through epidermis and is translocated to other parts of a plant, according to the studies of Tomizawa (1965), Moriya et al. (1965), and Araki et al. (1965). Moriya et al. also proved that PMA in the tissues changes into non-phenylmercuric compounds. Kitamura (0000) reported the existence of a very small amount of PM compound in unsprayed leaves, two weeks after the routine application of PMI wettable powder or dust, most of the applied PMI being changed into non-phenylmercuric compound. According to our tests, no free PM compound was detected in a 100 p.p.m. solution of PMA mixed at the ratio of one μg. of PMA: sap of 8.5×10^{-4} g. of rice fresh leaves and in a 200 p.p.m. solution mixed at the ratio of one μg. of PMA: sap of 4.3×10^{-4} g. of fresh rice leaves, both ten minutes after mixing. Tested on the inhibition of conidial germination by the hanging drop method, one to two μg./cc. solution of PMA inhibited conidial germination by 99.5 percent, and ten μg./cc. perfectly, whereas either two μg./cc. solution mixed at the ratio of PMA two μg.: sap of 0.05 g. of fresh rice leaves, or one μg./cc. solution mixed at the ratio of PMA two μg.: sap of 0.1 g. of fresh rice leaves lost the inhibitory action perfectly; ten μg./cc. solution of PMA mixed at the ratio of PMA two μg.: sap of 0.01 g. of fresh rice leaves inhibited the germination by merely 6.1 percent. These phenomena show that there exists no free PM compound or active mercuric compounds sufficient to inhibit the mycelial growth in the tissues. Adsorption of PMA by cell membranes or others may occur in the tissues. The growth of blast fungus contacted with the adsorbed PMA may be inhibited, to some extent, as on the surface of a leaf.

Referring to the soluble amino acids in the sap of rice leaves, Nasuda et al. (1957) reported that the quantities of glutamic acid, alanine, etc. which are recognized to be negatively correlated with blast resistance, decreased by PMA application. We also recognized the decrease of these substances in the fresh leaf sap of rice plants treated with PMA, compared with that of untreated plants.

As already mentioned, the inhibitory action of PM compound such as PMA against the three steps of infection of blast fungus was not recognized on unsprayed parts of a rice plant or on new parts emerged after its application. From these results, the decrease in number of lesions on the unsprayed leaves and ears without deposition of PMA is not attributed to the inhibition of infection of blast fungus, but attributed to the inhibition of mycelial growth in the tissues, affected by PMA absorbed on the other parts of the plant and translocated. We assumed that the inhibitory activity against mycelial growth of the blast fungus in the tissues of unsprayed parts of the rice plant will not be derived from the direct inhibitory action of free PMA in the tissues, but may be derived from the increase of physiological blast resistance introduced by the mercuric compounds originated from absorbed PMA, to some extent. Frankly speaking, it is true that there are a lot of unknown factors to explain the mechanism of this phenomenon. Therefore we have named the effectiveness of PMA compounds such as PMA, PMI, etc. on the unsprayed leaves and ears, without deposition of the chemical on the surface, "indirect effect" or "indirect action" of PM compounds to rice blast.

NAKAZAWA (1959) reported that organomercuric compounds permeated into the tissues inhibited the mycelial growth of blast fungus, and TATSUYAMA (1964) reported the increase of physiological blast resistance of rice plants induced by PMA and the fungicidal activity of PMA in the tissues. TERANAKA (1960) reported that carbon assimilation was strengthened by PMA application and also that the contents of glutamic acid, valine, aspartic acid, etc. in rice plants increased after PMA application. The increase of nitrogen in rice plants was reported by NASUDA (1957) and TERANAKA (1960). The results of the published reports do not always coincide with one another. It has not been confirmed whether the inhibition of mycelial growth in the tissues is due to the direct contact with free PM compounds (we do not agree with this) or due to the increase of physiological blast resistance induced by absorbed PM compounds (if so, by what mechanism?). Anyhow, further investigation will be required for the clarification of the mechanism of the indirect action of PM compounds against rice blast fungus.

IX. Resistance of phenylmercuric compounds to rainfall

PM compounds such as PMA are easily sorbed and translocated to other parts of the plant to act effectively for the control of rice blast. Therefore, PM compounds are resistant to the washing off by rainfall, different from mechanical resistance to it.

X. Inhibition of sporulation

Severe damage from blast is derived from the dispersion of conidia. Therefore, the inhibition of sporulation by chemicals is strongly required. PM compounds such as PMA and PMI inhibited sporulation strongly for a long period.

According to the observation in fields treated with PMA, two months' duration was recognized by the spore trap method. With the PM compounds strong inhibitory action against sporulation and the "indirect effect," PM compounds for the control of neck rot can be applied at the rooting stage when spraying is easily carried out.

Summary

Phenylmercuric acetate and iodide show high effectiveness for rice blast control when applied to rice plants as sprays. Their effectiveness is derived from the prevention of infection, inhibition of mycelial growth in the tissues of sprayed leaves, and inhibition of mycelial growth in the unsprayed leaves. Phenylmercuric compounds are easily adsorbed, absorbed, and translocated in the tissues. Adsorbed compounds on leaf surfaces do not show inhibition of germination of conidia, but do show high inhibition of the formation of appressoria and penetration, different from the phenomena on glass plates.

Résumé *

L'action fongicide des composés phénylmercuriques

L'acétate et iodure phénylmercurique, appliqués par pulvérisation sur les plantes de riz, sont très efficaces dans la lutte contre la piriculariose du riz. Ils ont une action préventive contre l'infection et une action inhibitrice contre la croissance du mycélium sur les feuilles traitées et aussi sur les feuilles non-exposées au traitement. Les composés phénylmercuriques sont facilement adsorbés, absorbés, et transportés dans les tissus. Les composés adsorbés sur la surface des feuilles n'arrêtent pas la germination des conidies mais causent une forte inhibition de la formation des appressoria et de la pénétration, une action différente de celle sur des plaques de verre.

Zusammenfassung **

Über die fungicide Wirkung von Phenylquecksilber-Verbindungen

Phenylquecksilberacetat und-jodid haben als Sprühmittel eine starke Wirksamkeit für die Kontrolle des Reisbrandes. Sie verhindern

---

* Traduit par H. Gordon.
** Übersetzt von W. Loher.

die Infektion und hemmen das Mycelwachstum in besprühten und unbesprühten Blättern. Phenylquecksilber-Verbindungen werden leicht von den Geweben adsorbiert, absorbiert und weitergeleitet. Adsorbierte Verbindungen an Blatteroberflächen können die Keimung der Conidien nicht verhindern, haben aber eine stark hemmende Wirkung auf die Bildung von Haftorganen und ihre Eindringungsfähigkeit; dies wurde an besprühten Glasplatten nicht beobachtet.

References

ARAKI, T.: Sampu-suiginzai no shokubutsutai niokeru dotai to zanryu. Tokyo: Association of Plant Protection, Japan (1965).

ISHIYAMA, T.: Yuki-suigin sakkinzai. Ann. Org. Syn. Chem. 18, 550 (1960).

—— Chemical structure and fungicidal activity of organo-mercurials. Ann. Phytopathol. Soc. Japan (The Golden Jubilee Issue) p. 81 (1965).

MORIYA, S.: Sampu-suiginzai no shokubutsutai niokeru dotai to zanryu. Tokyo: Association of Plant Protection, Japan (1965).

NAKAZAWA, G.: Fungicidal activity and its mechanism of organomercuric fungicides. Bull. Aichi Agr. Expt. Sta. 19, 1 (1959).

OKAMOTO, H.: Chugoku Agr. Research 12, 134 (1960 a).

—— Varietal difference of mercuric phytotoxicity in rice plants. Bull. Chugoku Agr. Expt. Sta. 4, 225 (1960 b).

—— and Y. SAITO: On the indirect effect of Ceresan-slaked lime. Chugoku Agr. Research 3, 32 (1953).

—— and T. YAMAMOTO: On the effect of mercuric compounds, Chugoku Agr. Research 14, 73, (1960).

TATSUYAMA, K.: Direct and resistance-intensifying effects of fungicides on the control of blast disease of rice plant. Special Rept. Pl. Pathol. Lab., Shimane Agr. College 3, 1 (1964).

TERANAKA, R.: Suiginzai tofu niyoru ine-tainai amino acid, amide nado no henkwa ni tsuite. Ann. Phytopathol. Soc. Japan 25, 8 (1960 a).

—— Suito no tainaiseibun ni oyobosu yukisuiginzai no eikyo. Ann. Phytopathol. Soc. Japan 25, 39 (1960 b).

TOMIZAWA, C.: Sampu-suiginzai no shokubutsutai ni okeru dotai to zanryu. Tokyo: Assoc. Pl. Protection, Japan (1965).

YAMAMOTO, T., and H. OKAMOTO: Studies on a simple laboratory assay method of the effectiveness of fungicides to rice blast. Chugoku Agr. Research 28, 19 (1964).

Fungicidal action of organophosphorus compounds

by

MASARU KADO * and EIICHI YOSHINAGA *

Contents

I. Introduction

In Japan, some organomercuric compounds and an antibiotic blasticidin S have been used for many years for the control of the rice blast which is caused by the infestation of *Pyricularia oryzae*. Recently remarkable controlling effectiveness of some benzyl phosphoric esters on the rice blast was found by the research group of the *Ihara Chemicals Co.* (KADO et al. 1965). Successively two other organophosphorus compounds have also been developed for the same purpose by other workers. As universally known, organophosphorus compounds have generally been used for the control of various insect pests, and their fungicidal activity has been almost left unknown. The fungicidal mechanisms of these compounds have not been elucidated in detail yet. The results of our experiments carried out on benzyl phosphoric esters are summarized in this paper.

II. Chemical structures effective on rice blast

Bioassay experiments of the rice blast controlling agents are generally carried out by the methods of spore germination, mycelial growth or sporulation inhibition in laboratory, or that of application to potted rice plants. The so-called bed test which is a small scale field test and

---

* Agricultural Chemicals Research Laboratory, Ihara Chemicals Co., Ltd., 100 Shibuka, Shimizu, Shizuoka, Japan.

large scale paddy field test are also made for the development of new controlling agents. The fungicidal activity of benzyl phosphoric esters was found by the screening method with application of chemicals to plants in greenhouses. That is, plants grown until the fourth or fifth leaf stage were inoculated with spore suspension of *P. oryzae;* the suspension of chemicals in water was sprayed on infested plants on the second or third day after inoculation, with effectiveness assessed on the seventh to tenth day after inoculation.

As shown in Figure 1, the effectiveness of *O,O*-diethyl S-benzyl

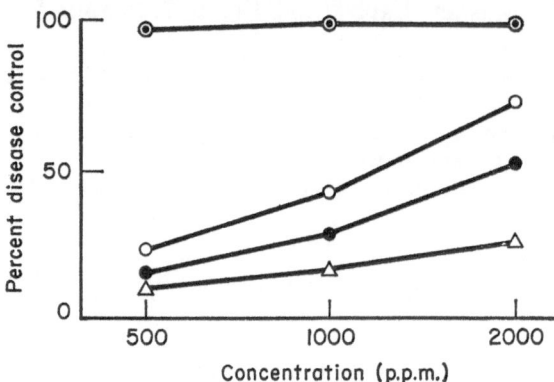

Fig. 1. Relation between the percent control of rice blast on potted plants and the concentrations of chemicals (Kado *et al.* 1965): ☉ = *O,O*-diethyl S-benzyl phosphorothiolate; ○ = *O,O*-diethyl S-benzyl phosphorodithioate; ● = *O,O*-diethyl O-benzyl phosphate; and △ = *O,O*-diethyl O-benzyl phosphorothionate

phosphorothiolate on the rice blast was higher than that of the corresponding *O*-benzyl phosphate and phosphorothionate and S-benzyl phosphorodithioate. Similar results were obtained in all cases of *O,O*-dialkyl derivatives, such as dimethyl, diethyl, and others. It is considered that *O,O*-dialkyl S-benzyl phosphorothiolates are the most effective group on the rice blast among benzyl phosphoric esters. With respect to alkyl radicals, ethyl, propyl, and butyl radicals give the higher effectiveness. It is considered that the alkyl radicals of benzyl phosphoric esters play an important role in the permeation of active ingredient into the fungus.

Kitazin (*O,O*-diethyl S-benzyl phosphorothiolate, EBP) and Kitazin-P (*O,O*-diisopropyl S-benzyl phosphorothiolate, IBP) are the rice blast controlling chemicals developed by the Ihara research group. Two other organophosphorus compounds, Hinozan (*O*-ethyl S,S-diphenyl phosphorodithiolate, EDDP) and Inezin (*O*-ethyl S-benzyl phenylphosphonothiolate, ESBP), are the chemicals developed commercially for the same purpose by the Bayer and the Nissan research groups, respectively. Mammalian and fish toxicities of these four chemicals are much lower than those of most organophosphorus insecticides.

Nearly all organophosphorus insecticides used practically are phosphorothionates or phosphorodithioates, and phosphorothiolates are not used so much. On the contrary, organophosphorus rice blast controlling chemicals mentioned above belong to a group of phosphorothiolates. It has not been elucidated, however, why phosphorothiolates are effective on the rice blast as controlling agents.

III. Fungicidal mechanisms

Kitazin or Kitazin-P inhibits the spore germination of *P. oryzae* completely at 20 to 30 p.p.m. on glass slides. Nevertheless, in the inoculation test on rice plants, this inhibition cannot be recognized as the main mechanism of fungicidal action.

Inactive dark green lesions with yellow zone circles outside are formed when Kitazin or Kitazin-P is applied to the rice plant infected with *P. oryzae* (Fig. 2). These inactive lesions are not functional, be-

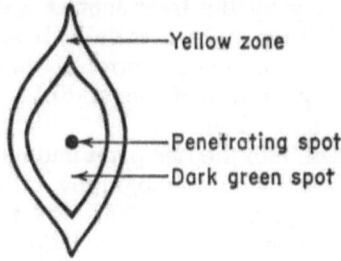

Fig. 2. Inactive lesion formed by Kitazin treatment at a concentration of 500 p.p.m. (Kado *et al.* 1965)

cause the inhibitions of sporulation and of mycelial growth are observed on these inactive lesions. The same lesion is also formed by the application of antibiotics such as blasticidin S and kasugamycin (Hirano *et al.* 1965, Ishiyama *et al.* 1965).

The relations between the formation of inactive lesions by controlling agents and the time elapsed after the inoculation of *P. oryzae* are shown in Figure 3. In this experiment, the potted plants were inoculated with suspension of spore beforehand; Kitazin solution at a concentration of 500 p.p.m. was then sprayed on them on the first, second, third, and fourth day after inoculation. Any lesion could not be observed on the leaves of untreated plants within one or two days after inoculation. Complete inhibitions of the mycelial growth in tissues and of the formation of lesions were obtained by the Kitazin spray in this period. High percent formation of inactive lesions was

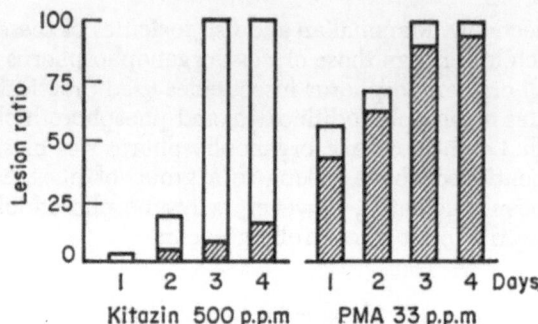

Fig. 3. Formation of inactive □ and active lesion ▨. Kitazin and PMA (phenyl
mercury acetate) were sprayed on the 1st, 2nd, 3rd, or 4th day after
inoculation with a spore suspension of *P. oryzae* (KADO *et al.* 1965)

also obtained by the Kitazin spray on the third or fourth day after
inoculation which is just the incipient or the development stage of
lesion. On the other hand, the sporulation on the lesions of rice blast
was inhibited completely by the treatment at a concentration of 350
p.p.m. of Kitazin or 300 p.p.m. of Kitazin-P. It is considered that the
main mechanisms of the rice blast control of these chemicals are the
inhibitions of mycelial growth in tissue, and of lesion formation or of
sporulation of fungus on rice plants.

Absorption of Kitazin into the rice plant and its translocation within
the plant were confirmed by autoradiography with P^{32}- or S^{35}-Kitazin.
Kitazin absorbed through a treated leaf was translocated to other
leaves and then accumulated at necrotic lesions. As rice blast controlling agents, this systemic action is one of the excellent characters.

Kitazin or Kitazin-P was also absorbed into the rice plant through
the root system and translocated to stems and leaves. The relations
between the percent control of rice blast obtained by the root dipping

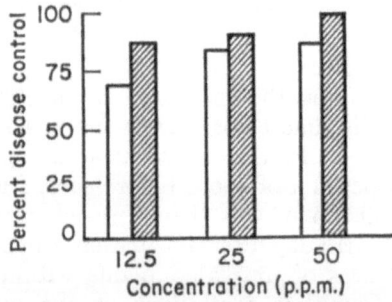

Fig. 4. Rice blast control obtained by the dipping of root systems in Kitazin
solution for 24 hours □ or 48 hours ▨ before inoculation with a spore
suspension of *P. oryzae* (YOSHINAGA *et al.* 1965)

method and the concentration of Kitazin solution are shown in Figure 4. As the result of this experiment, it was proved that the soil treatment with Kitazin or Kitazin-P is an excellent method for rice blast control in paddy field.

Organophosphorus fungicides used for rice blast control are also effective for other diseases of the rice plant such as the disease having symptoms of the "ear blight" caused by *Helminthosporium oryzae* and others, the stem-rot (*Helminthosporium sigmoideum* and *H. sigmoideum* var. *irregulare*), and the sheath blight (*Pellicularia sasakii*).

As mentioned above, fungicidal mechanisms of Kitazin and Kitazin-P have not been elucidated yet, and the investigations of organophosphorus fungicides including Kitazin and Kitazin-P have just started.

Summary

Organophosphorus fungicides developed for the control of rice blast (*Pyricularia oryzae*) in Japan were introduced. Among the benzyl phosphoric esters, *O,O*-dialkyl S-benzyl phosphorothiolates are more effective on the rice blast than the corresponding *O*-benzyl phosphates and phosphorothionates and S-benzyl phosphorodithioates. With respect to alkyl radicals, ethyl, propyl, and butyl have the highest effectiveness. Fungicidal mechanisms of Kitazin (*O,O*-diethyl S-benzyl phosphorothiolate) and Kitazin-P (*O,O*-diisopropyl S-benzyl phosphorothiolate) are considered to be the inhibition of mycelial growth in tissue, and of lesion formation or of sporulation of fungus on the rice plant. Both chemicals have a systemic action.

Résumé *

Action fongicide des composés organophosphorés

Des fongicides organo-phosphorés développés pour le traitement de la brunissure du riz (*Pyricularia oryzae*) ont été introduits au Japon. Parmi les esters benzyl-phosphoriques, les O,O-dialkyl S-benzyl thiophosphates sont plus efficaces sur ce parasite que les O-benzyl phosphates et thionophosphates correspondants ainsi que les S-benzyl dithiophosphates. Au sujet des radicaux alkyles, éthyl, propyl et butyl ont la plus grande efficacité. Kitazine (O,O-diéthyl S-benzyl thiophosphate) et kitazine (O,O-diisopropyl S-benzyl thiophosphates) doivent leur activité fongicide à l'inhibition de la croissance du mycelium dans les tissus, et à la formation d'une lésion ou d'une sporulation du champignon sur le plant de riz. Les deux produits ont une action endothérapique.

---

* Traduit par R. MESTRES.

Zusammenfassung *

Fungizidwirksamkeit von Organophosphorverbindungen

Organophosphorfungizide, welche zur Kontrolle des Reisbrands (*Pyricularia oryzae*) entwickelt wurden, wurden in Japan eingeführt. Von den Benzylphosphorsäureestern ist O,O-Dialkyl-S-benzylphosphorthiolat wirksamer gegen den Reisbrand als die entsprechenden O-Benzylphosphate und Phosphorthionate und S-Benzylphosphordithioate. Was die Alkylradikale anbetrifft, so haben Aethyl-, Propyl- und Butyl- die grösste Wirksamkeit. Fungizide Mechanismen des Kitazins (O,O-Diäthyl-S-benzylphosphorthiolat) und Kitazin-P (O,O-Diisopropyl-S-benzylphosphorthiolat) werden als die Hemmer des Myzelwachstums in Gewebe und der Schädigungsbildung oder der Sporenbildung des Pilzes auf der Reispflanze angesehen. Beide Chemikalien haben systemische Wirkung.

References

Hirano, K.: Inactivation of rice blast lesion on leaves by application of fungicides. Shokubutsu Boeki (Japan) 19, 373 (1965).

Ishiyama, T., and H. Umezawa: Studies on the rice blast control by Kasugamycin. 1. Effectiveness and phytotoxicity (Abstr.). Ann. Phytopathol. Soc. Japan 30, 111 (1965).

Kado, M., T. Maeda, E. Yoshinaga, T. Iwakura, and T. Uchida: Studies on the rice blast control. I. Chemical structure and fungicidal activities of benzyl phosphoric esters (Abstr.). Ann. Phytopathol. Soc. Japan 30, 109 (1965).

—— —— —— —— —— Studies on the rice blast control. II. Rice blast control by Kitazin (Abstr.). Ann. Phytopathol. Soc. Japan 30, 110 (1965).

Yoshinaga, E., T. Uchida, and T. Iwakura: Studies on the rice blast control. III. Formation of inactive lesion by application of Kitazin (Abstr.). Ann. Phytopathol. Soc. Japan 30, 110 (1965).

—— —— —— Studies on the rice blast control. IV. Rice blast control of Kitazin by root treatment (Abstr.). Ann. Phytopathol. Soc. Japan 30, 307 (1965).

---

* Übersetzt von A. Schumann.

Pentachlorobenzyl alcohol, a rice blast control agent

by

M. Ishida,* H. Sumi,* and H. Oku *

Contents

I. Introduction

Mercurial fungicides have been the target for considerable criticism, especially in recent years in Japan. Opponents have stressed possible human health hazards in their preparation, application, and as chemical residues in rice grains, the Japanese staple. A strong and healthy trend toward substitution of mercurials by much safer chemicals has accelerated the operation of extensive and improved screening programs. In 1964, our research people disclosed a very peculiar activity of a known and simple compound toward rice blast (Sumi *et al.* 1968 a): 2,3,4,5,6-pentachlorobenzyl alcohol, or Blastin.

Since this compound was first prepared by Beilstein and Kuhlberg (1869), it had been deeply sleeping in the literature for almost a century until British people (Carter *et al.* 1958) tried to awake it by evaluating it *in vitro* as a microbiocide. Their attempt, however, was in vain, and yet we had to wait for several years before its practical

---

* Sankyo Co., Ltd., Agricultural Chemicals Research Laboratories, No. 2, 1-Chome, Hiromachi, Shinagawaku, Tokyo, Japan.

use was discovered by Sumi and his collaborators (1968 a) with their *in vivo* tests on rice plants.

II. Physical and chemical properties

Blastin is colorless crystalline powder with a melting point of 197° to 198° C. It has poor solubilities in organic solvents at room temperature: 0.72 percent in acetone, 0.56 percent in ethanol, 5.8 percent in dimethylformamide, 13.8 percent in pyridine, and 0.0002 percent in water. It is stable to heat, acids, and alkalis. No degradation products have been detected after irradiation with an artificial sun light-mimic flux (Kondo 1966). Microquantities of the chemical are detectable with gas liquid chromatography with an electron-capture detector.

III. Mode of action

Blastin is virtually inactive in a Petri dish, but shows an astonishing effect against rice blast when applied on rice leaves (Sumi *et al.* 1968 a). The activity of Blastin seems to be highly specific to rice blast, since no other plant diseases have been found to be controllable by the chemical so far.

a) Inhibition of hyphal penetration

By spraying Blastin at various times in relation to inoculation with *Pyricularia oryzae*, Sumi and his coworkers (1968 a) found that Blastin exhibited the highest and long-lasting activity if applied before inoculation, while post-inoculation application gave rise to the reduction of effectiveness to some extent. Closer observation revealed that Blastin did have little effects on spore germination and appressorium formation, but specifically prevented the penetration of hyphae into leaves (Fig. 1). Their experiments were furthered to show that Blastin ex-

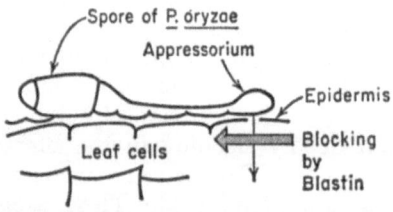

Fig. 1. Morphological mode of action of Blastin

hibited this peculiar activity in full strength only when applied before appressorium formation (Fig. 2).

Malformation of appressorium has been observed on a glass plate

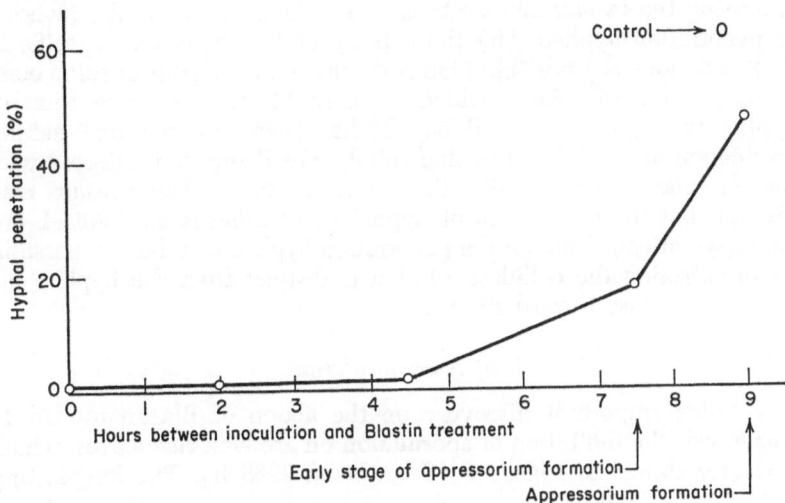

Fig. 2. Application-time-dependence of effectiveness of Blastin (100 μg./ml.)

at the concentration of 300 μg./ml., suggesting a possibility that a direct interaction between Blastin and the fungus had taken place prior to the formation of penetration hyphae in appressoria (SUMI et al. 1968 a).

The blocking of hyphal penetration by Blastin has been clearly demonstrated with an *in vitro* experiment (OKU and SUMI 1968): a drop of spore suspension is placed on a sheet of cellophane membrane (Seamless cellulose tubing, Visking Co., 1 x 1 cm.) which is floated on a boiling-water-extract of rice straw containing Blastin, incubated for 48 hours at 27° C. in a Petri dish, and then the pores in the cellophane membrane made by penetration of hyphae are counted under a microscope. The results shown in Table I indicate the following,

Table I. *Inhibitory effect of Blastin on hyphal penetration in vitro*

Blastin (p.p.m.)	Pore indices [a]		
	Pyricularia oryzae	*Cochliobolus miyabeanus*	*Alternaria citri*
0 (Control)	100	100	100
6	20	136	88
12.5	10	148	96
25	15	149	88
50	11	140	81

[a] Pore index = $\dfrac{\text{No. of pores in treated dish}}{\text{No. of pores in control dish}}$ x 100.

endorsing the *in vivo* observations: (a) Blastin prevents the invasion of penetration hyphae, (b) the activity of Blastin is very specific to *P. oryzae,* and (c) possible biotransformation of Blastin to some compound(s) in rice plants which actually blocks the penetration of hyphae is most likely ruled out. It has been reported that neither production nor activity of hyphal cellulase of *P. oryzae* is affected with Blastin (Oku and Sumi 1968, Sumi and Ito 1966). These results may suggest that the mechanism of hyphal penetration is mechanical, but not enzymatical, although the penetration hyphae may have a possibility of secreting the cellulase(s) that is distinct from the hyphal one in their response toward Blastin.

b) Antisporulation

Another important discovery on the action of Blastin toward *P. oryzae* was the inhibition of sporulation on growing rice leaves, which is clearly shown in Figure 3 (Sumi *et al.* 1968 b). The long-lasting

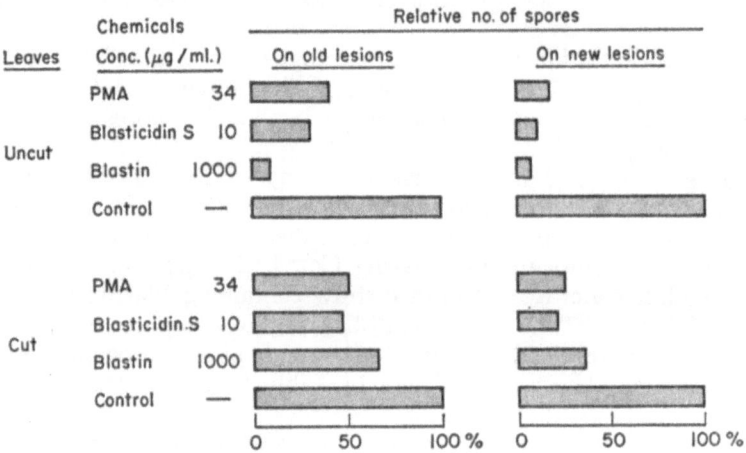

Fig. 3. Inhibition of sporulation by Blastin. Data are averages of eight replications. Cutting of leaves was carried out 24 hours after application of chemicals and the isolated leaves were kept in a moist chamber

effectiveness of Blastin observed in fields may be partially attributed to this activity. Since the inhibition was not observed on culture media (Sumi *et al.* 1968 b) and was remarkably reduced when the treated leaves were cut and kept in a moist chamber (Fig. 3), normal and complete physiological functions of the host seem to be required to bring Blastin into action. In this connection, it may be worthwhile to mention that foliage application of Blastin lowers the concentrations of glutamic and aspartic acids in rice leaves which are known to be preferable substrates for *P. oryzae* (Sumi *et al.* 1967).

After this seminar, Dr. J. G. HORSFALL (1967) kindly tested for us *in vitro* antisporulation activity of Blastin toward *Alternaria solani* and found that it was ineffective in contrast with its analogue, α-trichloromethylbenzyl alcohol, but produced a swollen spore which often eventually broke open.

IV. Metabolism

a) Metabolism in rats

It has been reported that Blastin has virtually no toxicity to mammals (*Sankyo Co., Ltd.* 1966). No fatal cases have been observed with acute oral, acute subcutaneous, and 90-day chronic toxicity tests on rats, mice, and guinea pigs at a higher level of dosage (1,450 to 3,600 mg./kg.). It, however, is still desirable to investigate the biotransformation and fate of Blastin in animals, because it may be contained in rice grains as a residue.

Elimination of radioactivity from rats orally treated with approximately ten microcuries (μc) of C^{14}-Blastin (prepared by FUJITA and ISHIDA 1968) or C^{14}-γ-BHC (both are uniformly labeled in rings, 653 μc/millimole and 650 μc/millimole, respectively) is shown in Figures 4 and 5 (ISHIDA 1968). Fecal metabolite was identified with Blastin

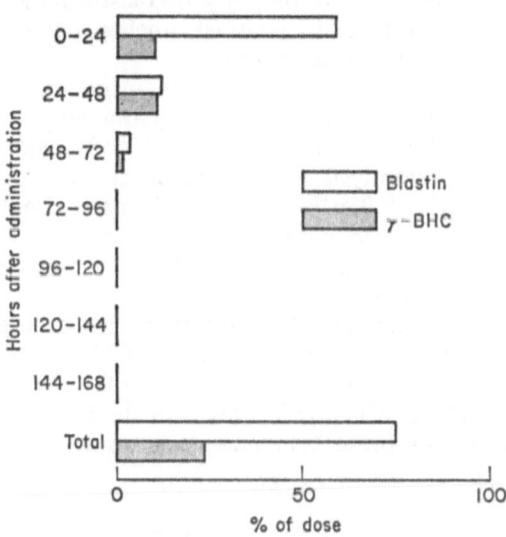

Fig. 4. Fecal excretion of radioactivity from rats orally treated with C^{14}-Blastin or C^{14}-γ-BHC

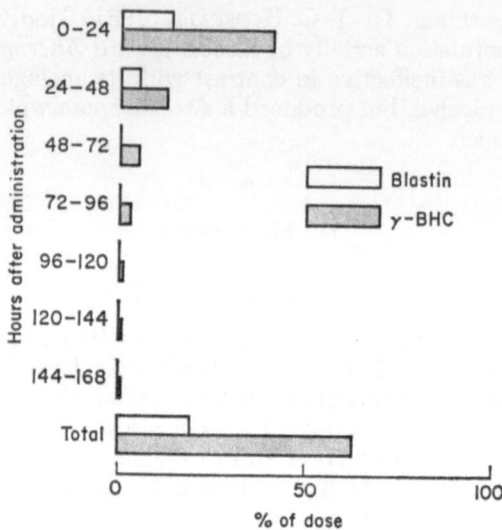

Fig. 5. Urinary excretion of radioactivity from rats orally treated with C^{14}-Blastin or C^{14}-γ-BHC

by gas liquid chromatography, thin-layer chromatography (TLC), and infrared spectroscopy. It seems that the orally administered Blastin is not easily absorbed through the intestines, but is excreted unchanged in the feces, and the non-toxic property of Blastin may be attributed in part to this poor uptake. In contrast, unchanged γ-BHC in feces did not exceed five percent of the dose.

Three water-soluble metabolites in the urine were detected by paper chromatography: the major one was chromatographically identified with pentachlorobenzoic acid, the second major one, which gave Blastin upon hydrolysis with β-glucuronidase, appeared to be the ether glucuronide of Blastin, and the minor metabolite was found to be easily transformed to Blastin when incubated in acetate buffer of pH 4.2 for three hours at 37° C. (Fig. 6). None of them have yet been isolated in good purity.

Experiments on retention of radioactivity in rat organs demonstrated that no significant accumulation of C^{14} was found in rats treated with Blastin, while C^{14} originated from C^{14}-γ-BHC was found high in kidney and fat, which coincides with the fact that measurable radioactivity was still present in one-week urine of γ-BHC-treated rats (Fig. 5).

b) Metabolism in rice plants

Three days after foliar application of C^{14}-Blastin (five μc), rice leaf blades (55 g.) were homogenized and extracted with ether. Most of the radioactivity in the ether extract was due to Blastin which ap-

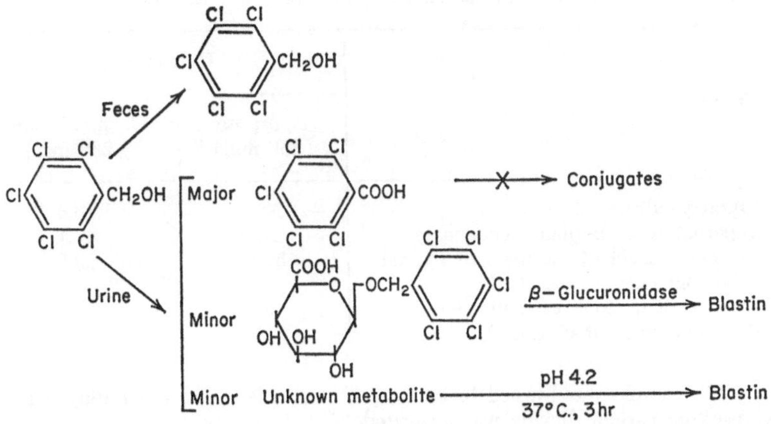

Fig. 6. Metabolism of Blastin in rats

peared to remain unchanged in and on the leaves. Three ether-soluble metabolites were detected with TLC. The major one, which amounted to approximately one percent of the dose, was chromatographically identified to be pentachlorobenzaldehyde (PCBAD). This was also confirmed by deriving the aldehyde to its 2,4-dinitrophenylhydrazone. Dehydrogenation of Blastin to PCBAD in rice plants has already been reported by KAKIGI and MISATO (1967). Identification of the other two metabolites is meeting difficulties mainly because of their poor production rates. KAKIGI and MISATO (1967) have reported that TLC of benzene-acetic acid extracts of rice leaves which absorbed Blastin through the roots give the spots that have the R_f values corresponding to those of pentachlorobenzoic acid and pentachlorophenol. Our experiment with labeled Blastin, however, gave negative results.

The aqueous layer separated from the ether extract contained a metabolite which was found to have a large molecular size by a gel filtration (Sephadex G-10) (NAKAGAWA and ISHIDA, unpublished results).

c) Metabolism in fungi

Each of sensitive (P. oryzae) and insensitive (C. miyabeanus) fungi was cultured in a C^{14}-Blastin containing medium with shaking, separated into four fractions, and then the distribution of added radioactivity was investigated. Table II shows that both of the fungi take the radioactivity up to a great extent and the difference in uptake between two fungi seems not to be able to interpret the specificity of Blastin toward P. oryzae. The same two minute metabolites were detected with TLC in all three liquid fractions from both of the fungi: one was identified to be PCBAD with TLC and the other neutral substance is under investigation (NAKAGAWA and ISHIDA, unpublished

Table II. *Uptake of radioactivity by fungi and its distribution*

Fractions	Uptake (%)	
	P. oryzae (250 mg.)	C. miyabeanus (280 mg.)
Filtered culture solution	20.5	10.4
Supernatant of hyphal homogenate	22.2	47.1
Acetone extract of sediment of hyphal homogenate	51.6	39.2
Residue of acetone extraction [a]	3.8	1.4
Recovery, percent of dose [b]	98.1	98.1

[a] The sample was burned in an oxygen atmosphere and the radioactivity of the resulting carbon dioxide was measured.

[b] Seven μc of C^{14}-Blastin.

results). The presence of a phenolic metabolite has been claimed by KAKIGI and MISATO (1967).

Studies on fungal metabolism of Blastin at around the stage of appressorium formation are in progress, expecting to disclose the mechanism of inhibition of hyphal penetration.

Summary

Phytopathological studies on the mode of action of 2,3,4,5,6-pentachlorobenzyl alcohol, Blastin, toward rice blast revealed that both inhibition of hyphal penetration of rice blast fungus into rice leaves and antisporulation on growing leaves were playing key roles in its action. Metabolic studies in rats demonstrated that a large portion of orally administered Blastin was excreted unchanged in the feces and the rest was eliminated in the urine as pentachlorobenzoic acid and the ether glucuronide of Blastin. Very minute portion of Blastin applied on rice leaves was transformed to pentachlorobenzaldehyde which was also found to be the major metabolite in fungi.

Résumé *

L'alcool pentachlorobenzylique, un agent de lutte contre la nielle du riz

Des études phytopathologiques sur le mode d'action de l'alcool 2,3,4,5,6-pentachlorobenzylique, Blastin, à l'égard de la nielle du riz ont montré que l'inhibition de la pénétration de l'hyphe du parasite

* Traduit par S. DORMAL-VAN DEN BRUEL.

dans le feuillage et l'inhibition de la sporulation sur les feuilles en croissance jouaient un rôle essentiel dans l'action du produit. Des études du métabolisme chez le rat ont démontré qu'une fraction importante du Blastin, administré par voie orale, était excrétée non modifiée dans les fèces et que le reste était éliminé dans l'urine sous forme d'acide pentachlorobenzoïque et d'éther glucuronique du Blastin. De très petites quantités de Blastin appliqué sur le feuillage du riz ont été transformées en pentachlorobenzaldéhyde; ce produit fut aussi trouvé comme étant le principal métabolite chez les cryptogames.

Zusammenfassung *

Pentachlorbenzylalkohol, ein Mittel zur Kontrolle des Reisbrands

Phytopathologische Studien über die Wirkungsweise von 2,3,4,5,6-Pentachlorbenzylalkohol, Blastin, auf Reisbrand zeigte, dass sowohl die Hemmung der Hypheninvasion des Reisbrandpilzes in die Reisblätter als auch die Antisporenbildung an wachsenden Blättern Schlüsselrollen bei seiner Wirkung spielen. Metabolische Studien an Ratten haben erwiesen, dass eine grosse Menge des oral eingegebenen Blastins unverändert in den Fäkalien und der Rest im Urin als Pentachlorbenzoesäure und als das Aetherglukoronid des Blastin ausgeschieden wurde. Ein sehr kleiner Anteil von Blastin, der auf Reisblätter appliziert worden war, wurde in Pentachlorbenzaldehyd überführt, das auch als der Hauptmetabolit im Pilz gefunden wurde.

References

BEILSTEIN, VON F., and A. KUHLBERG: Untersuchungen über Isomerie in der Benzoëreihe. Zehnte Abhandlung. Über Di- und Trichlorbenzoësäure. Ann. Chem. 152, 224 (1869).

CARTER, D. V., P. T. CHARLTON, A. H. FENTON, J. R. HOUSLEY, and B. LESSEL: The preparation and the antibacterial and antifungal properties of some substituted benzyl alcohols. J. Pharm. Pharmacol. 10, 149T (1958).

FUJITA, H., and M. ISHIDA: Synthesis of C14-2,3,4,5,6-pentachlorobenzyl alcohol, Blastin. Ann. Rept. Sankyo Research Lab., in press (1968).

HORSFALL, J. G.: Personal communication (1967).

ISHIDA, M.: The metabolism of pentachlorobenzyl alcohol (Blastin) in rats. Unpublished work, ms. submitted Agr. Biol. Chem. Tokyo (1968).

KAKIGI, K., and T. MISATO: The metabolism of Blastin in rice plants and rice blast fungus. Synopsis of paper presented Annual Meeting Kanto Branch Phytopathol. Soc. Japan (1967). Ann. Phytopathol. Soc. Japan 33, 319 (1967).

KONDO, M.: Personal communication (1966).

OKU, H., and H. SUMI: Mode of action of pentachlorobenzyl alcohol, a rice blast control agent.—Inhibition of hyphal penetration of *Pyricularia oryzae* through artificial membrane. Ann. Phytopathol. Soc. Japan, in press (1968).

Sankyo Co., Ltd.: Toxicity of Blastin. Shin-noyaku 20, 68 (1966).

---

* Übersetzt von A. SCHUMANN.

SUMI, H., Y. TAKAHI, K. NAKAGAMI, and Y. KONDO: Physiological effects of penta-chlorobenzyl alcohol on rice plants. Ann. Phytopathol. Soc. Japan **33**, 150 (1967).

———— ———— ———— ———— Controlling effect of pentachlorobenzyl alcohol on rice blast. Ann. Phytopathol. Soc. Japan **34**, 114 (1968 a).

———— ————, and Y. KONDO: Inhibitory effect of pentachlorobenzyl alcohol (PCBA) on sporulation of *Pyricularia oryzae* Cavara. Ann. Phytopathol. Soc. Japan **34**, 122 (1968 b).

Radiotracer studies on metabolism, degradation, and mode of action of insecticide chemicals *

by
JOHN E. CASIDA **

Contents

I. Introduction

The fields of intermediary metabolism and detoxication mechanisms were largely rewritten after the advent of radiotracer approaches. The same is true for the areas of insecticide metabolism and degradation, and, in small part, also for their mode of action. Radiotracer studies on metabolism and fate of pesticides are no longer just a means of satisfying academic curiosity on compounds which are already in use; on the contrary, they are now generally a necessary and critical step in understanding the persistence and action of a pesticide before it enters actual use. These investigations usually are designed to provide a total accounting, or balance sheet, of the physical and chemical fate of the compound, and frequently they are done under conditions simulating those involved in actual use. Thus, they involve studies on metabolic fate in plants, mammals, and pest species and on photodegradation, in addition to mode of action studies concerning combination or reaction at the site of physiological disruption. Radioanalysis usually can be achieved at nanogram levels of the

* Work reported herein was partially supported by AEC Contract AT(11-1)-34, Project Agreement No. 113, and by PHS Grants ES 00049 and GM 12248.
** Division of Entomology, University of California, Berkeley, California 94720.

labeled compound, and the specificity of analysis depends largely on the degree of separation and clean-up before scintillation counting or radioautography. Fortunately, these techniques are, in most instances, adequately developed so that the limiting factor is the experimental design.

This discussion concerns largely the strategy of recent studies with labeled insecticides done in the author's laboratory. The general steps involved are as follows: 1) select the site for labeling of the insecticide chemical; 2) perform the radiosynthesis; 3) introduce the labeled compound into an appropriate biological system or degradation situation; 4) determine the chemical and physical fate of the compound; and 5) interpret the results in relation to the mechanism, selectivity, and efficiency of action as an insecticide chemical.

II. Selection of site for labeling

Both theoretical and practical considerations enter into the selection of the site to be radiolabeled. Frequently the functional group critical for biological activity is known and is common to a group of pesticides. Labeling this biologically-active or toxophoric grouping is justified because an understanding of reactions occurring at this site will probably lead most directly to an understanding of the mode of action. In addition, by labeling at this position a common intermediate can be used for preparing several members of the group of related compounds for comparative study. Examples of such labeled intermediates for synthesis of insecticide and synergist chemicals are as follows: methyl isocyanate-C^{14} for methylcarbamate insecticides (KRISHNA et al. 1962), chrysanthemumic acid-C^{14} for pyrethroid insecticides (NISHIZAWA and CASIDA 1965 b), methylene-C^{14} iodide for methylenedioxyphenyl synergists (KUWATSUKA and CASIDA 1965), and a relatively few P^{32}-labeled intermediates for organophosphorus insecticides (CASIDA 1961).

When a mixture of isomers results from radiosynthesis, whether or not the isomers differ in biological activity, separation or resolution of these isomers is necessary because labeled isomers other than the desired labeled compound behave like impurities from the standpoint of interpreting the results. Examples of such radiosyntheses and separations include cis and trans isomers of organophosphates (ARTHUR and CASIDA 1959, CASIDA 1955), diastereoisomers of sulfoxide synergist (KUWATSUKA and CASIDA 1965), the d-trans-acid d-alcohol isomer of allethrin (NISHIZAWA and CASIDA 1965 b, YAMAMOTO and CASIDA 1966, YAMAMOTO et al. 1968), and the 6aβ,12aβ,5'β-configuration of natural rotenone (NISHIZAWA and CASIDA 1965 a).

Practical considerations lead to use of unlabeled fragments, derived from degradation of natural products or other relatively complex materials, to react with a labeled intermediate suitable for building again the original biologically-active molecule, but in a labeled form. In pyrethrin I-C^{14} synthesis, d-pyrethrolone derived from pyrethrum extract was used (YAMAMOTO *et al.* 1968), for rotenone-6a-C^{14}, derritol was used to preserve the $5'\beta$-configuration in the final product (NISHIZAWA and CASIDA 1965 a), and sulfoxide synergist was easily labeled by demethylenation of the unlabeled compound with aluminum trichloride followed by methylenation of the resulting unlabeled catechol with methylene-C^{14} iodide (KUWATSUKA and CASIDA 1965).

The subsequent fate of fragments liberated on metabolism or degradation must also be considered. Labeling of groups easily removed from the molecule on metabolism can be used with advantage in some cases, but labeling at such a position is of great disadvantage, in other cases. Groups frequently removed are, among others, ester moieties, and O-alkyl and N-alkyl groups, the latter taking place if mixed-function oxidase enzymes are present. Labeled fragments which are liberated and enter general metabolic pools often cause complications because a great variety of metabolites are formed; for example, oxidation of O-methyl-C^{14}, N-methyl-C^{14}, and methylene-C^{14}-dioxyphenyl compounds gives products entering the one-carbon metabolic pool and the nucleic acid pool, while initial O-dealkylation of organophosphate-P^{32} insecticides ultimately yields orthophosphate^{32} which enters the phosphate pool. For this reason, there is often justification for making the more difficult syntheses involved in labeling of groups which are more central in the molecule, such as aromatic rings or ring substituents. Tagging with tritium has the advantages of ease of labeling and the high specific activities attainable; disadvantages are the special procedures necessary in analysis of the weak beta radiation, the frequent unspecific labeling of the molecule, and particularly the possibility of the loss of the label through simple exchange reactions with hydrogen in the course of the experiment. These disadvantages of tritium are greatly minimized in *in vitro* studies or when used in conjunction with other labeled compounds, such as those incorporating C^{14} or P^{32}.

Multiple-labeled samples, particularly with the labels on different potential "leaving groups" from the molecule, greatly facilitate a variety of studies on degradation and mode of action of insecticides. This is convenient, for example, with combinations of any two of the three radiolabels, H^3, C^{14}, and P^{32}, or by the use of separate but identical experiments either with C^{14} samples separately labeled at different positions in the molecule or with C^{14} and S^{35} or C^{14} and Cl^{36} samples. This procedure has been particularly useful with carbamates, involving carbonyl, N-methyl, and ring or ring-substituent labels (CASIDA 1963, DOROUGH and CASIDA 1964, DOROUGH *et al.* 1963, KRISHNA and CASIDA 1966, KUHR and CASIDA 1967, LEELING and CASIDA 1966, OONNITHAN and CASIDA 1966 and 1968), with pyrethroids separately labeled in the acid and alcohol moieties (YAMAMOTO and CASIDA 1966, YAMA-

MOTO *et al.* 1968), with organophosphates labeled at the phosphorus, *O*-alkyl and/or phenolic or enolic moiety positions (BERENDS *et al.* 1959, CASIDA *et al.* 1962, HODGSON and CASIDA 1962, MENZER and CASIDA 1965, ROGER *et al.* 1964 and 1968, SMITH *et al.* 1967, STIASNI *et al.* 1967), and with methylenedioxyphenyl synergists labeled in a ring-substituent position in addition to the methylene-C^{14}-dioxy moiety (CASIDA *et al.* 1966 a, ESAAC *et al.* 1968, KUWATSUKA and CASIDA 1965, SCHMIDT and DAHM 1956).

III. Identification of labeled degradation products and metabolites

Isolation in a radiochemically pure state is accomplished when the degradation product cannot be resolved, by chromatography or other separation means, into more than one labeled compound, and in a pure state free of unlabeled impurities when, in addition, its specific activity is equivalent to that of the original labeled compound used. When isolated in a state of high chemical purity in amounts of 20 μg. or more, characterization then becomes largely a matter of analysis by spectral means such as ultraviolet, infrared, nuclear magnetic resonance spectroscopy, and mass spectroscopy, combined with degradation or derivative studies. Use of enzyme systems or unicellular test organisms facilitates isolation of metabolites in a pure state, because fewer interfering materials are generally involved, in those cases where such systems are known to produce the same products as the higher living organisms.

Radiochemical purity is the most easily achieved, and degradation products frequently can be tentatively characterized at this stage either by derivative formation, by carrier crystallization, or cochromatography with authentic compounds from synthesis. The problem is—what compounds need to be synthesized for comparison? If the choice is not obvious, then degradation studies on the labeled product and characterization of the labeled fragments, by cochromatography or labeled-derivative formation, might suggest where and what chemical changes have taken place. Frequently, these labeled derivatives can be formed directly on the thin-layer chromatography (TLC) plates by use of appropriate reagents which, of course, vary with the functional group to be tested.

Additional characterization evidence is provided by bioassay of the labeled metabolites compared with known compounds suspected to be of identical structure; examples are dimetilan-C^{14} metabolites as assayed for toxicity to houseflies (ZUBAIRI and CASIDA 1965), carbamate metabolites as assayed *in situ* on TLC plates for cholinesterase-inhibiting potency (OONNITHAN and CASIDA 1966 and 1968), and rotenone-C^{14} metabolites as assayed for inhibition of reduced nicotinamide adenine dinucleotide (NADH) oxidase (FUKAMI *et al.* 1967).

Metabolite conjugates frequently can be labeled for characterization studies by administering either the foreign compound or the conjugating portion, or both, in a labeled form as has been done with labeled synergists and C^{14}-labeled amino acids which combine to give various amino acid conjugates of piperonylic acid in houseflies (ESAAC and CASIDA 1968).

IV. Active sites of enzymes and reaction kinetics

Radiolabeled substrates and inhibitors are useful in studies on cholinesterase activity and on the active site of this important enzyme. In a mole-for-mole reaction, acetylcholine, carbamates, and organophosphates acylate the active site. The stability of the acyl intermediate increases and the turnover number of the enzyme decreases in the order indicated. A radiolabel on the acylating moiety (such as the acetyl, carbamoyl, or phosphoryl group, respectively) serves to radiolabel the esteratic site at the hydroxyl function of a serine residue; a label at another position in the molecule enables following the reaction rate by analysis for the "leaving group". Acetyl-C^{14}-choline is very rapidly hydrolyzed by cholinesterase to yield acetate-C^{14}; since acetate-C^{14} but not acetyl-C^{14}-choline is acid volatile, radioanalysis for acid-volatile product gives a quantitative measure of acetyl-C^{14}-choline hydrolysis and, therefore, of cholinesterase activity. This method of analysis is particularly useful in considering the kinetics of carbamate inhibition of cholinesterases (WINTERINGHAM and DISNEY 1964). 3,5-Diisopropylphenyl-H^3 methylcarbamate has been used in kinetic studies on the release of 3,5-diisopropylphenol-H^3 on reaction with cholinesterase, a reaction which is biphasic as expected for the initial carbamoylation phase followed by subsequent turnover of the enzyme, and which is blocked by organophosphate inhibitors (O'BRIEN et al. 1966).

The low turnover number of organophosphates means that, under usual conditions, only one molar equivalent of leaving group is liberated, at which time a stable dialkylphosphoryl enzyme is formed. Dialkylphosphoryl-P^{32} cholinesterase has served as the basis for the analytical method for determining the amino acid sequence about the serine group which ultimately is phosphorylated (COHEN and OOSTERBAAN 1963). Dialkylphosphoryl cholinesterase undergoes an "aging process", both in vivo and in vitro, and when this reaction is studied in vitro with diisopropyl-C^{14}-phosphoryl and diisopropylphosphoryl-P^{32} cholinesterases, it is found that isopropanol-C^{14} is liberated from the inhibited enzyme as it undergoes the aging process to form monoisopropylphosphoryl cholinesterase; so, this process is one of spontaneous O-dealkylation on the enzyme surface (BERENDS et al. 1959).

On reaction of certain insecticide chemicals with microsomal-reduced nicotinamide adenine dinucleotide phosphate (NADPH) enzymes, the insecticides are oxidized and fragments from the insecticides bind to the "microsomes". Methyl- and dimethylcarbamate insecticide chemicals and/or certain of their oxidation products carbamoylate "microsomes", the N-methyl-C^{14} and carbonyl-C^{14} samples yielding bound radiocarbon, but the label is not bound when using ring- or ring-substituent-C^{14}-labeled preparations (Oonnithan and Casida 1968); methylcarbamoylation of proteins is a plausible explanation of the persistence of certain of the fragments from carbamates in mammals (Krishna and Casida 1966). Parathion-S^{35} undergoes oxidation to paraoxon, but the S^{35} remains bound to the "microsomes" (Nakatsugawa and Dahm 1967). In future studies, this binding of fragments should be considered in more detail in order to find out whether a specific or an unspecific binding site is involved with microsomal systems, particularly where labeled synergists are used as substrates.

Radiolabeled respiratory inhibitors (rotenone-C^{14} or piericidin A-C^{14} and triphenyltin^{113}, which act at different sites in the electron transport pathway) and uncouplers of respiratory chain phosphorylation (2-sec-butyl-4,6-dinitrophenol-C^{14} and its esters, and substituted-2-trifluoromethyl-C^{14}-benzimidazoles) should prove useful in further studies on the nature of the active sites involved on the enzymes with which these compounds interact or react. A beginning in this direction has been made with rotenone-C^{14} in binding experiments using submitochondrial particles (electron transport particles, or ETP) prepared from beef heart mitochondria (Horgan et al. 1968). One of the major problems was to differentiate binding at the enzymatically active site, or specific binding, from unspecific binding. This was accomplished by the observation that piericidin A, amytal, and a number of rotenoids appear to react at the same site as rotenone because they compete with rotenone-C^{14} for binding at the specific site, and the competition with rotenone is proportional to the inhibitory power of the compounds tested for NADH oxidase (Fig. 1).
Studies are continuing in the hope of defining the nature of the binding site and of the interactions involved. Inhibitors of this enzyme system, such as rotenone and piericidin A, are potent insecticides, so this information is important.

V. Illustrative studies

Comparative studies on 10 methyl- and dimethylcarbamates with various positions of C^{14}-labeling were made to determine their fates in rats (Krishna and Casida 1966), in plants (Abdel-Wahab et al. 1966, Kuhr and Casida 1967), on incubation with microsome-NADPH systems from housefly abdomens (Tsukamoto and Casida 1967 a and b), and rat liver (Oonnithan and Casida 1966 and 1968), and on exposure to sunlight on plant surfaces (Abdel-Wahab et al. 1966). By the use of dimethyl-H^3-amino- and carbonyl-C^{14}-labeled samples of

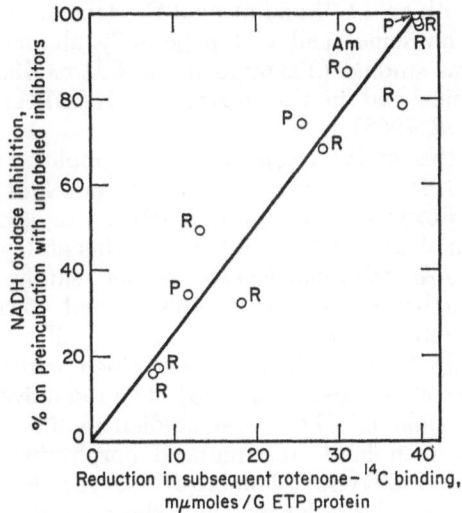

Fig. 1. Effect of various rotenoids (R), piericidin A (P), and amytal (Am) on NADH oxidase activity and on rotenone-C^{14} binding

4-dimethylamino-3-cresyl and 4-dimethylamino-3,5-xylyl methylcarbamates, the photooxidation to N-demethylate the dimethylamino moiety was followed on plant surfaces (ABDEL-WAHAB and CASIDA 1967). Metabolites and photodecomposition products of carbamates were synthesized (ABDEL-WAHAB and CASIDA 1967, BALBA and CASIDA 1968), to yield structures suggested by degradation and derivative-formation studies on the labeled compounds. C^{14}-Labeled carbamates were also used in studies on the plant enzymes that are involved in the degradation of carbamates and in cleaving the conjugates formed from hydroxylated metabolites of carbamates in plants (KUHR and CASIDA 1967). Studies were initiated on plant and animal metabolism, and on photodegradation, using labeled preparations of other carbamates, rotenone, pyrethroids, dinitrophenolic compounds, and chloro-substituted-2-trifluoromethylbenzimidazoles.

More than 20 labeled insecticides and synergists were used in studies on the properties, substrate-specificity, and biochemical genetics of the housefly microsome-NADPH enzyme system (TSUKAMOTO and CASIDA 1967 a and b, TSUKAMOTO et al. 1968). Ten methylene-C^{14}-dioxyphenyl compounds were used in investigating the relation between the mechanism of synergistic action and of the hydroxylation of the methylene-C^{14} group in living mice and houseflies, and in enzyme systems derived from mice and houseflies (CASIDA et al. 1966 a, ESAAC et al. 1968). [These studies point out the importance of selective trapping of volatile products because safrole and dihydrosafrole volatilize rapidly from houseflies in an unchanged form, following injection, and so does acetone resulting from O-depropylation of isopropoxy-compounds (CASIDA et al. 1966 a, SHRIVASTAVA et al. 1968).]

By the use of both acid-C^{14} and alcohol-C^{14}-labeled esters, the fate of pyrethroids in houseflies and in the housefly abdomen-NADPH enzyme system was studied (YAMAMOTO and CASIDA 1966), as was also their fate in mice and in the mouse liver-NADPH enzyme system (YAMAMOTO et al. 1968).

Several of the major rotenone-C^{14} metabolites formed in the microsome-NADPH systems and in living houseflies and rats were characterized (FUKAMI et al. 1967). A labeled organophosphate was used to confirm evidence from other experimental approaches that adult helminths, Ascaris lumbricoides, do not readily oxidize or activate organophosphorus toxicants (KNOWLES and CASIDA 1966). The chemistry of dimethoate degradation in methyl cellosolve formulations to yield a highly toxic product was examined with dimethoate-P^{32} (CASIDA and SANDERSON 1961 and 1963). The reductive dechlorination of DDT-C^{14} to yield DDD-C^{14} was studied in the presence of lake water, bovine rumen fluid, and reduced porphyrins (MISKUS et al. 1965). Labeled 3',5'-diesters of 5-fluoro-2'-deoxyuridine (FUDR) were used in determining the order in which the ester groups are cleaved in vivo and in vitro to free FUDR, which is a carcinostatic agent and insect chemosterilant (CASIDA et al. 1966 b, NISHIZAWA et al. 1965). In studies on organophosphate-induced teratogenesis in chick embryos, labeled samples of Bidrin (the teratogen) and nicotinamide (the alleviating agent) were used in addition to many radioactive biochemical intermediates in labeled pool approaches to the mechanism of teratogenic action (ROGER et al. 1964 and 1968, UPSHALL et al. 1968).

Summary

Radiotracer approaches have played an important role in the past, and will undoubtedly continue to play an important role in the future, in studies on the chemistry, biochemistry, and mode of action of insecticide chemicals. The general steps involved in studies with labeled insecticides are as follows: select the site for labeling and perform the radiosynthesis; determine the chemical and physical fate of the compound in an appropriate biological system or degradation situation; and interpret the results in relation to the mechanism, selectivity, and efficiency of action as an insecticide chemical.

Résumé *

Etudes, à l'aide de traceurs radioactifs, sur le métabolisme, la dégradation et le mode d'action des insecticides chimiques

Les approches à l'aide de traceurs radioactifs ont joué, dans le passé, un rôle important et continueront, assurément, à assumer le même rôle dans le futur, en ce qui concerne les études sur la chimie,

* Traduit par S. DORMAL-VAN DEN BRUEL.

la biochimie et le mode d'action des insecticides chimiques. Les étapes générales impliquées dans les études sur les insecticides marqués sont les suivantes: choisir l'emplacement du marquage et réaliser la radiosynthèse; déterminer l'évolution chimique et physique du composé dans un système biologique approprié ou un état de dégradation; interpréter les résultats en relation avec le mécanisme, la sélectivité et l'efficacité de l'action d'un insecticide chimique.

Zusammenfassung *

Radiotracer-Studien über Metabolismus, Abbau und Wirkungsweise von insektiziden Chemikalien

Radiotracer-Methoden haben in der Vergangenheit eine wichtige Rolle gespielt und werden zweifellos weiterhin eine wichtige Rolle in der Zukunft für Untersuchungen über Chemie, Biochemie und Wirkungsweise von insektiziden Chemikalien spielen. Die folgenden Schritte sind bei Studien mit markierten Insektiziden üblicherweise einbegriffen: Wahl des Markierungsortes und Durchführung der Radiosynthese, Bestimmung des chemischen und physikalischen Verhaltens der Verbindung in einem entsprechenden biologischen System oder einer Abbausituation, Interpretation der Ergebnisse in Bezug auf den Mechanismus, Selektivität und Nutzwert der Wirkung als insektizides Chemikal.

References

ABDEL-WAHAB, A. M., and J. E. CASIDA: Photooxidation of two 4-dimethylaminoaryl methylcarbamate insecticides (Zectran and Matacil) on bean foliage and of alkylaminophenyl methylcarbamates on silica gel chromatoplates. J. Agr. Food Chem. 15, 479 (1967).
——, R. J. KUHR, and J. E. CASIDA: Fate of C14-carbonyl-labeled aryl methylcarbamate insecticide chemicals in and on bean plants. J. Agr. Food Chem. 14, 290 (1966).
ARTHUR, B. W., and J. E. CASIDA: Biological activity and metabolism of Hercules AC-528 components in rats and cockroaches. J. Econ. Entomol. 52, 20 (1959).
BALBA, M. H. M., and J. E. CASIDA: Synthesis of possible metabolites of methylcarbamate insecticide chemicals. Hydroxyaryl and hydroxyalkylphenyl methylcarbamates. J. Agr. Food Chem. 16, 561 (1968).
BERENDS, F., C. H. POSTHUMUS, I. VAN DER SLUYS, and F. A. DEIERKAUF: The chemical basis of the "ageing process" of DFP-inhibited pseudocholinesterase. Biochim. Biophys. Acta 34, 567 (1959).
CASIDA, J. E.: Isomeric substituted-vinyl phosphates as systemic insecticides. Science 122, 597 (1955).
—— Metabolism of organophosphate insecticides in plants: A review. In: Radioisotopes and radiation in entomology, pp. 49-64. Internat. Atomic Energy Agency, Vienna (1961).

* Übersetzt von A. SCHUMANN.

—— Radiotracer approaches to carbamate insecticide toxicology. In: Radiation and radioisotopes applied to insects of agricultural importance, pp. 223-239. Internat. Atomic Energy Agency, Vienna (1963).

——, J. L. ENGEL, E. G. ESAAC, F. X. KAMIENSKI, and S. KUWATSUKA: Methylene-C^{14}-dioxyphenyl compounds: Metabolism in relation to their synergistic action. Science 153, 1130 (1966 a).

—— ——, and Y. NISHIZAWA: 3',5'-Diesters of 5-fluoro-2'-deoxyuridine and thymidine: Hydrolysis by esterases in human, mouse, and insect tissue. Biochem. Pharmacol. 15, 627 (1966 b).

——, L. MCBRIDE, and R. P. NIEDERMEIER: Metabolism of 2,2-dichlorovinyl dimethyl phosphate in relation to residues in milk and mammalian tissues. J. Agr. Food Chem. 10, 370 (1962).

——, and D. M. SANDERSON: Toxic hazard from formulating the insecticide dimethoate in methyl 'cellosolve'. Nature 189, 507 (1961).

—— —— Reaction of certain phosphorothionate insecticides with alcohols and potentiation by breakdown products. J. Agr. Food Chem. 11, 91 (1963).

COHEN, J. A., and R. A. OOSTERBAAN: The active site of acetylcholinesterase and related esterases and its reactivity towards substrates and inhibitors. In: Cholinesterases and anticholinesterase agents, pp. 299-373. G. B. Koelle, Sub-Ed. Berlin: Springer-Verlag (1963).

DOROUGH, H. W., and J. E. CASIDA: Nature of certain carbamate metabolites of the insecticide Sevin. J. Agr. Food Chem. 12, 294 (1964).

——, N. C. LEELING, and J. E. CASIDA: Nonhydrolytic pathway in metabolism of N-methylcarbamate insecticides. Science 140, 170 (1963).

ESAAC, E. G., and J. E. CASIDA: Piperonylic acid conjugates with alanine, glutamate, glutamine, glycine, and serine in living house flies. J. Insect Physiol., 14, 913 (1968).

——, F. X. KAMIENSKI, and J. E. CASIDA: Unpublished results (1968).

FUKAMI, J-I., I. YAMAMOTO, and J. E. CASIDA: Metabolism of rotenone in vitro by tissue homogenates from mammals and insects. Science 155, 713 (1967).

HODGSON, E., and J. E. CASIDA: Mammalian enzymes involved in the degradation of 2,2-dichlorovinyl dimethyl phosphate. J. Agr. Food Chem. 10, 208 (1962).

HORGAN, D. J., T. P. SINGER, and J. E. CASIDA: Studies on the respiratory chain-linked reduced nicotinamide adenine dinucleotide dehydrogenase. XIII. Binding sites of rotenone, piericidin A, and amytal in the respiratory chain. J. Biol. Chem. 243, 834 (1968).

KNOWLES, C. O., and J. E. CASIDA: Mode of action of organophosphate anthelmintics: Cholinesterase inhibition in Ascaris lumbricoides. J. Agr. Food Chem. 14, 566 (1966).

KRISHNA, J. G., and J. E. CASIDA: Fate in rats of the radiocarbon from ten variously labeled methyl- and dimethylcarbamate-C^{14} insecticide chemicals and their hydrolysis products. J. Agr. Food Chem. 14, 98 (1966).

——, H. W. DOROUGH, and J. E. CASIDA: Synthesis of N-methylcarbamates via methyl isocyanate-C^{14} and chromatographic purification. J. Agr. Food Chem. 10, 462 (1962).

KUHR, R. J., and J. E. CASIDA: Persistent glycosides of metabolites of methylcarbamate insecticide chemicals formed by hydroxylation in bean plants. J. Agr. Food Chem. 15, 814 (1967).

KUWATSUKA, S., and J. E. CASIDA: Synthesis of methylene-C^{14}-dioxyphenyl compounds: Radioactive safrole, dihydrosafrole, myristicin, piperonyl butoxide, and diastereoïsomers of sulfoxide. J. Agr. Food Chem. 13, 528 (1965).

LEELING, N. C., and J. E. CASIDA: Metabolites of carbaryl (1-naphthyl methylcarbamate) in mammals and enzymatic systems for their formation. J. Agr. Food Chem. 14, 281 (1966).

MENZER, R. E., and J. E. CASIDA: Nature of toxic metabolites formed in mammals, insects, and plants from 3-(dimethoxyphosphinyloxy)-N,N-dimethyl-ciscrotonamide and its N-methyl analog. J. Agr. Food Chem. **13**, 102 (1965).

MISKUS, R. P., D. P. BLAIR, and J. E. CASIDA: Conversion of DDT to DDD by bovine rumen fluid, lake water, and reduced porphyrins. J. Agr. Food Chem. **13**, 481 (1965).

NAKATSUGAWA, T., and P. A. DAHM: Microsomal metabolism of parathion. Biochem. Pharmacol. **16**, 25 (1967).

NISHIZAWA, Y., and J. E. CASIDA: Synthesis of rotenone-6a-C^{14} on a semimicro scale. J. Agr. Food Chem. **13**, 522 (1965 a).

—— —— Synthesis of d-*trans*-chrysanthemumic acid-C^{14} and its antipode on a semimicro scale. J. Agr. Food Chem. **13**, 525 (1965 b).

—— ——, S. W. ANDERSON, and C. HEIDELBERGER: 3′,5′-Diesters of 5-fluoro-2′-deoxyuridine: Synthesis and biological activity. Biochem. Pharmacol. **14**, 1605 (1965).

O'BRIEN, R. D., B. D. HILTON, and L. GILMOUR: The reaction of carbamates with cholinesterase. Mol. Pharmacol. **2**, 593 (1966).

OONNITHAN, E. S., and J. E. CASIDA: Metabolites of methyl- and dimethylcarbamate insecticide chemicals as formed by rat liver microsomes. Bull. Environmental Contamination and Toxicol. **1**, 59 (1966).

—— —— Oxidation of methyl- and dimethylcarbamate insecticide chemicals by microsomal enzymes and anticholinesterase activity of the metabolites. J. Agr. Food Chem. **16**, 28 (1968).

ROGER, J-C., H. CHAMBERS, and J. E. CASIDA: Nicotinic acid analogs: Effects on response of chick embryos and hens to organophosphate toxicants. Science **144**, 539 (1964).

——, D. G. UPSHALL, and J. E. CASIDA: Unpublished results (1968).

SCHMIDT, C. H., and P. A. DAHM: The synthesis of C^{14}-labeled piperonyl butoxide and its fate in the Madeira roach. J. Econ. Entomol. **49**, 729 (1956).

SHRIVASTAVA, S. P., M. TSUKAMOTO, and J. E. CASIDA: Unpublished results (1968).

SMITH, G. N., B. S. WATSON, and F. S. FISCHER: Investigations on Dursban insecticide. Uptake and translocation of [^{36}Cl] O,O-diethyl O-3,5,6-trichloro-2-pyridyl phosphorothioate and [^{14}C] O,O-diethyl O-3,5,6-trichloro-2-pyridyl phosphorothioate by beans and corn. J. Agr. Food Chem. **15**, 127 (1967).

STIASNI, M., D. REHBINDER, and W. DECKERS: Absorption, distribution, and metabolism of O-(4-bromo-2,5-dichlorophenyl)-O,O-dimethylphosphorothioate in the rat. J. Agr. Food Chem. **15**, 474 (1967).

TSUKAMOTO, M., and J. E. CASIDA: Albumin enhancement of oxidative metabolism of methylcarbamate insecticide chemicals by the house fly microsome-NADPH$_2$ system. J. Econ. Entomol. **60**, 617 (1967 a).

—— —— Metabolism of methylcarbamate insecticides by the NADPH$_2$ - requiring enzyme system from house flies. Nature **213**, 49 (1967 b).

——, S. P. SHRIVASTAVA, and J. E. CASIDA: Biochemical genetics of house fly resistance to carbamate insecticide chemicals. J. Econ. Entomol. **61**, 50 (1968).

UPSHALL, D. G., J-C. ROGER, and J. E. CASIDA: Biochemical studies on the teratogenic action of Bidrin and other neuroactive agents in developing hen eggs. Biochem. Pharmacol. **17**, 1529 (1968).

WINTERINGHAM, F. P. W., and R. W. DISNEY: A radiometric study of cholinesterase and its inhibition. Biochem. J. **91**, 506 (1964).

YAMAMOTO, I., and J. E. CASIDA: O-Demethyl pyrethrin II analogs from oxidation of pyrethrin I, allethrin, dimethrin and phthalthrin by a house fly enzyme system. J. Econ. Entomol. **59**, 1542 (1966).

——, E. C. KIMMEL, J. L. ENGEL, and J. E. CASIDA: Unpublished results (1968).

ZUBAIRI, M. Y., and J. E. CASIDA: Detoxication of dimetilan in cockroaches and houseflies J. Econ. Entomol. **58**, 403 (1965).

Mode of action of natural insecticides

by
Izuru Yamamoto *

Contents

I. Nicotinoids

Nicotine is an autonomic blocking agent, and acts like acetylcholine but only at ganglia and neuromuscular junctions, initially stimulating then depressing them. Those symptoms caused by applied acetylcholine which resemble those in nicotine poisoning are called "nicotinic". In search for the structural feature for nicotine-like activity of acetylcholine, it has been postulated that the electron density at the carbonyl and/or the ethereal oxygen and the distance between either oxygen and the quaternary nitrogen are the important factors in determining the activity of the acetylcholine analogs (BARLOW 1955). Nicotine is nearly 90 percent ionized at the physiological pH, and the ionized form has been assumed *á priori* to be an active form by analogy to quaternary nitrogen of acetylcholine.

Quite contrast to the assumption in pharmacology, nicotine has been described as acting as the free base for insects, because free nicotine is more toxic than its salt when applied in usual practice. In fact, superiority of free form is attributable to the preferential cuticular permeability. The study of structure-activity relationships for insects by YAMAMOTO et al. (1962) reveals that they should be fairly strong bases but not quaternarized, and that they should mimic acetylcholine with respect to their molecular dimensions and electron make-up; specifically, they should contain the following essential moiety:

---

* Department of Agricultural Chemistry, Tokyo University of Agriculture, Tokyo, Japan.

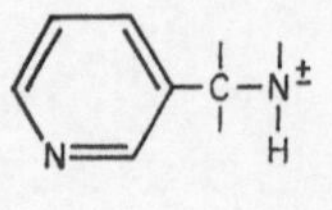

All nicotinoids of high toxicity have this essential moiety, whereas all nicotinoids not provided with the essential moiety have little or no toxicity. There are considerable evidences that ions very slowly penetrate insect ganglia and nerves, and it can be expected that ionic toxicants have much less effect upon insect nervous systems than their non-ionic analogs. We can see this case in Amiton and Tetram (O'Brien 1959). In this and other cases, the ionizable moiety is an auxiliary part, but in the case of the nicotinoids, it is an essential part of the molecule. At a glance it seems to be strange that more ionizable nicotinoids are more insecticidal. However, when we assume that the ionized form interacts with the receptor, the difficulty can be overcome. Based on this assumption, we proposed the scheme as shown in Figure 1. (Yamamoto et al. 1962, Yamamoto 1965).

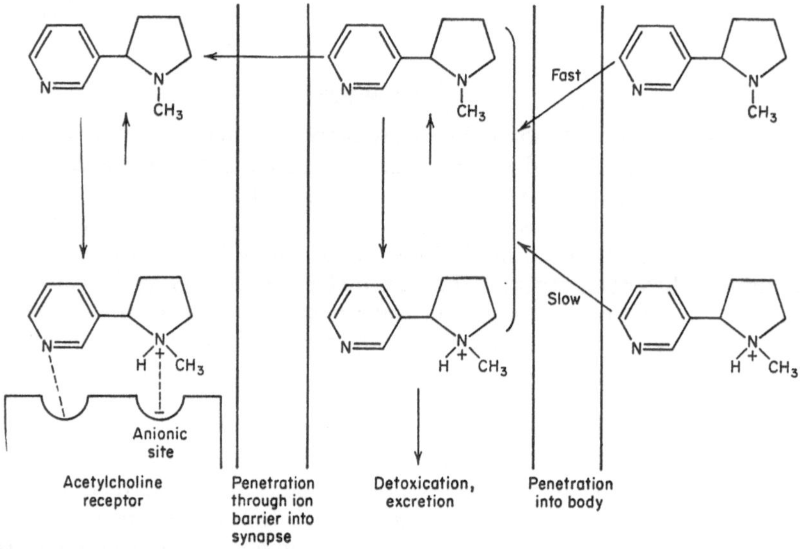

Fig. 1. Behavior of nicotine in insects. Yamamoto et al. (1962)

A direct approach to verify the interaction between a toxicant and a receptor is needed. However, because of the difficulty of isolating the receptor from nerve, we used house fly head cholinesterase as a model of the receptor. First it was shown that nicotine inhibited cholinesterase rather competitively at its lower concentrations and pH 7.4 (Soeda and Yamamoto 1968). Change of the enzyme surface at pH 7.4 and at 8.4 is small as indicated from the Michaelis constants (3.5 x

$10^{-4}M$ and $4.1 \times 10^{-4}M$, respectively), whereas the affinity between the enzyme and nicotine at pH 7.4 when nicotine is 76 percent ionized is nearly twice greater than that at pH 8.4 when nicotine is only 24 percent ionized as indicated from the inhibition constants ($1.0 \times 10^{-3}M$ and $2.3 \times 10^{-3}M$, respectively). Nicotine monomethiodide showed a complete competitive inhibition. Considering also the pH-dependency of cholinesterase inhibition, it was concluded that the cationic head of nicotinium ion interacts with the anionic site in the active center of cholinesterase. Assuming that other nicotinoids and 3-pyridylmethyl-amines also inhibit cholinesterase competitively, YAMAMOTO et al. (1968) compared their inhibitory power in relation to their toxicity and ionization. We used four types of compounds for the study: (I) nicotinoids, (II) N,N-disubstituted 3-pyridylmethylamines, (III) un- or mono-substituted 3-pyridylmethylamines, and (IV) other 3-pyridyl-alkylamines. Figure 2 shows a relationship between relative toxicity

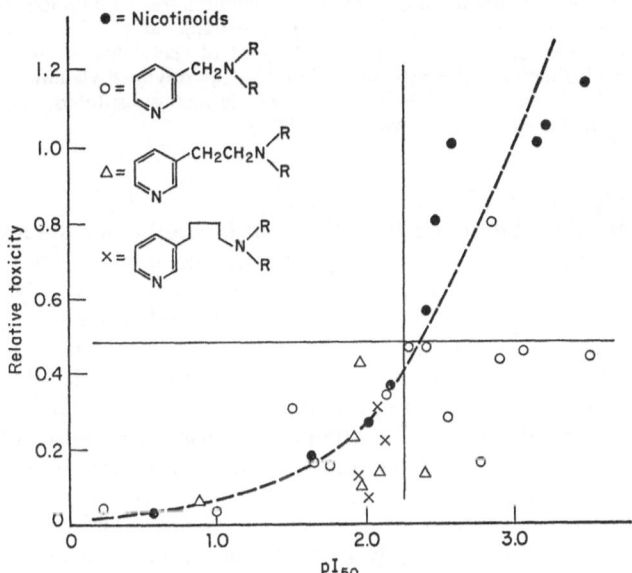

Fig. 2. Relationship between house fly head cholinesterase inhibition (as pI_{50}) and relative toxicity (1-nicotine = 1.0: house fly, topical LD_{50}); pI_{50} at 2.0 mM of acetylcholine at pH 7.4 (YAMAMOTO et al. 1968)

and cholinesterase inhibition. The highly toxic compounds showed the pI_{50} higher than a certain level, whereas the compounds having pI_{50} lower than the level showed low toxicity. It is noticeable that type IV compounds which are not provided with the essential moiety showed low toxicity even though they showed higher pI_{50}. Figure 3 shows the

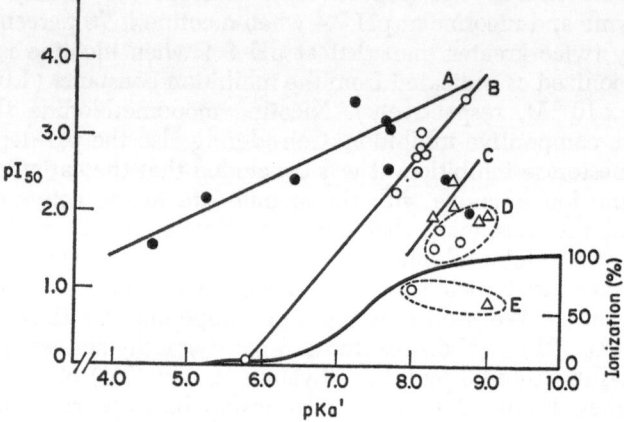

Fig. 3. Relationship among house fly head cholinesterase inhibition (as pI_{50}), basicity (as pKa'), and ionization percentage at pH 7.4 in nicotinoids and 3-pyridylalkylamines. pI_{50} at 2.0 mM of acetylcholine and at pH 7.4. A = nicotinoids, B = N,N-disubstituted 3-pyridylmethylamine, C = N,N-disubstituted 3-pyridylethylamine, D = N-monosubstituted, and E = N-unsubstituted (YAMAMOTO et al. 1968).

relationship among pKa', ionization at pH 7.4, and pI_{50}. General tendencies are that the higher the pKa', that is, the higher the ionization, the higher the pI_{50} becomes. Primary or secondary amine type of compounds show lowered pI_{50}, even though they are highly ionized. WILSON (1952) demonstrated that in an ethanolamine series, acetylcholinesterase inhibition became low when the number of methyl group on the nitrogen decreased. Therefore, the anionic site binding on cholinesterase seems to require some steric factors which reinforce the coulombic attraction. The anomalous position of nornicotine can be explained by its secondary amine nature. However, nornicotine is a rather highly toxic compound, so that steric differences of sites in the cholinesterase and in the nervous receptor might be present. Generally speaking, the highly basic (hence highly inhibitory) nicotinoids and 3-pyridylmethylamines are of high toxicity, whereas low basic (hence low inhibitory) ones are of low toxicity.

A preliminary attempt was made to correlate the structures of nicotinoids and 3-pyridylmethylamines with their effect on house fly nerve activities (YAMAMOTO and ISHIDA 1965). Details about experimental procedures were described by NARAHASHI (1964). The EC_{50} values for fly nerves were evaluated by counting the action potential outbreak frequencies with a spike counter, when nicotine and other analogs were applied. Highly basic compounds (highly toxic) gave more frequencies, whereas low basic or unsubstituted compounds (low toxicity) gave fewer frequencies.

From the above discussion, the toxicity of nicotinoids and 3-pyridylmethylamines is best explained by postulating the receptor carrying the anionic site which is electrostatically very similar to the anionic site of cholinesterase and, as YAMAMOTO (1965) postulated, the essential moiety of nicotinoids can be regarded as consisting of three parts: the highly basic nitrogen, which is protonated in the insect body and anchors the molecule to the anionic site of the receptor; a carbon bridge, which arranges the pyridine ring and the nitrogen in a definite position; and the pyridine ring, which effects some unidentified influence on the nerve membrane.

II. Pyrethroids

These are defined as axonic nerve poisons. Elucidation of the complete stereochemistry gave the basis for discussing structure-activity relationship. Four natural pyrethrins and synthetic d-*trans*-d-allethrin are represented by the same absolute configuration. It is a remarkable feature that all active pyrethroids contain a planar or pseudoplanar ring system in the alcohol moiety. The planar ring might fit the planar part of the receptor. The hydroxyl function should not be coplanar with the planar ring, as phenol esters lose toxicity. The eight possible isomers of allethrin has been prepared by LAFORGE *et al.* (1954 and 1956) and their insecticidal activities against house flies were studied by GERSDORFF (1947). ELLIOTT (1954) assigned rough relative potencies to the stereoisomers from toxicity data to typical insect species. Taking these values, we can correlate the change of toxicity to the change of configuration of the molecule (Fig. 4). The natural acid

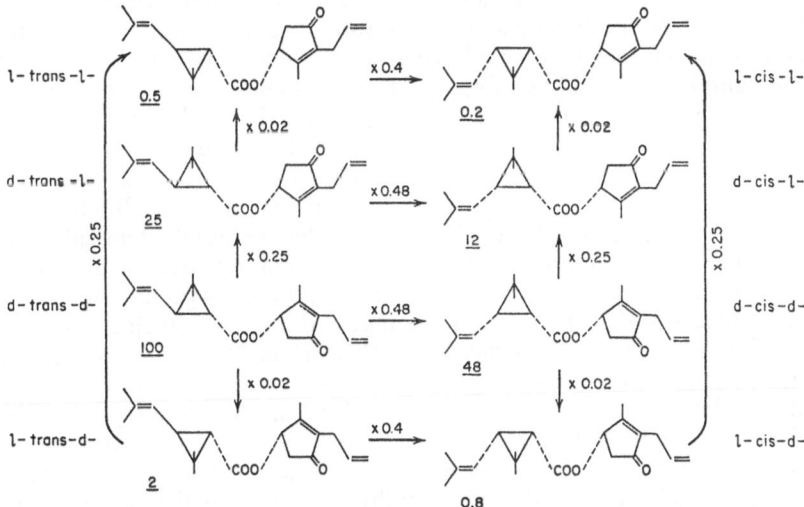

Fig. 4. Isomerization and toxicity changes of allethrin to typical insect species (from data of ELLIOTT 1954)

has the d-*trans* configuration, and this has been shown to be more insecticidally active than any of the other isomers when combined with the alcohol moiety. Again, the natural configurations of the ketoalcohol in the esters are insecticidally more active. We can assume that the most active d-*trans*-d-allethrin is provided with the most suitable arrangement of indispensable groups to fit on the receptor, and the deviation from this configuration will cause various degree of poor fit to the receptor, resulting in decreased toxicity. Transfer of the isobutenyl side chain from *trans* to the *cis* position on the cyclopropane ring decreases toxicity by 0.4 to 0.48. Change of d-allethrolone to l-allethrolone decreases toxicity by 0.25. The most significant change of toxicity occurs, when the d-*trans* acid is replaced by l-*trans* acid or d-*cis* acid is replaced by l-*cis* acid: the toxicity is decreased by 0.02. An understanding is made for this toxicity change by postulating that (1) the absolute configuration of $C_{(1)}$ on the cyclopropane ring is the most important factor, which determines the fitness of the gem-dimethyl group on the receptor, (2) when d-allethrolone is converted to l-allethrolone, the planarity of the ketoalcohol moiety on a hypothetical planar part of the receptor does not change, but the relative position of substituents on the ring is reversed, and (3) the contribution of the isobutenyl side chain is of secondary importance.

Next we will discuss the molecular feature involved in pyrethroid metabolism studied by Yamamoto and Casida (1966). Acid moiety-labeled d-*trans*-d-allethrin, d-*trans*-d-pyrethrin I, d-*trans*-dimethrin, d-*trans*-phthalthrin, and the alcohol-moiety-labeled d-*trans*-dl-allethrin and d-*trans*-phthalthrin were synthesized by them. As shown in Figure 5, allethrin-^{14}C was converted to ten or more metabolites by using house fly abdomen homogenate-NADPH$_2$ as metabolizing system. Each of these metabolites had the ester group intact, because the same products were detected when acid-labeled and alcohol-labeled compounds were used as substrate. The metabolite mixture was further resolved by an ammonia-containing solvent. The solvent minimized the movement of acidic metabolites. While chrysanthemumic acid was not found and all metabolites were esters, the major metabolite also appeared to be acidic. On saponification, the major metabolite derived from acid-labeled allethrin yielded a labeled material identical with known chrysanthemum dicarboxylic acid. On reaction with diazomethane, the major metabolite yielded a neutral ester product identical with allethrin II. This evidence suggested that the major allethrin metabolite might be O-demethyl allethrin II or "allethrin-ω-oic acid", and so this compound was synthesized. The synthetic compound was identical with the major metabolite and both are converted to allethrin II by methylation. The unmodified nature of the allethrolone moiety in the structure of the major metabolite was further demonstrated: when treated with 2,4-dinitrophenylhydrazine in methanolic sulfuric acid, it gave the dinitrophenylhydrazone of allethrolone methyl ether.

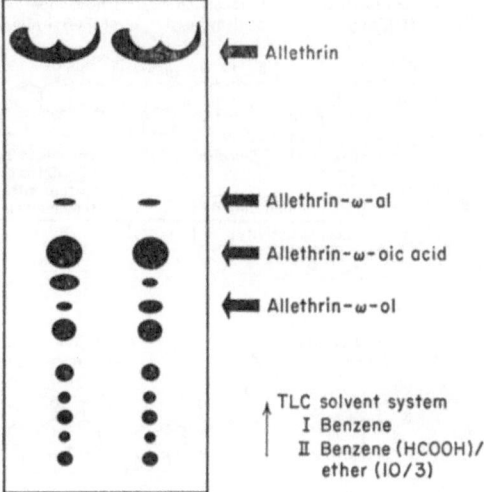

Allethrin

Allethrin-ω-al

Allethrin-ω-oic acid

Allethrin-ω-ol

TLC solvent system
I Benzene
II Benzene (HCOOH)/
ether (10/3)

Fig. 5. Allethrin metabolites for acid- and alcohol-labeled preparations incubated with R_{Hokota} house fly abdomen-NADPH$_2$ enzyme system (YAMAMOTO and CASIDA 1967)

Three neutral metabolites designated as *A*, *A'*, and *B* were also produced by metabolism. By using same type of experiments as used for the major metabolite, metabolite *A* was characterized as "allethrin-ω-ol" and metabolite *B* was characterized as "allethrin-ω-al". Metabolite *A'* gave an uncharacterized acid by hydrolysis, but its allethrolone

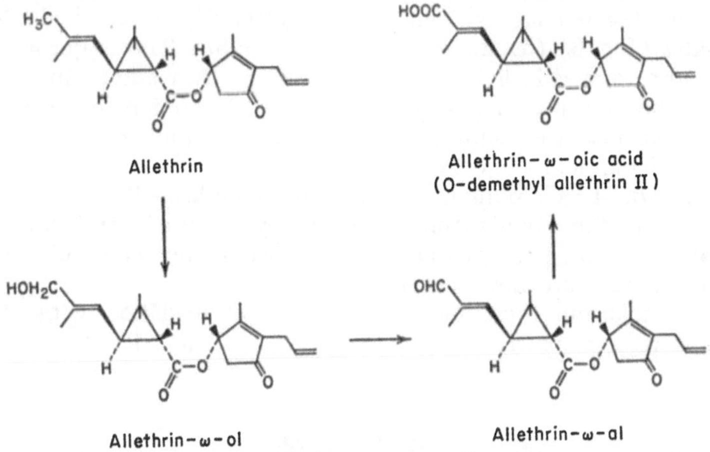

Allethrin

Allethrin-ω-oic acid
(O-demethyl allethrin II)

Allethrin-ω-ol

Allethrin-ω-al

Fig. 6. Metabolic pathway for allethrin in house flies (YAMAMOTO and CASIDA 1967)

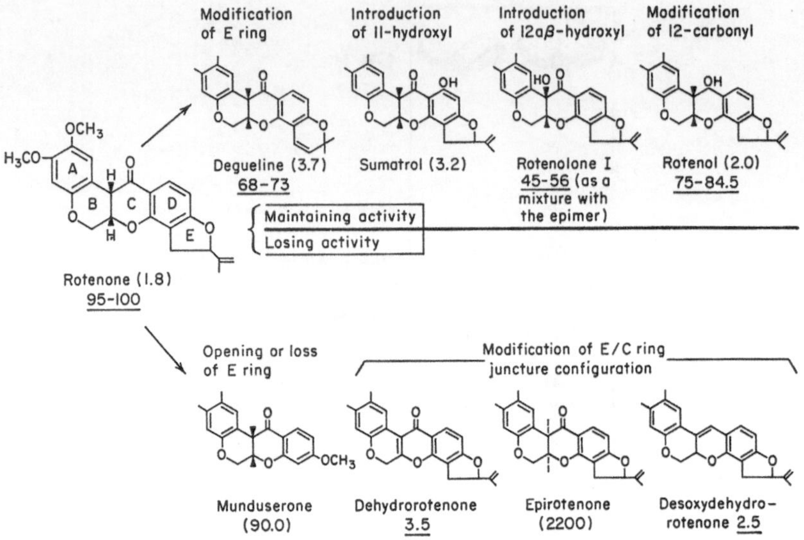

Fig. 7. Structural modifications and activity changes of rotenone. The figures underlined are percent inhibition of "glutamic dehydrogenase" (inhibitor, $10^{-5}M$) (Fukami et al. 1959) and the figures in parentheses are concentration for 50 percent inhibition of $NADH_2$ oxidase (μM/mg. protein x 10^5) (Burgos and Redfearn 1965)

moiety was not modified. When we examined the allethrin metabolite mixture as a whole, it was found that almost no modification occurred on the allethrolone moiety and chrysanthemumic acid was not formed; this means that allethrin is not hydrolyzed enzymatically. The major allethrin metabolite, allethrin-ω-oic acid, is thus formed in the fly abdomen homogenate system through alcohol and aldehyde intermediates (Fig. 6) (Yamamoto and Casida 1967). This major metabolite is also formed in living house flies and by liver-microsomal oxidation in the presence of NADPH₂. The overall reaction involves a detoxification because allethrin is greater than 30-fold more toxic than allethrin-ω-oic acid (injection, house flies).

Pyrethrin I-¹⁴C, phthalthrin-¹⁴C, and dimethrin-¹⁴C are also metabolized in the same manner; one methyl group in the isobutenyl moiety is oxidized to a carboxyl group, while the ester linkage and alcohol linkage are not modified.

In conclusion, this study establishes that oxidation without hydrolysis is a major mechanism for pyrethroid detoxication in house flies.

III. Rotenoids

The mechanism of action was defined as inhibiting specifically the coupled oxidation of $NADH_2$ and reduction of cytochrome b (LINDAHL and ÖBERG 1961). At first we will discuss the structure-activity relationship (data taken from FUKAMI et al. 1959, BURGOS and REDFEARN 1965) (Fig. 7). The presence of A, B, C, and D rings with *cis* fusion of B and C rings as in rotenone is essential. Spatial conformation at B/C rings seems to be important, as epirotenone, (+)-isorotenone, dehydrorotenone, and rotenolone II are inactive. Slight modification at B/C rings, as far as it does not affect the conformation of B/C fusion, is permissible, as in the case of rotenolone I and rotenolol. Various types of modification at the E ring occur without serious loss of potency as seen in dihydrorotenone, (−)-isorotenone, elliptone, degueline, and rotenone hydrochloride, whereas munduserone and tetrahydrorotenone having no E ring are of slight or no potency. Introduction of the hydroxyl group at $C_{(11)}$ to the active rotenoids reduces the potency to a half. As (±)-6a, 12a-dihydro-6H-rotoxen-12-one provided with the essential A, B, C, and D rings showed only 1/2,222th of potency (1/1,111th for the active enantiomorph), two methoxyl groups on the A ring and the presence of the E ring may constitute a sufficient condition.

The metabolism of rotenone has been recently elucidated by FUKAMI et al. (1967). Rotenone-6a-^{14}C yielded eight metabolites *in vitro* by the microsome-$NADPH_2$ system as derived from liver or house fly abdomens. Each metabolite is more polar than rotenone, and designanted as A through H. The solid spots are metabolites, A, C, D, F, and G. The sequence of metabolite formation was first determined as follows:

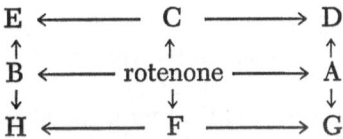

At least four initial site of attack on the molecule are involved. Four additional metabolites, D, E, G, and H, are subsequently formed from the initial metabolites, so they must involve at least two modifications (Fig. 8).

The major metabolite is F. Ultraviolet and infrared spectra indicated a great similarity to rotenone except for the hydroxyl function. The hydroxyl is probably of an aliphatic nature, as it reacted with methylisocyanate but not with diazomethane. A mass spectrum gave a molecular weight of 428, comparable to introducing two hydroxyl groups into the molecule of rotenone; this rules out the replacement of hydrogen with a hydroxyl group, for example ring hydroxylation. The A/B and B/C/D rings are unmodified, but the D/E ring portion and specifically the side chain are modified. The fragmentation patterns of rotenoids were studied by REED and WILSON (1963). Metabo-

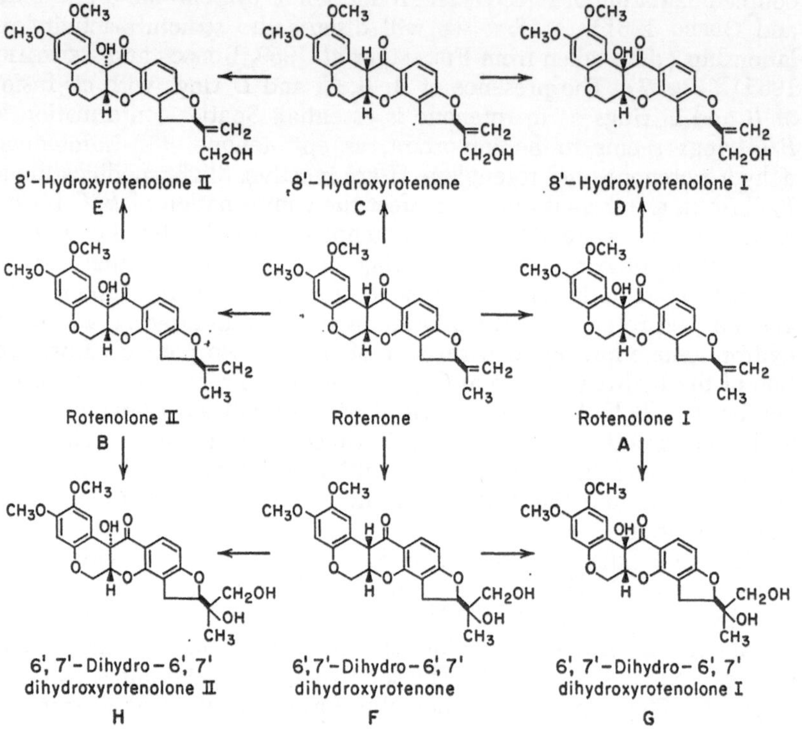

Fig. 8. Metabolic pathway of rotenone (FUKAMI *et al.* 1967)

lite *F* gave the ions indicating the unmodified nature of its *A/B/C* ring system. These observations suggested dihydrodihydroxyrotenone as *F*. 6′,7′-Dihydro-6′,7′-dihydroxyrotenone synthesized by osmium tetroxide oxidation of rotenone was identical with metabolite *F*. Metabolites *A* and *B* were obtained chemically by alkaline oxygenation of rotenone (as spots). This oxygenation results in introduction of a hydroxyl group at the 12a-position of the *B/C* ring juncture to form a mixture of rotenolone I and rotenolone II; this characteristic reaction also occurs with related compounds. By the same treatment, metabolite *C* gave *D* and *E*, and metabolite *F* gave *G* and *H*. (Rotenolones I and II are only general names because the products formed on alkaline oxygenation consists of a mixture of four diastereoisomers. Rotenolone I is a mixture of $6a\beta$, $12a\beta$- and $6a\alpha$, $12a\alpha$ isomers having the *cis B/C* fusion. Rotenolone II is a mixture of $6a\beta$, $12a\alpha$- and $6a\alpha$, $12a\beta$-isomers having the *trans B/C* fusion. Rotenolone I and II are resolvable by TLC, but the two components of each of these rotenolones are usually not resolved by this means. The assumption was made that the rotenolones formed by enzymatic hydroxylation are of the $6a\beta$, $12a\beta$-configuration (rotenolone I) and $6a\beta$, $12a\alpha$-configuration (rotenolone II), because only with these two of the four possible diastereoisomers is the configuration retained at positions other than the 12a position involved in the hydroxylation.)

Therefore, metabolite G probably is dihydrodihydroxyrotenolone I, and metabolite H probably is dihydrodihydroxyrotenolone II. These structures were synthesized, and identification with metabolites G and H was made. The structure of G was further supported by its mass spectrogram. Metabolite C was found to be 8'-hydroxyrotenone (amorphigenin), so metabolites D and E are probably the 12aβ-hydroxyl and 12aα-hydroxyl derivatives of metabolite C, respectively. Rotenone is therefore susceptible to attack at the 12a-carbon to give the isomeric β-hydroxyl compound (rotenolone I) and the α-hydroxyl compound (rotenolone II). The isopropenyl side chain is metabolized to yield 8'-hydroxyrotenone and 6',7'-dihydro-6',7'-dihydroxyrotenone. Various combinations of these sites of attack yield eight metabolites in all. Additional, more polar metabolites are subsequently formed, and these may be in part conjugates of the initial hydroxylation products.

These initial reactions are not necessarily detoxification, because the toxicity of certain of the metabolites is of the same order as that of rotenone. When injected into male mice, LD_{50} values as mg./kg. were observed as follows: rotenone = 2.8, 8'-hydroxyrotenone (C) = 2.6, dihydrodihydroxyrotenone (F) = 10, rotenolone I (A) = 4.1, and rotenolone II (B) = > 25.

Further studies on rotenone toxicology and biochemistry should take these metabolites into account.

Summary

Historical natural insecticides, pyrethrins, nicotine, and rotenone, exert their toxic action through different mechanisms. *Nicotine* is a synaptic blocking agent and the physiological data made it likely that nicotine is active by its similarity to acetylcholine. Recent study on the structural requirement of nicotine analogs for insecticidal activity showed that they should be fairly strong bases but not quaternized and their ionized form should mimic acetylcholine with respect to molecular dimensions and electron make-up. Although more precise study is needed for defining the nature of the receptor, the detailed analysis of inhibition by nicotine and the related compounds of house fly head cholinesterase as a model of the nerve receptor gave some insights to the interaction between the cationic head of the toxicant and the hypothetical anionic site in the receptor. *Pyrethroids* are defined as an axonic nerve poison. Modification of the configuration at each asymmetric center in the allethrin molecule results in stepwise changes of insecticidal activity. This relationship indicates that the absolute $C_{(1)}$ configuration on the cyclopropane ring is the most important factor for toxicity and that the isobutenyl side chain on the ring is of secondary importance. However, the recent study on the molecular features involved in pyrethroid metabolism in the insect revealed that the isobutenyl side chain is a point of metabolic attack. In common to allethrin, pyrethrin I, dimethrin, and phthalthrin, the chrysanthemumic acid moiety of the molecule is oxidized to chrysanthemum dicarboxylic acid form via the corresponding alcohol and aldehyde, while the ester linkage is intact. *Rotenone* exerts its toxic action by inhibiting the coupled oxidation of NADH$_2$ and reduction of cytochrome b. Rather rigid configuration of the ring system of the rotenoid molecule is required for the inhibitory action. Recently the structures of eight products from *in vitro* and *in vivo* metabolisms of rotenone were elucidated. Hydroxylation occurs at the 12a, 6'-7', and 8' positions. As certain of the metabolites are as toxic as rotenone, they need to be considered in the discussion of the mode of action and the selective toxicity.

Résumé *

Mode d'action des insecticides naturels

Les premiers insecticides naturels: pyréthrines, nicotine et roténone exercent leur effet toxique grâce à des mécanismes divers. La *nicotine* bloque les synapses et les données physiologiques semblent indiquer que son activité est dûe à sa ressemblance avec l'acétylcholine. Une étude récente des configurations structurales nécessaires pour que les analogues de la nicotine aient une activité insecticide, a montré que ces produits devaient être des bases fortes mais non quaternaires et que la forme ionisée devait ressembler à l'acétylcholine au point de vue de la dimension moléculaire et de la configuration électronique. Bien qu'une étude plus précise soit nécessaire pour définir la nature du récepteur, l'analyse détaillée de l'inhibition par la nicotine et les composés voisins de la cholinestérase de la tête de la mouche domestique—pris comme modèle de récepteur nerveux—a donné quelques aperçus de l'intéraction entre la tête cationique du toxique et l'hypothétique site anionique du récepteur. Les *pyréthrines* sont définies comme des poisons nerveux. Des modifications de la configuration de chacun des centres asymétriques dans la molécule d'alléthrine produisent un changement graduel de l'activité insecticide. Cette relation indique que la configuration absolue (C_1) sur le noyau cyclopropanique est le facteur le plus important pour la toxicité et que la chaine latérale isobuténique sur le cycle a une importance secondaire. Cependant, l'étude récente sur la caractéristique moléculaire impliquée dans le métabolisme des pyréthrines chez l'insecte a révélé que la chaine latérale isobuténique était un point de l'attaque métabolique. Dans l'alléthrine, la pyréthrine I, la diméthrine comme dans la phthalthrine, la partie acide chrysanthémique de la molécule est oxydée en acide chrysanthème-dicarbonique avec passage par la forme alcool et aldé-

---

* Traduit par R. Mestres.

hyde, tandis que la liaison ester demeure intacte. La *roténone* exerce son action toxique par inhibition de l'oxydation de $NADH_2$ et de la réduction du cytochrome b. Une configuration plutôt rigide des cycles de la molécule est nécessaire pour cette action inhibitrice. Les structures de huit produits de métabolisme in vitro et in vivo de la roténone ont été récemment élucidées. L'hydroxylation survient aux positions 12a, 6'-7' et 8'. Comme certains de ces métabolites sont aussi toxiques que la roténone, ils doivent être pris en considération dans la discussion du mode d'action et de la toxicité sélective.

Zusammenfassung *

Wirkungsweise von natürlichen Insektiziden

Historische, natürliche Insektizide wie die Pyrethrine, Nicotin und Rotenon üben ihre toxische Wirkung durch verschiedene Mechanismen aus. *Nicotin* ist ein synaptisches Blockierungsmittel, und die physiologischen Daten machten es wahrscheinlich, dass Nicotin wegen seiner Aehnlichkeit mit Acetylcholin wirksam ist. Neuere Forschung über die strukturellen Bedingungen von Nicotinanalogen für insektizide Wirksamkeit zeigten, dass sie ziemlich starke Basen sein sollten, aber keine quaternären Gruppen haben, und ihre ionisierte Form sollte Acetylcholin ähneln in Bezug auf Molekulargrösse und Elektronenanordnung. Obwohl genauere Untersuchungen nötig sind, um die Natur des Rezeptors zu definieren, gab die genaue Untersuchung der Hemmung (durch Nicotin und verwandte Verbindungen) von Hausfliegenkopf-Cholinesterase als Modell für den Nervenrezeptor einige Einsicht in die Wechselwirkung zwischen dem kationischen Kopf der toxischen Verbindung und der hypothetischen anionischen Seite am Rezeptor. *Pyrethroide* werden als axonische Nervengifte definiert. Modifizierung der Konfiguration an jedem der asymmetrischen Zentren des Allethrinmoleküls hat die stufenweise Aenderung der insektiziden Aktivität zur Folge. Diese Beziehung zeigt an, dass die absolute Konfiguration $C_{(1)}$ am Cyclopropanring der wichtigste Faktor für die Toxizität ist und dass die Isobutenylseitenkette am Ring von sekundärer Bedeutung ist. Jedoch hat neue Forschung über die molekularen Merkmale, welche mit dem Pyrethroidmetabolismus im Insekt verwickelt sind, offenbart, dass die Isobutenylseitenkette ein Angriffspunkt für Metabolismus ist. Wie beim Allethrin, Pyrethrin I, Dimethrin und Phthalthrin, wird der Chrysanthemumsäureanteil des Moleküls zur Chrysanthemum-dicarboxysäureform über den entsprechenden Alkohol und das Aldehyd oxidiert, während die Esterbindung intakt bleibt. *Rotenon* übt seine toxische Wirkung durch Hemmung der gekuppelten Oxidation von $NADH_2$ und Reduktion von Cytochrom b aus. Eine ziemlich starre Konfiguration des Ringsystems des Rotenoidmoleküls ist für eine

---

* Übersetzt von A. Schumann.

Hemmwirkung erforderlich. Kürzlich sind die Strukturen von 8 Produkten von *in vitro* und *in vivo* Metabolismen von Rotenon aufgeklärt worden. Hydroxylierung findet an den 12a-, 6'-7'- und 8'-Positionen statt. Da einige der Metaboliten ebenso toxisch sind wie Rotenon, sollten sie in jeder Diskussion über Wirkungsweise und selektive Toxizität berücksichtig werden.

References

BARLOW, R. B.: Introduction to chemical pharmacology. London: Methuen (1955).

BURGOS, J., and E. R. REDFEARN: The inhibition of mitochondrial reduced nicotinamide-adenine dinucleotide oxidation by rotenoids. Biochem. Biophys. Acta 116, 475 (1965).

ELLIOTT, M.: Allethrin. J. Sci. Food Agr. 5, 505 (1954).

FUKAMI, J., T. NAKATSUGAWA, and T. NARAHASHI: The relation between chemical structure and toxicity in rotenone derivatives. Japan J. Applied Entomol. Zool. 3, 259 (1959).

——, I. YAMAMOTO, and J. E. CASIDA: Metabolism of rotenone in vitro by tissue homogenates from mammals and insects. Science 155, 713 (1967).

GERSDORFF, W. A.: Toxicity to house flies of the pyrethrins and cinerins, and derivatives, in relation to chemical structure. J. Econ. Entomol. 40, 878 (1947).

LaFORGE, F. B., N. GREEN, and M. S. SCHECHTER: Allethrin. Resolution of dl-allethrolone and synthesis of the four optical isomers of trans-allethrin. J. Org. Chem. 19, 457 (1954) and 21, 455 (1956).

LINDAHL, P. E., and K. E. ÖBERG: The effect of rotenone on respiration and its point of attack. Expt. Cell Research 23, 228 (1961).

NARAHASHI, T.: Insecticide resistance and nerve sensitivity. Jap. J. Med. Sci. Biol. 17, 46 (1964).

O'BRIEN, R. D.: Effect of ionization upon penetration of organophosphates to the nerve cord of the cockroach. J. Econ. Entomol. 52, 812 (1959).

REED, R. I., and J. M. WILSON: Electron impact and molecular dissociation. Part XII. The cracking patterns of some rotenoids and flavones. J. Chem. Soc., p. 5949 (1963).

SOEDA, Y., and I. YAMAMOTO: Studies on nicotinoids as an insecticide. Part V. Inhibition of house fly head cholinesterase by nicotine. Agr. Biol. Chem. 32, 568 (1968).

WILSON, I. B.: Acetylcholinesterase. XII. Further studies of binding forces. J. Biol. Chem. 197, 215 (1952).

YAMAMOTO, I.: Nicotinoids as insecticides. Adv. Pest Control Research 6, 231 (1965).

——, and J. E. CASIDA: O-Demethyl pyrethrin II analogs from oxidation of pyrethrin I, allethrin, dimethrin, and phthalthrin by a house fly enzyme system. J. Econ. Entomol. 59, 1542 (1966).

—— —— Metabolism of pyrethroids by mammals and insects. Proc. 153rd Ann. Meeting Amer. Chem. Soc., Miami Beach (1967).

——, and K. ISHIDA. Unpublished (1965).

——, H. KAMIMURA, R. YAMAMOTO, S. SAKAI, and M. GODA: Studies on nicotinoids as an insecticide. Part I. Relation of structure to toxicity. Agr. Biol. Chem. 26, 709 (1962).

——, Y. SOEDA, H. KAMIMURA, and R. YAMAMOTO: Relationship between cholinesterase inhibition and insecticidal activity of nicotinoids. Proc. Ann. Meeting Agr. Chem. Soc. Japan, Nagoya (1968).

Selective toxicity of systemic insecticides

by

TETSUO SAITO [*]

Contents

I. Introduction

Systemic action of plant tissues may offer a possibility of developing ecological selectivity: once the insecticide is absorbed and translocated in the plant tissues, only phytophagous pests will consume it directly, and it reduces the population of plant-feeding pests without killing beneficial organisms. Therefore, systemic activity is a very desirable property of both insecticides and fungicides. The usage of systemic compounds may decrease the spray coverage, making possible the use of lighter application equipment, and decrease the cost of application. Also, pests whose habitats prevent intoxication by contact insecticide, e.g., apple leaf curling aphid, Myzus malisuctus Matsumura, may be controlled by the systemic insecticide.

There have been some trials of applying chemical compounds to roots or stems for controlling the pest since the late nineteenth century. For example, MONZEN (1933 and 1957) found that nicotine sulfate applied to the soil of potted apple trees was absorbed and translocated in the apple plant, and eventually killed the woolly apple aphid, Erisoma lanigerum Hausmann. He showed, also, that dipping of rice seedlings in the nicotine sulfate were successful for controlling the rice leaf miner, Agromyza oryzae Munakata, and proposed the concept of

* Laboratory of Applied Entomology and Nematology, Faculty of Agriculture, Nagoya University, Chikusa, Nagoya, Japan.

"innertherapy" in pest control. Oya (1950) determined by chemical analysis that the nicotine content in the rice plant tissues which were soaked in 0.1 percent aqueous nicotine sulfate solution for 24 hours was 400 p.p.m., and found that this concentration was maintained in the plant body for four to five days.

Systemic activity of sodium selenate (Hurd-Karrer and Poos 1936), rotenone (Fulton and Mason 1937), and sodium fluoroacetate (David and Gardiner 1951) has been observed. However, the well-known discovery by Schrader and Kükenthal of the systemic insecticidal properties of organophosphorus compounds greatly stimulated subsequent research in this field, and their study has contributed much to practical methods of pest control (Geary 1953). Many efforts have been carried out since then to find better insecticides having systemic activity and some of them have been proved to be satisfactory practically. However, it must be recognized that most of them are ineffective against chewing insects and some compounds even exhibit a high mammalian toxicity. Since the first review on systemic insecticides by Geary (1953), various aspects of the subject have been reviewed by many workers.

II. Selective toxicity of systemic insecticides

Mechanism of the selective toxicity of systemic insecticides may be broadly classified into two groups, one ecological and the other physiological. Although these two mechanisms operate concurrently in plant, insect, and animal bodies, let us discuss the mode of action of insecticides based on these two factors.

1. Ecological factors: Differences in the feeding habit of insects.
2. Physiological factors: (a) differences in distribution patterns of insecticides in the insect body, (b) differences in the metabolisms of insecticides, and (c) differences in the sensitivity of the sites of action to insecticides.

Generally, systemic insecticides are not distributed uniformly in the plant body. They are absorbed by the plant tissues and transported to other parts in the transpiration stream of the xylem of the shoot axis and leaves. The upward translocation toward the apex is more intensive than the downward translocation. Furthermore, translocation of the toxicant toward the younger leaves and the peripheral growing areas of the upper leaves was found to be more intensive than that to the aged tissues.

Feeding behaviors of pests differ among different species. Hemipterus pests are known to suck the plant sap, while beetles and Lepidopterus larvae usually eat the tissues of leaves, stems, or fruits. Certain non-selective systemic compounds, however, can be used to achieve ecological selectivity by the following application methods: seeds, soil or bark treatment, or trunk implantation in order to minimize the unfavourable effects on natural enemies. This phenomenon was demonstrated for dimefox and mipafox (RIPPER et al. 1951). RIDGWAY et al. (1966) have shown that significant mortalities of boll weevils were observed by the application of a systemic insecticide and a feeding stimulant, although the concentration of insecticide was lower in the bolls, which are attackable by weevils.

Physiological selectivity has been proved only for schradan. Schradan is effective against sucking insects but not effective against chewing insects. Schradan was activated and detoxicated by the insects. The metabolic rates of schradan in the different pests varied considerably and definite relationships were not found between metabolism and toxicities. Definite relationship was not observed between the sensitivity of cholinesterase of insects and toxicities, but relationship appears to exist in the distribution pattern of toxicants in insect bodies between sucking and chewing insects. More toxicants accumulated in the central nervous systems of susceptible pests than in those of non-susceptible pests. Electron microscopic examination on the central nerve sheath showed differences in thickness and structure of the sheath between susceptible and non-susceptible insects. Therefore, the distribution of toxicants in pests and the character of the nerve sheath which acts as a barrier against the penetration of toxicants may be the most important factors responsible for the selective toxicity of schradan (CASIDA and STAHMANN 1953, O'BRIEN and SPENCER 1953 and 1955, SAITO 1960 a, and 1960 b, SAITO and MATSUI 1960). Experimental results using schradan analogues have suggested that O-p-nitrophenyl N,N-dimethyl phosphorodiamidate has a toxicity pattern similar to that of schradan (METCALF and MARCH 1949). However, dimefox and mipafox failed to show any selective activity, unless they are applied to the soil to avoid the direct contact with the insect's body (RIPPER et al. 1951).

Schradan is one of the selective mammalicides. Insect nerve has been shown to have a sheath impermeable to sodium, potassium, acetylcholine, and Tetram. This sheath is absent in the mammal. So far, these substances are shown to be highly toxic to mammalian as compared to most other insects (HOYLE 1952, TOBIAS et al. 1946, TWAROG and ROEDER 1956, YAMASAKI and NARAHASHI 1958). Schradan is an interesting compound, which is not ionized yet shows a toxic pattern somewhat similar to that of an ionized organophosphates, e.g., Tetram, which was toxic to mammals, aphids, and mites. The active metabolite of schradan, the hydroxymethyl derivative, is a very polar compound which does not penetrate along with other ionized compounds (O'BRIEN 1961). O'BRIEN (1961) has stated "It is not anticipated that a fervent need for such selective mammalicides will arise. However, the generalization can be immediately useful in suggesting which compounds the insecticide designer should avoid".

Dimethoate [O,O-dimethyl S-(N-methyl carbamoylmethyl) phosphorodithioate] is an interesting compound, because of its low mammalian toxicity and its high systemic activity in animals and plants. It is broken down much more rapidly in the mouse than in the housefly or American cockroach (Krueger *et al.* 1960). Carboxyesterase or carboxyamidase are thought to be responsible for the degradation of this compound, analogous to the case of malathion breakdown (Dauterman *et al.* 1959).

Thioether systemics, demeton, methyl demeton, disulfotone, and phorate are very effective against sucking insects, Hemipterus pests, but not so effective against chewing insects, army worm, and rice stem borer. However, there has been no detailed study in attempting to elucidate the factors involved in this phenomenon. Whereas many workers studied the metabolism of these compounds and found that many related metabolites were produced in the insects, plants, and animals from the compounds but that the rate of metabolism differed by different organisms (Bull 1965, Bowman and Casida 1958, March *et al.* 1955).

Monofluoroacetate is more toxic to mammals than its amide, anilide, or *p*-bromoanilide derivatives. These derivatives are known to be converted to the acetate by insect enzymes, which have fairly different properties compared with the enzymes present in the warm-blooded animals. These findings thus suggest that some selective insecticides may be discovered among the derivatives of monofluoroacetic acid (Ando and Nakamura 1966, Matsumura and O'Brien 1963).

Carbamate insecticides are broken down by hydrolysis, methyl hydroxylation, and hydroxylation of the aromatic ring and the conjugation before or after hydrolysis of the carbaryl grouping. These reactions are known to be the major pathway for the metabolism of biologically active carbamates. Biological significance of these phenomena has been discussed in relation to the species specificity of insecticides (Casida 1963, Weiden and Moorefield 1965).

III. Toxicological aspects of *O,O*-dimethyl dichlorohydroxyethyl phosphonate and trichlorfon

We have investigated the systemic action of several commercially established systemic insecticides against the adult of the green peach aphid, *Myzus persicae* L. and the larvae of the tobacco cutworm, *Prodenia litura* Fab., in cabbage plants. It was demonstrated that these compounds are highly toxic to the aphid but less toxic to the worm (Table I).

Various phosphate esters (thiophosphate, phosphate, phosphonate, and pyrophosphate) were examined to discover the wormcidal compounds. The final results of our examination in the series are shown in Table II. Schradan was effective to the aphid but ineffective to the worm, while some new phosphonate compounds, NS 2662, *O,O*-

Table I. *Systemic insecticidal activity of some systemic insecticides to the green peach aphid and to the tobacco cutworm*

Insecticide	Mortality (%)			
	Aphid p.p.m.		Worm p.p.m.	
	1,000	100	1,000	100
Thiol methyl demeton	100	90	60	0
Thiono methyl demeton	100	95	60	10
Thiol ethyl demeton	100	85	70	0
Thiono ethyl demeton	100	100	30	0
Phorate	100	50	20	0
Disulfotone	80	85	40	0
Schradan	90	70	0	0
Control	—	(0)	—	(0)

dimethyl dichlorohydroxyethyl phosphonate, and its acetyl derivative (NS 2664), and trichlorfon and its ethyl methyl homologue (NS 2666) were found to be more toxic to the worm compared to the aphid (SAITO and HONDA 1966). The same insecticidal activities of these compounds against the larvae of rice stem borer, *Chilo suppressalis* Walk., and the female adult of the green rice leafhopper, *Nephotettix cincticeps* Uhler, in the rice plant were observed (SATO and SAITO unpublished).

Table II. *Systemic insecticidal test of some organophosphorus esters to the green peach aphid and to the tobacco cutworm*

Compound	Mortality (%)			
	Aphid p.p.m.		Worm p.p.m.	
	100	10	100	10
Schradan	70	15	0	0
Trichlorfon	25	0	100	30
NS 2523 $(CH_3O)_2P(O)O(Br)HC(Br)Cl_2$	10	0	60	30
NS 2662 $(CH_3O)_2P(O)CH(OH)CHCl_2$	15	0	100	20
NS 2664 $(CH_3O)_2P(O)CH(OCOCH_3)CHCl_2$	5	0	90	20
NS 2666 $(CH_3O)(C_2H_5O)P(O)CH(OH)CCl_3$	25	0	100	30

Several factors may control the selective toxicities of NS 2662 and trichlorfon as mentioned above. The absorption and translocation of ^{32}P-NS 2662 and ^{32}P-trichlorfon from the root and the feeding amounts of insecticides with the four insect species were observed. Considerable amounts of chloroform-partitioned radioactivities were detected in the aerial parts of the plants and there were no relationships between the feeding amounts and the systemic toxicities for those insects (SATO and SAITO unpublished).

Physiological factors, distribution, metabolism, and *in vitro* cholinesterases inhibition of those insecticides were studied for the larvae of the rice stem borer and the tobacco cutworm, the adult of the American cockroach, the housefly, the green rice leafhopper, the black rice bug (*Scotinophara lurida* Bur.), and the corbett rice bug (*Leptocorixa corbetti* China). No relationships were observed between the susceptible and non-susceptible insects regarding the permeabilities of cuticle, distribution in the various tissues and penetration in the dissected thoracic ganglions of ^{32}P-NS 2662 and ^{32}P-trichlorfon (Tables III, IV, and V).

Table III. *Amounts of ^{32}P-NS 2662 and ^{32}P-trichlorfon in the various parts of insect bodies after topical application*

Insect	Amounts found (μg./g.)				
	Ganglions	Gut	Fat	Coxa	Others
NS 2662					
Rice stem borer	95 μg/g	8	4	—	12
Tobacco cutworm	138	33	23	—	7
American cockroach	84	164	34	42	41
Housefly	49	39	65	30	22
Green rice leafhopper	20	9	12	21	15
Black rice bug	169	52	69	154	78
Corbett rice bug	25	58	19	36	21
Trichlorfon					
Rice stem borer	12	6	3	—	1
Tobacco cutworm	62	25	29	—	2
American cockroach	43	95	26	16	17
Housefly	75	36	122	36	19
Green rice leafhopper	27	6	11	9	8
Black rice bug	52	27	51	23	14
Corbett rice bug	20	19	37	50	16

Table IV. *The absorption and excretion of ^{32}P-NS 2662 and ^{32}P-trichlorfon by various insects after topical application of insecticides*

Insect	Hours	Absorption or excretion (%)					
		NS 2662			Trichlorfon		
		Outer	Internal	Excreta	Outer	Internal	Excreta
Rice stem	0.5	3.4	1.1	95.5	10.9	3.2	85.9
borer	1.0	2.9	5.1	92.0	4.3	1.9	93.8
	2.0	2.9	6.9	90.2	0.4	2.9	96.8
Tobacco	0.5	31.6	32.0	36.4	21.2	42.0	36.4
cutworm	1.0	24.0	35.6	40.4	11.9	15.3	72.8
	2.0	10.6	34.4	55.0	6.0	14.6	79.4
American	0.5	31.3	51.9	16.8	14.7	31.7	53.6
cockroach	1.0	28.8	35.0	36.2	13.1	29.7	57.2
	2.0	2.0	15.8	79.2	11.0	48.1	40.9
Housefly	0.5	42.0	41.5	10.5	26.4	14.1	59.5
	1.0	17.3	7.8	74.9	27.8	6.4	65.8
	2.0	34.0	12.9	53.1	22.6	15.1	62.3
Green rice	0.5	15.5	31.9	52.6	1.4	6.4	92.6
leafhopper	1.0	14.4	35.8	51.8	1.3	8.8	89.9
	2.0	12.4	45.2	42.4	0.9	5.7	93.4
Black rice	0.5	25.3	20.8	53.9	22.5	9.2	68.3
bug	1.0	18.5	27.4	54.1	16.8	9.7	73.5
	2.0	23.6	13.4	63.1	11.3	6.1	82.6
Corbett	0.5	37.0	19.8	43.2	5.2	4.7	90.1
rice bug	1.0	14.4	25.7	59.7	6.4	6.4	87.2
	2.0	6.6	7.5	85.9	4.5	8.9	86.6

Table V. *Penetration of ^{32}P-NS 2662 and ^{32}P-trichlorfon into dissected thoracic ganglions of various insects immersed in Ringer solution*

Insect	Penetration (μg./g.)			
	NS 2662 Minutes		Trichlorfon Minutes	
	10	20	10	20
Rice stem borer	0.31	1.10	1.39	2.34
Tobacco cutworm	1.95	2.33	5.61	6.12
American cockroach	0.23	0.34	0.74	0.79
Housefly	1.93	2.17	1.73	3.71
Green rice leafhopper	0.55	0.64	0.53	0.87
Black rice bug	0.86	1.28	2.77	3.59
Corbett rice bug	1.57	1.71	3.23	4.42

Results of the *in vivo* metabolism of ^{32}P-NS 2662 and ^{32}P-trichlorfon and their topical toxicities are shown in Figure 1. A relationship appears to exist between the metabolism of these compounds and the susceptibility of insects to the compounds (SATO and SAITO unpublished).

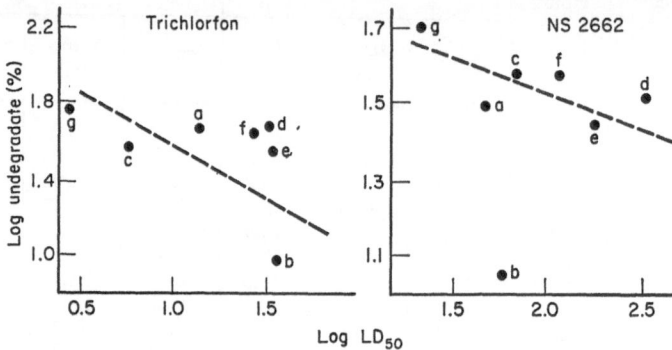

Fig. 1. The relationship for amounts of undegradate and topical toxicity of trichlorfon and NS 2662: (*a*) rice stem borer, (*b*) tobacco cutworm, (*c*) American cockroach, (*d*) housefly, (*e*) green rice leafhopper, (*f*) black rice bug, and (*g*) corbett rice bug

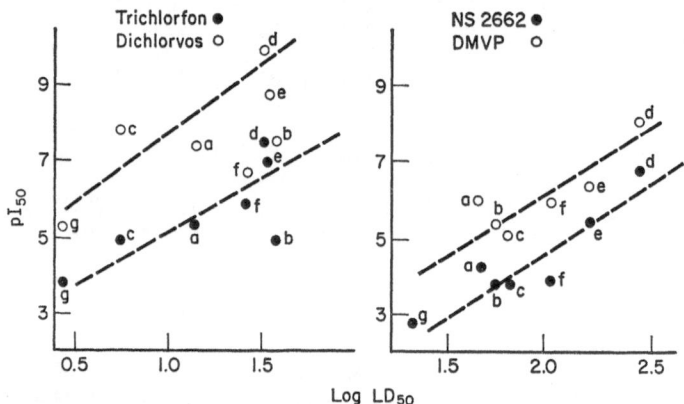

Fig. 2. The relationship between *in vitro* cholinesterase inhibition and topical toxicity of trichlorfon, NS 2662, and their vinyl phosphate. See legend of Figure 1 for explanation of *a-g*

Sensitivity of the cholinesterase of these insects to NS 2662, trichlorfon, and their vinyl phosphate derivatives was examined, and a negative correlation was found to exist with topical toxicities (Fig. 2).

The different rates of degradation of the insecticides in insect bodies may account for the selective toxicities of NS 2662 and trichlorfon. In order to evaluate the detoxification metabolism, water-soluble metabolites were analysed by paper chromatography. Table

VI shows that much greater amounts of the desmethyl compounds were found in sucking insects, *e.g.*, green rice leafhopper and black rice bug, as compared in the other insects (SATO and SAITO unpublished).

Table VI. *Amounts of degradates of* ^{32}P-NS 2662 *and* ^{32}P-trichlorfon *by various insects*

Compound	R_f	Amounts (%) [a]					
		a	b	c	d	e	f
NS 2662							
$(HO)_2P(O)OH$	0.0	3.7	4.4	12.8	2.6	8.3	0.9
$(CH_3O)P(O)(OH)_2$	0.10	0.8	0.6	4.7	0.5	6.7	1.2
$(CH_3O)(HO)P(O)OCH=CHCl$	0.46	1.1	1.7	17.0	0.6	3.3	0.5
$(CH_3O)_2P(O)OH$	0.64	58.9	61.6	51.4	59.6	54.8	52.8
$(CH_3O)(HO)P(O)CH(OH)CHCl_2$	0.76	6.7	2.8	3.2	4.7	10.6	12.7
Unknown	0.89	28.9	28.8	10.9	32.0	16.3	31.9
Trichlorfon							
$(HO)_2P(O)OH$	0.0	4.5	36.6	27.1	9.2	10.6	5.3
$(CH_3O)P(O)(OH)_2$	0.10	0.8	0.9	13.3	4.1	8.7	1.3
$(CH_3O)(HO)P(O)OCH=CCl_2$	0.46	32.9	1.9	23.5	34.2	6.0	2.0
$(CH_3O)_2P(O)OH$	0.64	37.3	49.9	37.3	15.3	44.7	55.5
$(CH_3O)(HO)P(O)CH(OH)CCl_3$	0.77	0.7	0.5	2.8	2.7	17.6	12.8
Unknown	0.90	24.8	10.2	16.0	34.5	12.4	25.1

[a] See the legend of Figure 1 for explanation of *a-g.*

NS 2662 has low mammalian toxicity (the oral LD_{50} is 2,632 mg./kg.) and it shows characteristic symptoms to mammals different from those of other organophosphorus compounds. NS 2662 poisoning of mice produces no marked hyperactivity or convulsion which is the characteristic symptom of other organophosphorus compounds such as parathion, trichlorfon, and dichlorvos. In roaches, however, it shows hyperactivity, chronic and tonic convulsions, paralysis, and death. But mice poisoned with the dehydrochlorinated NS 2662 (*O,O*-dimethyl 2-monochlorovinyl phosphate or DMVP, oral LD_{50} 248 mg./kg.) showed hyperactivity and convulsions. The metabolism and cholinesterase inhibition were studied *in vivo* and *in vitro* on roach and mice. Poisoning of NS 2662 in the roach was the result of cholinesterase inhibition caused by both NS 2662 and its metabolite DMVP, but the mode of action in mice and the reason for its low mammalian toxicity remain obscure.

Acknowledgments

The writer thanks Dr. Kisabu Iyatomi, Professor of the Nagoya University, for his valuable suggestions to this work. He also wishes to express his gratitude to Dr. Teruo Yamasaki, Professor of the Tokyo University, Dr. Tomijiro Oya, Professor of the Iwate University, and Dr. Tyoji Kusano, Professor of the Tottori University, who provided valuable literature. The contributions of the following coworkers are gratefully acknowledged: Hachiro Honda, Osamu Morikawa, Kimihiko Sato, Tadashi Miyata, and Minoru Yamada. This work was supported in part by grants from the IAEA contract no. 236/RB.

Summary

The mechanism of selective toxicity of trichlorfon and NS 2662 (O,O-dimethyl dichlorohydroxyethyl phosphonate), which are more toxic to chewing insects than to sucking insects for systemic application, were studied and a relationship appears to exist between the metabolism of these compounds and the susceptibility of insects to the compounds; the amounts of desmethyl compounds were larger in sucking insects than in chewing insects.

Résumé *

Toxicité sélective des insecticides endothérapiques

Le mécanisme de la toxicité sélective du trichlorfon et du NS 2662 (O,O-diméthyl dichlorohydroxyéthyl phosphonate) plus toxiques pour les insectes broyeurs que pour les insectes suceurs en application endothérapique, a été étudié. Une relation paraît exister entre le métabolisme de ces composés et la susceptibilité des insectes vis à vis d'eux; les teneurs en composés desméthyl étaient plus grandes chez les insectes suceurs que chez les insectes broyeurs.

Zusammenfassung **

Selektive Toxizität von systemischen Insektiziden

Der Mechanismus der selektiven Toxizität von Trichlorphon und NS 2662, O,O-Dimethyldichlorhydroxyäthylphosphonat, welche bei systemischer Applizierung toxischer für die kauenden als für die saugenden Insekten sind, wurden untersucht, und eine Beziehung scheint zwischen dem Metabolismus dieser Verbindungen und der

---

* Traduit par R. MESTRES.
** Übersetzt von A. SCHUMANN.

Empfindlichkeit der Insekten auf diese Verbindungen zu existieren; die Mengen an Demethylverbindungen waren grösser in saugenden als in kauenden Insekten.

References

ANDO, K., and T. NAKAMURA: Enzymatic hydrolysis of monofluoroacetanilides in insects. Botyu-Kagaku 31, 157 (1966).

BOWMAN, J. S., and J. E. CASIDA: Further studies on the metabolism of thimet by plants, insects, and mammals. J. Econ. Entomol. 51, 838 (1958).

BULL, D. L.: Metabolism of Di-Syston by insects, isolated cotton leaves, and rats. J. Econ. Entomol. 58, 249 (1965).

CASIDA, J. E.: Mode of action of carbamates. Ann. Rev. Entomol. 8, 39 (1963).

——, and M. A. STAHMANN: Metabolism and mode of action of schradan. J. Agr. Food Chem. 1, 883 (1953).

DAUTERMAN, W. C., J. E. CASIDA, J. B. KNAAK, and T. KOWALCZYK: Bovine metabolism of organophosphorus insecticides. Metabolism and residues associated with oral administration of dimethoate to rats and three lactating cows. J. Agr. Food Chem. 7, 188 (1959).

DAVID, W. A. L., and B. O. C. GARDINER: Investigations on the systemic insecticidal action of sodium fluoroacetate and of three phosphorus compounds on Aphis fabae Scop. Ann. Applied Biol. 38, 91 (1951).

DE PIETRI-TONELLI, P.: Penetration and translocation of rogor applied to plants. Advances Pest Control Research 6, 31 (1965).

FULTON, R. A., and H. C. MASON: The translocation of derris constituents in bean plants. J. Agr. Research 55, 903 (1937).

GEARY, R. J.: Development of organic phosphates as systemic insecticides. J. Agr. Food Chem. 1, 880 (1953).

HOYLE, F.: High blood potassium in insects in relation to nerve conduction. Nature 169, 281 (1952).

HURD-KARRER, A. M., and F. W. POOS: Toxicity of selenium-containing plants to aphids. Science 84, 252 (1936).

KRUEGER, H. R., R. D. O'BRIEN, and W. C. DAUTERMAN: Relationship between metabolism and differential toxicity in insects and mice of diazinon, dimethoate, parathion, and acethion. J. Econ. Entomol. 53, 25 (1960).

MATSUMURA, F., and R. D. O'BRIEN: A comparative study of the mode of action of fluoroacetamide and fluoroacetate in the mouse and American cockroach. Biochem. Pharmacol. 12, 1201 (1963).

MARCH, R. B., R. L. METCALF, T. R. FUKUTO, and M. G. MAXON: Metabolism of systox in the white mouse and American cockroach. J. Econ. Entomol. 48, 355 (1955).

METCALF, R. L., and R. B. MARCH: Studies of the mode of action of parathion and its derivatives and their toxicity to insects. J. Econ. Entomol. 42, 721 (1949).

MONZEN, K.: Some experiments on the innertherapy for injurious insects (Preliminary report). Tottori Soc. Agr. Sci. 4, 271 (1933).

—— Studies on the innertherapy with special reference to the root dipping. Ann. Report Gakugei Fac., Iwate Univ. 12, 19 (1957).

O'BRIEN, R. D.: Selective toxicity of insecticides. Advances Pest Control Research 4, 75 (1961).

——, and E. Y. SPENCER: Metabolism of octamethylpyrophosphoramide by insects. J. Agr. Food Chem. 1, 946 (1953).

—— —— Further studies on the insect metabolism of octamethylpyrophosphoramide. J. Agr. Food Chem. 3, 56 (1955).

OYA, T.: Studies of the quantities of nicotine absorbed by rice-seedling by means of the root dipping method (Report I.). Bull. Morioka College Agr. Forest., Iwate Univ. No. 26, p. 58 (1950).

RIDGWAY, R. L., S. L. JONES, and L. J. GORZYCKI: Tests for boll weevil control with a systemic insecticide and a boll weevil feeding stimulant. J. Econ. Entomol. 59, 149 (1966).

RIPPER, W. E., R. M. GREENSLADE, and G. S. HARTLEY: Selective insecticides and biological control. J. Econ. Entomol. 44, 448 (1951).

SAITO, T.: Distribution of ^{32}P-labeled schradan in various insects. Botyu-Kagaku 25, 64 (1960 a).

—— Cholinesterase inhibition and metabolism of schradan in various insects. Botyu-Kagaku 25, 163 (1960 b).

——, and H. HONDA: Systemic insecticidal properties of certain organic phosphorus compounds to the green peach aphid, *Myzus persicae* Sulzer, and the tobacco cutworm, *Prodenia litura* Fab. Botyu-Kagaku 31, 77 (1966).

——, and C. MATSUI: Electron microscopy of the ganglionic sheath of insect. Botyu-Kagaku 25, 71 (1960).

SATO, K., and T. SAITO: Selective toxicity of NS 2662 and trichlorfon. Unpublished data (1968).

TOBIAS, J. M., J. J. KOLLROS, and J. SAVIT: Relation of absorbability to the comparative toxicity of DDT for insects and mammals. J. Pharmacol. 86, 287 (1946).

TWAROG, B. W., and K. D. ROEDER: Properties of the connective tissue sheath of the cockroach abdominal nerve cord. Biol. Bull. 111, 278 (1956).

WEIDEN, M. H. J., and H. H. MOOREFIELD: Synergism and species specificity of carbamate insecticides. J. Agr. Food Chem. 13, 200 (1965).

YAMASAKI, T., and T. NARAHASHI: Synaptic transmission in the cockroach. Nature 182, 1805 (1958).

Specificity and mechanism
in the action of saligenin cyclic phosphorus esters

by
MORIFUSA ETO *

Contents

I. Introduction

The chemical structure of saligenin cyclic phosphorus esters is shown by the general formula I. Cyclic phosphates of saligenin were first discovered as the active metabolite of tri-*o*-tolyl phosphate (known as tri-*o*-cresyl phosphate or TOCP) (CASIDA *et al.* 1961, ETO *et al.* 1962 a). Their biological activities are remarkably affected by an exocyclic substituent group on phosphorus. Some of them have a deleterious biological activity to cause ataxia in higher animals as TOCP has (CASIDA *et al.* 1963). On the contrary, some others do not show such a harmful activity, but a high insecticidal activity (ETO *et al.* 1963 c). Moreover, another group shows a fungicidal activity.

In this review the mechanism of the metabolic formation and the specificity in the biological activity of the saligenin cyclic phosphorus esters are discussed.

---

* Department of Agricultural Chemistry, Kyushu University, Fukuoka, Japan.

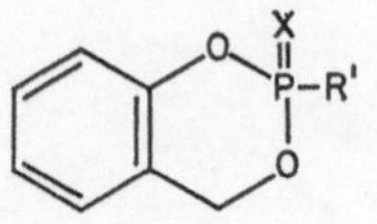

I

X = O, S
R'= R, OR, SR, NHR, NR$_2$
R = Alkyl, Aryl

II. Structural specificity in the metabolic formation of saligenin cyclic phosphates

TOCP is an undesirable impurity in tri-tolyl phosphates prepared for a variety of industrial uses. The ortho-isomer is a neurotoxic poison and causes paralysis in some animals (SMITH and ELVOVE 1930) and potentiates the toxicity of malathion [O,O-dimethyl S-(1,2-dicarbethoxyethyl) phosphorodithioate] (MURPHY et al. 1959). It is metabolized to form potent esterase inhibitors (ALDRIDGE 1954). The author, CASIDA, and their coworkers isolated and characterized the active metabolites of TOCP (CASIDA et al. 1961, ETO et al. 1962 a). The main active metabolite was saligenin cyclic o-tolyl phosphate, i.e., 2-o-tolyloxy-4H-1,3,2-benzodioxaphosphorin-2-oxide   (I,   X = O;   R' = o-tolyloxy). The biological activities of the cyclic metabolite and TOCP are compared in Table I. The in vitro antiesterase activity of the metabolite is ten million times as high as that of TOCP. The metabolite is much more active than TOCP and causes paralysis in the chicken and potentiates the toxicity of malathion (CASIDA et al. 1963).

Table I. Biological activities of TOCP and its active metabolite

Activity	TOCP	Metabolite
Antiesterase activity (units/mg.)	10^{-2}	1.2×10^5
Paralysis in chicken (mg./kg.)	250-500	4-8
Potentiation of malathion (times/dose in mg./kg.)	4/100	100/20

Rat liver microsomes and NADH$_2$, in the presence of air, are effective in activation of TOCP in vitro (ETO et al. 1962 a). Liver microsome systems, in the presence of NADH$_2$ or NADPH$_2$, are known to be active in oxidation or hydroxylation of organic compounds foreign to the body. Thus, the first step in TOCP activation is probably hydroxylation to yield di-o-tolyl o-(αhydroxy)tolyl phosphate.

The conversion of this hydroxylated intermediate to the corresponding cyclic phosphate was demonstrated both in vivo and in vitro.

Diaryl o-(α-hydroxy)tolyl phosphates are rather unstable and spontaneous cyclization proceeds slowly with liberation of an aryl group. ETO et al. (1967) found that the cyclization was greatly accelerated by various plasmata and that the cyclizing activity of plasma was associated with the albumin component when this fraction was prepared either by salting-out with ammonium sulfate or by Cohn's method with ethanol (fraction V). The activity of the cyclizing enzyme could not be separated from albumin either by Sephadex G-100 or by DEAE-cellulose column chromatography. No other protein such as plasma globulin or egg albumin was active for the cyclization.

Some esters related to the hydroxylated intermediate were examined for the substrate specificity in the plasma-catalyzed cyclization reaction. Diaryl o-(α-hydroxy)tolyl phosphates are good substrates. However, no cyclic phosphate was produced from diethyl o-(α-hydroxy)tolyl phosphate. It appears, therefore, that alkyl ester group participates but little in the cyclization.

Thus, the metabolic activation mechanism for tri-o-tolyl phosphate and related compounds is now more completely understood, as indicated in Figure 1. A ring methyl group is hydroxylated at the first step by the action of the liver microsome hydroxylation system, and the product cyclizes at the second step, mainly owing to the action of the plasma albumin, to yield the corresponding saligenin cyclic phosphate through intramolecular transphosphorylation with liberation of one aryl group.

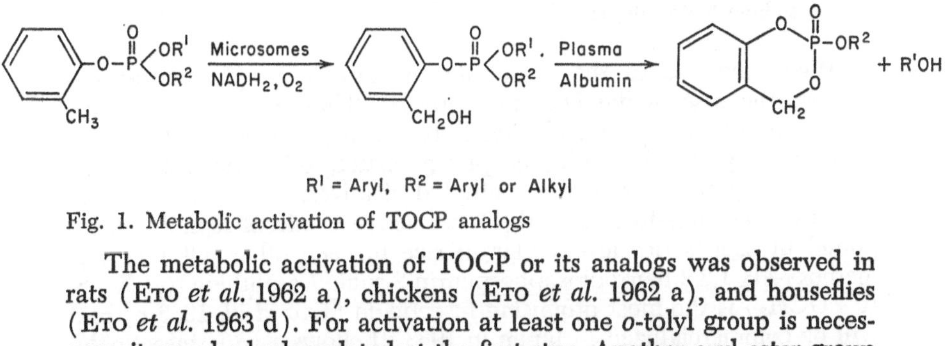

R^1 = Aryl, R^2 = Aryl or Alkyl

Fig. 1. Metabolic activation of TOCP analogs

The metabolic activation of TOCP or its analogs was observed in rats (ETO et al. 1962 a), chickens (ETO et al. 1962 a), and houseflies (ETO et al. 1963 d). For activation at least one o-tolyl group is necessary; it may be hydroxylated at the first step. Another aryl ester group is also required for the cyclization reaction; it may be liberated in the course of intramolecular transphosphorylation. Diethyl o-tolyl phosphate shows only little effect against houseflies. Thus, as shown in Table II, no phosphotriester which has no o-tolyl group or has only one aryl group shows such a distinct biological property as does TOCP.

Furthermore, another remaining ester group determines the physiological property of the tertiary phosphate ester: diaryl o-tolyl phosphates are ataxic to chickens (ALDRIDGE and BARNES 1961, HENSCHLER 1959, HINE et al. 1956) and synergistic with malathion (CASIDA 1961,

ETO *et al.* 1963 d), whereas alkyl aryl *o*-tolyl phosphates are insecticidal (ETO *et al.* 1963 d). This very interesting specificity in the biological action of phosphate triesters should be attributed to the produced cyclic phosphates.

III. Specificity in the biological activity of saligenin cyclic phosphorus esters

Many kinds of cyclic phosphorus esters were prepared by the means shown in Figure 2 (ETO *et al.* 1962 a, ETO and OSHIMA 1962 a, KOBAYASHI *et al.* 1966), and their insecticidal activities were tested (ETO *et al.* 1962 b, 1963 b, 1965, and 1966 a, KOBAYASHI *et al.* 1966). Aryl saligenin cyclic phosphorus esters as TOCP-metabolites were almost non-insecticidal, whereas alkyl homologs were insecticidal. This is true for all the analogous series of saligenin cyclic phosphorus esters, including phosphates (ETO *et al.* 1962 b and 1963 b), phosphoramidates (ETO *et al.* 1965), phosphorothioates (KOBAYASHI *et al.* 1966), phosphonates (ETO *et al.* 1966 a), and their thiono analogs (Table III).

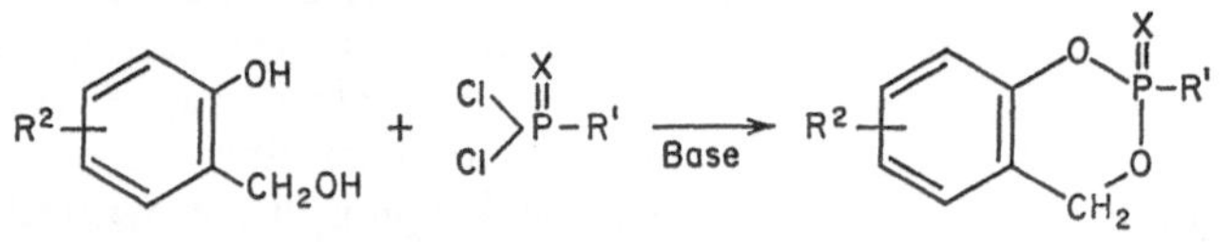

Fig. 2. Syntheses of saligenin cyclic phosphorus esters. $R^1 = R$, OR, SR, NHR, NR_2 and R = Alkyl, Aryl; $R^2 = CH_3$, Cl, etc.; X = O, S; Base = Pyridine, sodium hydroxide, etc.

When the size of the exo-cyclic substituent on phosphorus decreases, the insecticidal activity increases. Thus, methyl phosphate, methylphosphoramidate, and their thiono analogs show a high insecticidal activity. In the series of phosphonates, ethyl derivatives are exceptionally superior to the methyl derivatives. It is interesting to note that the substituents of the most active cyclic phosphorus compound in each series are similar in size to each other. Of these, 2-methoxy-4H-1,3,2-benzodioxaphosphorin-2-sulfide (salithion)(I, X = S; R' = OCH_3) is the most promising insecticide (ETO *et al.* 1963 c) and will be commerciallized in Japan in 1968. It shows a wide insecticidal spectrum.

Furthermore, some phosphorothiolate derivatives have activity to protect the rice plant from rice blast disease caused by the infection of *Pyricularia oryzae*.

On the other hand, aryl derivatives of saligenin cyclic phosphorus esters have quite different biological properties. They are not insecticidal, but are highly synergistic with malathion (ETO *et al.* 1966 b). They cause paralysis in chickens (CASIDA *et al.* 1963, ALDRIDGE and BARNES 1966) (Table IV).

Table II. Biological activities of phosphotriesters

Triester [a]	Insecticidal activity	Potentiation of malathion (LD_{50} in mg./kg.) [b]	Ataxia in chicken	Possible metabolite (I, X=O) [a]	
				R'¹	Detected
(ToO)₃P=O ‖ O	−	+(M) 190	+	OTo	yes (M)
(ToO)₂POPh ‖ O	−	+(H)	+	OTo OPh	yes (H) yes (H)
ToOP(OPh)₂ ‖ O	?	+(M) 180	+	OPh	−
(ToO)₂POMe ‖ O	+	−	−	OTo OMe	no (H) yes (H)
(ToO)₂POEt ‖ O	+	−	?	OTo OEt	no (H) yes (H)
ToOP(OEt)₂	−		−	OEt	?
(PhO)₃P=O	−	(M) 450	−	no	−

[a] Abbreviations. Et = ethyl, Me = Methyl, Ph = phenyl, To = o-tolyl; H = housefly, M = mouse.
[b] Malathion LD_{50}: mice treated with 100 mg. of test compound/kg. 24 hours prior to treatment with malathion.

Table III. Correlation between substituent and insecticidal activity (LD$_{50}$ μg./fly) of I

R' X R	R		OR		SR		NHR		NR$_2$	
	O	S	O	S	O	S	O	S	O	S
CH$_3$	0.13	0.31	0.04	0.05	0.09	0.18	0.05	0.04	0.3	0.38
C$_2$H$_5$	0.17	0.08	0.33	0.30	0.23	0.9	0.66	0.48	(0)[d]	0.63
n-C$_3$H$_7$	0.33[a]	0.09[a]	7.1	—	2.34	2.2	1.50	—	—	—
n-C$_4$H$_9$	7.0[b]	—	(40)[d]	—	4.37	10	(54)[d]	—	—	—
C$_6$H$_5$	(0)[c,d]	0.3	(3)[d]	2.0	—	(0)[d]	(5)[d]	—	—	—

[a] Isopropyl.
[b] Sec-butyl.
[c] Tert-butyl.
[d] Figures in parentheses are percentage mortality at 10 μg./fly.

Table IV. *Effects of substituent on biological activities of I* ($X = O$)

R'	Synergism with malathion against houseflies (cotoxicity coefficient)		LD_{50} of malathion to mice in equitoxic mixture (mg./kg.)	Minimum ataxic dose to chickens (mg./kg.)
	Sus.[a]	Res.[b]		
OCH$_3$	0.6	4.7	200	n. a.[c]
OC$_2$H$_5$	1.7	3.1	—	—
C$_2$H$_5$	—	—	260	n. a.[c]
OC$_6$H$_5$	2.3	9.2	85	1.5-2
OC$_6$H$_4$-o-CH$_3$	3.4	7.8	45	2-5
C$_6$H$_5$	2.5	8.0	40	200

[a] Sus. = susceptible strain.
[b] Res. = resistant strain.
[c] n. a. = no ataxia.

IV. Specificity in enzyme inhibition

It is very reasonable to presume that saligenin cyclic phosphorus esters manifest their biological activities through their inhibitory action against certain esterases. Thiono analogs are weak cholinesterase-inhibitors *in vitro*, but are strong *in vivo* (ETO *et al.* 1963 a). They may be converted *in vivo* to oxo analogs as are ordinary insecticides of the thiono type. The cyclic oxo esters react readily with chymotrypsin to inactivate its esteratic activity and liberate a free phenolic hydroxyl group (ETO *et al.* 1962 a). The major part of the hydroxyl group is not extractable with chloroform; this means it combines to the protein. The inactivated chymotrypsin contains phosphate equimolar with the protein, indicating that a stoichiometric reaction occurs between the enzyme and the inhibitor. Thus, it appears that the initial reaction with chymotrypsin or other esterases is a phosphorylation involving opening of the cyclic phosphate structure at the P-O-C (aryl) bond (Fig. 3).

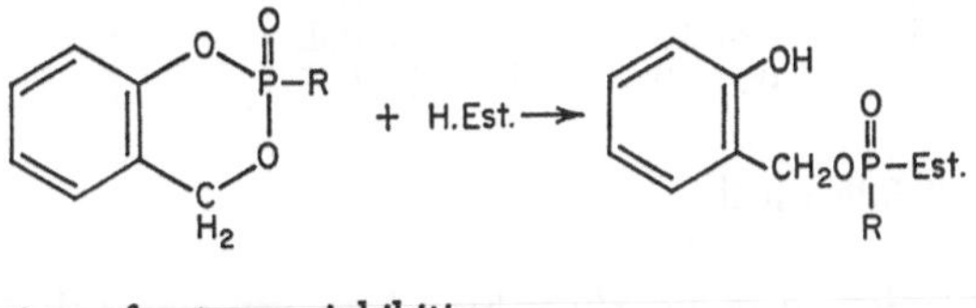

Fig. 3. Mechanism of esterase inhibition

In general, highly insecticidal organophosphates are shown by a structure $\frac{B}{B'} > P\frac{=X}{-A}$, where A is an acidic group (SCHRADER 1952) whose pK value is about seven or less (METCALF 1955). The P-A bond is very reactive. Thus, parathion (O,O-diethyl O-p-nitrophenyl phosphorothionate) is highly insecticidal, while its phenyl analog is noninsecticidal. On the contrary, saligenin cyclic phosphates are highly insecticidal, in spite of the fact that the pK value of saligenin is almost the same with that of phenol. Moreover, the introduction of a nitro group at the para position to the phenolic hydroxyl of saligenin diminishes the insecticidal activity of the cyclic esters. It appears, therefore, that saligenin cyclic phosphorus esters are quite different from ordinary organophosphorus insecticides in the mechanism of reaction.

Many neutral cyclic phosphorus esters have been prepared by several investigators. Cyclic esters, except halogenides (LANHAM 1960), derived from alkanediols lack insecticidal properties (EDMUNDSON and LAMBIE 1966) and anti-esterase activity (FUKUTO and METCALF 1965), even if they have a p-nitrophenyl phosphate linkage. TICHÝ et al. (1957) prepared cyclic phosphorothionates from catechol. Some of them inhibit blood cholinesterases but have only weak insecticidal activity. Cyclic esters of saligenin are characterized by a ring including an enol and a benzyl ester linkages, which may be responsible for the high chemical and biological activities of the esters.

The reactivity of saligenin cyclic phosphorus esters with nucleophiles is affected by virtue of the electronic characteristics of the substituent on phosphorus. The relative reaction rate is in the order; $NR_2 < NHR < OR < R$. The hydrolysis rate constants of some cyclic esters are shown in Table V. In the alkyl phosphate series (OR), methyl ester is the most active one in reactivity as well as in insecticidal activity. On the contrary, aryl derivatives are not insecticidal, in spite of their higher reactivity than the methyl ester.

Table V. *Hydrolysis rate constants of some saligenin cyclic phosphorus esters* (I, X = O) *in M/15 phosphate buffer* (pH 7.7) *at 25° C.*

R'	$K_{hyd.}$ min.$^{-1}$	R'	$K_{hyd.}$ min.$^{-1}$
C_6H_5	1.28×10^{-2}	$OC_3H_7(n)$	3.79×10^{-4}
C_2H_5	4.25×10^{-3}	$OC_4H_9(n)$	3.29×10^{-4}
OC_6H_5	6.30×10^{-3}	NHC_6H_5	2.40×10^{-4}
OCH_3	1.42×10^{-3}	$NHCH_3$	1.54×10^{-4}
OC_2H_5	5.00×10^{-4}	$N(CH_3)_2$	n. s.[a]

[a] n. s. = negligibly small.

This contradiction is clearly solved by comparing their specificities in enzyme inhibition. Table VI shows the correlation between the size of the substituent and the specificity in housefly esterase inhibition (ETO et al. 1963 a). When the size of the substituent increases, the ester becomes a more specific inhibitor of aliesterase. In contrast, the ester carrying a small substituent is a more specific inhibitor of cholinesterase. There is a significant difference in the behavior against insect esterases between the highly insecticidal methyl phosphate, salioxon (2-methoxy-4H-1,3,2-benzodioxaphosphorin-2-oxide), and the non-insecticidal phenyl phosphate.

Table VI. *Specificity in housefly esterase inhibition of I* (X = O)

R'	OC_6H_5	OC_4H_9	OC_3H_7	OC_2H_5	OCH_3
I_{50} ChE x 10^8	155	37.5	50.7	13.2	7.6
I_{50} AliE x 10^8	1.4	2.3	3.0	2.1	8.4
Ratio	116	16.3	16.9	6.2	0.9
LD_{50} μg./fly	(3)[a]	(40)[a]	7.1	0.33	0.035

[a] Figures in parentheses are percentage mortality at 10 μg./fly.

Similar results are obtained by using cholinesterases of human blood. Aryl derivatives are more specific to pseudo-cholinesterase, and alkyl derivatives are less specific to the esterase. These phenomena may be explained by the steric effect of the substituent. It is an analogous phenomenon that both true and pseudo-cholinesterases hydrolyze acetylcholine, whereas only pseudo-cholinesterase hydrolyzes benzoylcholine (MENDEL *et al.* 1943). The active centers of true cholinesterase and housefly cholinesterase may be sterically protected, while aliesterase and pseudo-cholinesterase may not be.

The above-mentioned effects of substituents on chemical and biological activities of saligenin cyclic phosphorus esters are summarized in Table VII. Thus, the reactivity is mainly attributable to the special cyclic structure and is somewhat affected by the substituent on phosphorus in virtue of its electronic characteristics. On the other hand, the biological specificity is remarkably influenced by the substituent by virtue of its steric characteristics. Therefore the difference in biological activities between methyl and phenyl derivatives is very distinct and appears in kind more than in degree.

Table VII. *Effects of substituents on chemical and biological activities of I* (X = O)

R'	Reactivity	$\dfrac{I_{50}ChE}{I_{50}AliE}$	Insecticidal activity	Synergism with malathion	Paralysis in chicken
OCH_3	100	0.9	100	–	–
OC_2H_5	35	6.2	10.6	±	?
OC_6H_5	444	116	<0.3	++	++
OC_6H_4-o-CH_3	–	128	<0.3	++	++

Many esterases are present in the insect body and were separated on agar electrophoresis according to the method of OGITA and KASAI (1965). β-Naphthyl acetate being used as a substrate, six to eight housefly-esterases were visualized by diazo coupling of the liberated naphthol. By previous incubation of housefly homogenate with an inhibitor, the specificity of the inhibitor was clearly demonstrated. A zymogram treated with saligenin cyclic phenyl phosphate is in high contrast with that of the insecticidal methyl derivative as shown in Figure 4 (OHKAWA et al. 1968). The phenyl derivative inhibits com-

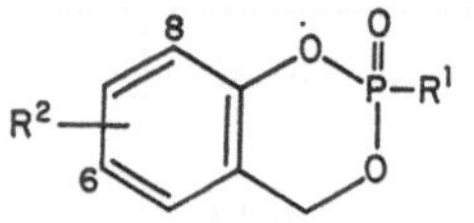

II

Fig. 4. Zymogram of housefly esterases inhibited by II

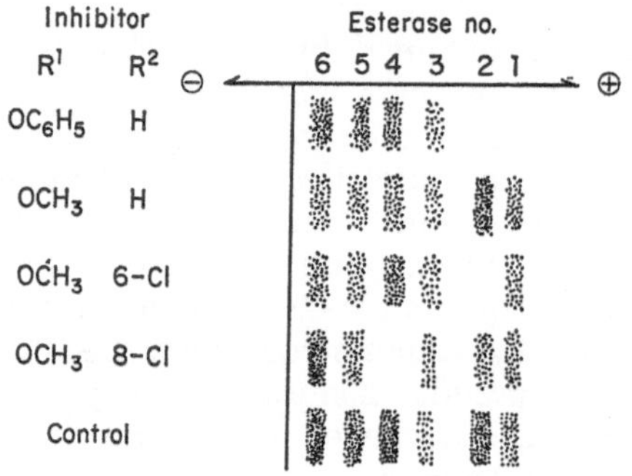

pletely bands 1 and 2, while the methyl derivative inhibits slightly the bands 5 and 6. The esterase activity of bands 1 and 2 is very strong in organophosphate-resistant strains of houseflies such as Hokota and G. Hokota is diazinon-resistant and shows cross-resistance to malathion. Strain G is highly resistant to malathion. The resistant strains have also a characteristic but not distinct band which moves toward the cathode. Malathion-degrading activity was observed in both those portions. The latter portion was much higher in activity. Saligenin cyclic phenyl phosphate, which is synergistic with malathion, inhibited completely the malathion-degrading activity of both portions (OHKAWA et al. 1967).

Substituents on the benzene ring also affect the specificity of esterase inhibition of saligenin cyclic phosphorus esters, though the effects are not so extensive as those of the substituent on phosphorus. Zymograms of housefly-esterases treated with these inhibitors are characteristic of the applied inhibitors. 6-Chloro derivatives inhibit specifically and band 2 and 8-chloro derivatives the band 4 of Figue 4. with these inhibitors are characteristic of the applied inhibitors. 6-Chloro derivatives inhibit specifically the band 2 and 8-chloro derivatives specifically the band 4 of Figure 4.

V. Degradation of saligenin cyclic phosphorus esters

Salithion is oxidized to its oxo form and rapidly decomposed *in vivo*. One hour after oral administration to the mouse, 86 percent of the salithion became water-soluble and 97.6 percent after three hours. Mouse liver homogenates have a high ability to hydrolyze the cyclic esters.

The oxo analogs of saligenin cyclic phosphorus esters are so reactive that they decompose gradually at an enolic ester linkage in the hetero ring in phosphate buffer (ETO and OSHIMA 1962 a) and on a reaction with certain oximes such as pyridine aldoxime methiodide (PAM) and monoisonitrosoacetone (ETO and OSHIMA 1962 b). These oximes and some mercaptans including glutathione react with the cyclic esters, probably after partial hydrolysis, to give their *o*-hydroxybenzyl derivatives (Fig. 5) (OHKAWA and ETO 1968). The

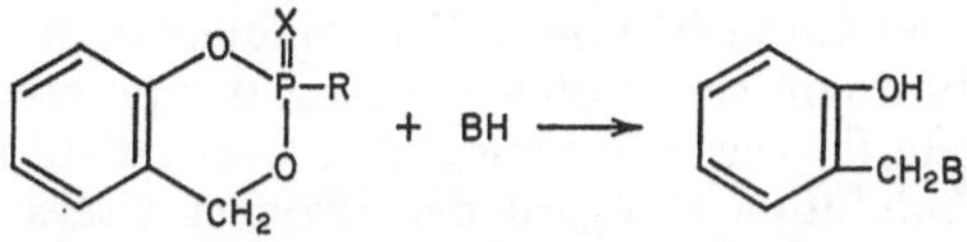

Fig. 5. *o*-Hydroxybenzylation of nucleophiles by saligenin cyclic phosphates. BH = oximes or mercaptans

reaction with glutathione occurs also enzymatically by incubation with liver homogenate. Both hydrolysis and glutathione-conjugation appear to be important mechanisms for detoxication of saligenin cyclic phosphorus esters.

VI. Conclusions

A study of the metabolic activation mechanisms of a poison disclosed a new biologically active skeleton. The results of the investigation on the activity-structure relationship suggest that biologically active substances may have two important sites in the molecule in order to manifest biological activities: one reacts actually with a target and another decides the specificity in the biological activity. A possibility to make poisons change to more useful substances by the structural modification of the latter is also shown in this study.

Summary

The physiological activities of phosphotriesters carrying an *o*-tolyl group and another aryl group are owing to saligenin cyclic phosphates produced *in vivo*. The reactive site of the cyclic esters is the enolic ester portion in the hetero ring. Substituent groups on phosphorus and on the benzene ring affect the chemical and biological properties of the cyclic ester by virtue of their electronic and steric characteristics.

Especially important is the size of the substituent on phosphorus to decide the specificity in biological activity. This is due to the fact that the inhibitor specificity against target enzymes is remarkably influenced by the steric factor.

Résumé *

Spécificité et mécanisme d'action des esters phosphoriques cycliques saligéniques

L'activité physiologique des triesters phosphoriques portant le groupement O-tolyl et un autre groupement aryl est due aux phosphates cycliques saligéniques produits *in vivo*. Le groupement actif des esters cycliques est la fraction ester enolique de l'hétérocycle. Des groupements substituants sur le phosphore et sur le cycle benzénique affectent les propriétés chimiques et biologiques de l'ester cyclique, en raison de leurs caractéristiques électroniques et stériques. La dimension du substituant sur le phosphore est d'importance particulière dans la spécificité de l'activité biologique. Ceci est dû au fait que la spécificité de l'inhibition à l'égard des enzymes visées est influencée d'une façon remarquable par le facteur stérique.

Zusammenfassung **

Spezifizität und Mechanismus in der Wirkung von cyklischen Saligeninphosphorsäureestern

Die physiologische Aktivität von Phosphorsäuretriestern, welche eine o-Tolylgruppe und eine andere Arylgruppe haben, beruht auf der *in vivo* Produktion von cyklischen Saligeninphosphaten. Die Reaktionsseite der cyklischen Ester ist der enolische Esteranteil im Heteroring. Substituierte Gruppen am Phosphor und am Benzolring beeinflussen die chemischen und biologischen Eigenschaften des cyklischen Esters durch ihre elektronischen und sterischen Eigenschaften. Besonders wichtig ist die Grösse des Substituenten am Phosphor, um über die Spezifizität bei der biologischen Aktivität zu entscheiden. Dies beruht auf der Tatsache, dass die Hemmungs-spezifizität gegenüber "target"-Enzymen merkbar von dem sterischen Faktor beeinflusst wird.

References

Aldridge, W. N.: Tricresyl phosphates and cholinesterase. Biochem. J. 56, 185 (1954).

---

* Traduit par S. Dormal-van den Bruel.
** Übersetzt von A. Schumann.

——, and J. M. BARNES: Neurotoxic and biochemical properties of some triaryl phosphates. Biochem. Pharmacol. 6, 177 (1961).
—— —— Further observations on the neurotoxicity of organophosphorus compounds. Biochem. Pharmacol. 15, 541 (1966).
CASIDA, J. E.: Specificity of substituted phenyl phosphorus compounds for esterase inhibition in mice. Biochem. Pharmacol. 5, 332 (1961).
——, R. L. BARON, M. ETO, and J. L. ENGEL: Potentiation and neurotoxicity induced by certain organophosphates. Biochem. Pharmacol. 12, 73 (1963).
——, M. ETO, and R. L. BARON: Biological activity of a tri-o-cresyl phosphate metabolite. Nature 191, 1396 (1961).
EDMUNDSON, R. S., and A. J. LAMBIE: Cyclic organophosphorus compounds as possible pesticides. II. 1,3,2-Dioxaphosphorinans. J. Chem. Soc. (C), p. 2001 (1966).
ETO, M., J. E. CASIDA, and T. ETO: Hydroxylation and cyclization reactions involved in the metabolism of tri-o-cresyl phosphate. Biochem. Pharmacol. 11, 337 (1962 a).
——, T. ETO, and Y. OSHIMA: Pesticidal activities of some cyclic phosphorus esters. Agr. Biol. Chem. (Tokyo) 26, 630 (1962 b).
——, K. HANADA, Y. NAMAZU, and Y. OSHIMA: The correlation between antiesterase activity and chemical structure of saligenin cyclic phosphates. Agr. Biol. Chem. (Tokyo) 27, 723 (1963 a).
——, Y. KINOSHITA, T. KATO, and Y. OSHIMA: Saligenin cyclic alkyl phosphates and phosphorothionates with insecticidal activity. Agr. Biol. Chem. (Tokyo) 27, 789 (1963 b).
—— —— —— —— Saligenin cyclic methyl phosphate and its thiono analogue: New insecticides related to the active metabolite of tri-o-cresyl phosphate. Nature 200, 171 (1963 c).
——, K. KISHIMOTO, K. MATSUMURA, N. OSHITA, and Y. OSHIMA: Studies on saligenin cyclic phosphorus esters with insecticidal activity. IX. Derivatives of phosphonic and phosphonothionic acids. Agr. Biol. Chem. (Tokyo) 30, 180 (1966 a).
——, K. KOBAYASHI, T. KATO, K. KOJIMA, and Y. OSHIMA: Saligenin cyclic phosphoramidates and phosphoramidothionates as pesticides. Agr. Biol. Chem. (Tokyo) 29, 243 (1965).
——, S. MATSUO, and Y. OSHIMA: Metabolic formation of saligenin cyclic phosphates from o-tolyl phosphates in house flies, Musca domestica. Agr. Biol. Chem. (Tokyo) 27, 870 (1963 d).
——, and Y. OSHIMA: Synthesis and degradation of cyclic phosphorus esters derived from saligenin and its analogues. Agr. Biol. Chem (Tokyo) 26, 452 (1962 a).
—— —— The reaction of cyclic phosphorus esters with some oximes. Agr. Biol. Chem. (Tokyo) 26, 834 (1962 b).
—— ——, and J. E. CASIDA: Plasma albumin as a catalyst in cyclization of diaryl o-(a-hydroxy)tolyl phosphates. Biochem. Pharmacol. 16, 295 (1967).
—— ——, S. KITAKATA, F. TANAKA, and K. KOJIMA: Studies on saligenin cyclic phosphorus esters with insecticidal activity. X. Synergism of malathion against susceptible and resistant insects. Botyu-Kagaku 31, 33 (1966 b).
FUKUTO, T. R., and R. L. METCALF: Reactivity of some 2-p-nitrophenoxy-1,3,2-dioxaphospholane 2-oxides and -dioxaphosphorinane 2-oxides. J. Med. Chem. 8, 759 (1965).
HENSCHLER, D.: Beziehungen zwischen chemischer Struktur und Lähmungswirkung von Triarylphosphaten. Arch. Expt. Pathol. Pharmakol. 237, 459 (1959).
HINE, C. H., M. K. DUNLAP, E. G. RICE, M. M. COURSEY, R. M. GROSS, and H. H. ANDERSON: The neurotoxicity and anticholinesterase properties of some substituted phenyl phosphates. J. Pharmacol. Expt. Therap. 116, 227 (1956).

KOBAYASHI, K., M. ETO, S. HIRAI, and Y. OSHIMA: Studies on saligenin cyclic phosphorus esters with insecticidal activity. XI. An improved method for preparation of 2-substituted 4H-1,3,2-benzodioxaphosphorin-2-sulfides. J. Agr. Chem. Soc. Japan **40**, 315 (1966).

LANHAM, W. M.: Bicyclo heterocyclic phosphorus compounds. U. S. Pat. 2,910,499; through Chem. Abstr. **54**, 3465 (1960).

MENDEL, B., D. B. MUNDELL, and H. RUNDNEY: Studies on cholinesterase. 3. Specific tests for true cholinesterase and pseudo cholinesterase. Biochem. J. **37**, 473 (1943).

METCALF, R. L.: Organic insecticides, their chemistry and mode of action. New York-London: Interscience (1955).

MURPHY, S. D., R. L. ANDERSON, and K. P. DUBOIS: Potentiation of toxicity of malathion by triorthotolyl phosphate. Proc. Soc. Expt. Biol. Med. **100**, 483 (1959).

OGITA, Z., and T. KASAI: Genetic control of multiple esterase in *Musca domestica.* Japan J. Genetics **40**, 1 (1965).

OHKAWA, H., and M. ETO: On the reaction of saligenin cyclic phosphates with SH-compounds. Ann. Meeting Agr. Chem. Soc. Japan (1968).

—— ——, and Y. OSHIMA: On the esterases of malathion-resistant insects. Ann. Meeting Agr. Chem. Soc. Japan (1967).

—— —— —— Comparative study on the specificity of cyclic phosphates and triphenyl phosphate in esterase inhibition. Botyu-Kagaku **33**, 21 (1968).

SCHRADER, G.: Die Entwicklung neuer Insektizide auf Grundlage organischer Fluorund Phosphor-Verbindungen. Weinheim: Verlag Chemie (1952).

SMITH, M. I., E. ELVOVE, and W. H. FRAZIER: Pharmacological action of certain phenol esters, with special reference to the etiology of so-called ginger paralysis. U. S. Pub. Health Rept. **45**, 2509 (1930); through Chem. Abstr. **25**, 348 (1931).

TICHÝ, V., V. RATTAY, J. JANOK, and I. VALENTINOVA: Zmiešané estery kyseliny fosforečnej a tiofosforečnej odvodené od pyrokatechínu. Chem. Zvesti **11**, 398 (1957).

Mechanisms of pesticide interactions in vertebrates *

by
SHELDON D. MURPHY **

Contents

I. Introduction

In 1957, FRAWLEY et al. reported that combinations of two organophosphorus insecticides, malathion [1] and EPN, were markedly synergistic in their toxicity to rats and dogs. These observations led to the requirement by the *U.S. Food and Drug Administration* that the toxicity of all new anticholinesterase insecticides proposed for use on food crops should be tested in combination with other insecticides of the same class for which food residue tolerances were established. Previous to that time, it had been customary to conduct toxicity tests and safety evaluation studies on single pesticidal chemicals independently.

The observation of synergism between malathion and EPN was

---

* The author gratefully acknowledges financial support for travel provided by the U.S. National Science Foundation. Recent and previously unreported research conducted by the author and contained herein was supported by Research grants ES 00084 and ES 00002 from the Division of Environmental Health Sciences, U.S. Public Health Service.

** Department of Physiology, the Kresge Center for Environmental Health, Harvard University School of Public Health, 665 Huntington Avenue, Boston, Massachusetts 02115, U.S.A.

[1] Chemical names of pesticides are shown in Table VIII.

not, however, the first demonstration of a markedly altered susceptibility of mammals to poisoning resulting from interactions between
the biologic effects of two different insecticides. Thus, BALL *et al.*
(1954) had reported that rats pretreated with several organochlorine
insecticides were much less susceptible to acute poisoning with parathion and tetraethyl pyrophosphate.

In recent years, the toxicity or the duration and intensity of phar-
macologic actions of several drugs have been shown to be altered by
preexposure of animals to organochlorine insecticides. Several examples of these interactions and the biochemical mechanisms involved
have recently been reviewed (CONNEY 1965 and 1967, CONNEY *et al.*
1967). Examples of interactions between pairs of organochlorine insecticides which resulted in altered storage of these compounds in
fatty tissues of mammals have also been reported (STREET *et al.* 1966 a
and b, STREET and CHADWICK 1967). Details concerning the mechanisms of these interactions can be found in the references cited, but
it is worthy of comment here that in most, if not all, cases they can be
explained on the basis of the capacity of organochlorine insecticides
to increase the concentration or activity of liver microsomal enzyme
systems that catalyze the metabolism of many drugs and other foreign
organic chemicals. DURHAM (1967) has recently reviewed the entire
subject of pesticide interactions. Although some duplication is un-
avoidable, I shall in this brief review restrict my discussion to some
of the known and probable biochemical mechanisms by which the
susceptibility of mammals to acute poisoning by organophosphorus
insecticides is altered by exposure to other insecticides or drugs. The
implications of knowledge gained through research on the mechanisms
of these toxicological interactions to problems of safety evaluation
studies and to interactions in vertebrate species, other than mammals,
are discussed in relation to recent findings in this and other laboratories.

II. Metabolism and biochemical action of organophosphorus insecticides

Several of the organophosphorus insecticides currently in use are
dialkyl, aryl esters of thiophosphoric acid. Therefore, the basic reaction scheme of PLAPP and CASIDA (1958), modified to show the relationship of metabolism to target enzymes, can be used for reference
to discussion of several reports of metabolic mechanisms of toxicologic interactions (Fig. 1). This general metabolic scheme was developed from isolation of metabolites in tissues of animals following
injections of ^{32}P-labelled insecticides.

Most toxicological studies of interactions of organophosphorus compounds in vertebrates have been conducted using rats, and reactions
I to V have been shown to occur in rats *in vivo*, although in this
species reactions II and IV appear to be significant only with certain

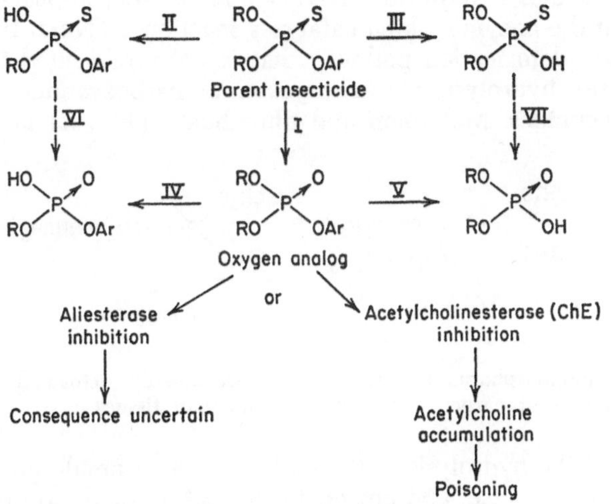

Fig. 1. General scheme of metabolism and mechanism of toxic action of dialkyl, aryl phosphorothioates

dimethyl substituted compounds. Reaction I yields a potent cholinesterase inhibitor and is catalyzed by a pyridine nucleotide-linked enzyme system(s) in liver microsomes (DAVISON 1955, MURPHY and DU BOIS 1957 a, NAKATSUGAWA and DAHM 1967, NEAL 1967) which resembles the enzyme systems that catalyze the metabolism of a large number of other drugs and foreign chemicals (CONNEY 1965). The liver has the greatest specific activity, but with parathion as the substrate, enzymes in kidneys, lungs, small intestines, and brains of rats also catalyzed reaction I (KUBISTOVA 1959, NEAL 1967) to some degree. Reactions II, III, IV, and V are detoxication reactions yielding products which do not inhibit acetylcholinesterase. For several years, and largely as the result of in vitro studies it was felt that reaction V, catalyzed by paraoxonase or the A-esterase of ALDRIDGE (1953), was the major pathway of detoxication of parathion. This enzyme is widely distributed among several tissues in rats and other mammals. It does not require addition of cofactors for measurements of activity in vitro and probably hydrolyzes several other organophosphates (P = O compounds), but apparently does not hydrolyze the P = S compounds directly. Several attempts to explain the enzymatic mechanisms of toxicologic interactions of other chemicals with phosphorothioate insecticides have been based on the concept that hydrolytic detoxication (reaction V) follows the formation of the oxygen analogs (reaction I). However, recent in vitro studies using ^{32}P-labeled parathion have shown that the aryl-phosphorus bond can be cleaved (Reaction III) without prior oxidation to paraoxon (NEAL 1967, NAKATSUGAWA and DAHM 1967). This reaction is catalyzed by a microsomal enzyme which, although it requires NADPH and oxygen, appears to be distinct from the enzyme which catalyzes reaction I (Neal 1967).

Another detoxication pathway that is not represented in Figure 1 involves the hydrolysis of carboxyester or carboxyamide linkages in some insecticides. Malathion and dimethoate (Fig. 2) are examples.

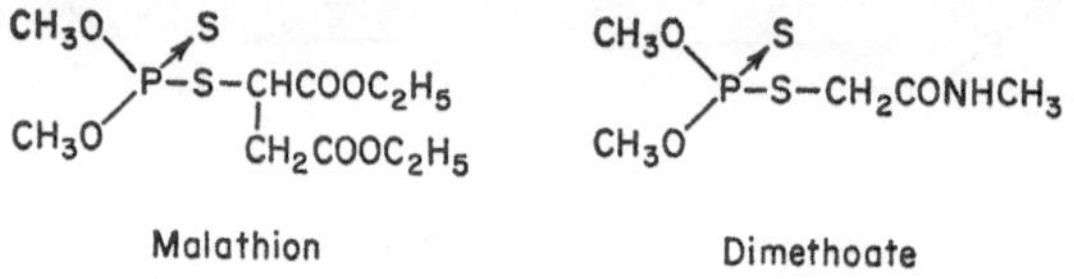

Malathion   Dimethoate

Fig. 2. Organophosphorus insecticides that are readily detoxified by carboxyesterases or carboxyamidases in mammalian tissues

Products of the hydrolysis of the carboxyester or amide groups do not inhibit cholinesterase, and enzymatic formation of these products has been demonstrated in *in vivo* and *in vitro* studies (March *et al.* 1956, Cook and Yip 1958, Dauterman *et al.* 1959, Uchida *et al.* 1964). In several species of mammals, this appears to be the major pathway of detoxication for these insecticides, and their selective insecticidal action is due to a relative lack of these hydrolytic enzymes in insects (Krueger and O'Brien 1959, Krueger *et al.* 1960).

III. Interactions that reduce organophosphorus toxicity

Ball *et al.* (1954) reported that three to four days after administration of the organochlorine insecticide, aldrin, the toxicity of parathion to female rats decreased by about seven-fold. Main (1956) showed that the oral but not the intravenous toxicity of paraoxon was decreased by aldrin pretreatment, while both the oral and the intravenous toxicity of parathion was decreased. He also showed that aldrin pretreatment doubled liver A-esterase activity and decreased serum A-esterase by 50 percent. Main concluded that the mechanism by which aldrin reduced parathion's toxicity was by stimulation of the hydrolysis of paraoxon in the liver after it was formed from parathion within this organ. He suggested that the net effect of the liver on parathion might be a detoxifying one. Recently Triolo and Coon (1966 a and b) confirmed the protective action of aldrin against parathion in mice and extended these studies to include several other organophosphorus insecticides and some anticholinesterase drugs (Table I). Dieldrin and chlordane also protected against parathion. They also showed that ethionine prevented both the aldrin-induced elevation of liver A-esterase and the protection against parathion, which suggested that aldrin acts to stimulate synthesis of the liver A-esterase. They speculated that, since mice were not protected against compounds such as OMPA and neostigmine, which do not readily penetrate the blood-

Table I. *Oral toxicity of anticholinesterase insecticides and drugs in mice four days after aldrin (16 mg./kg.) pretreatment* [a]

Anticholinesterase	Ratio, aldrin-pretreat./control LD_{50}'s
Parathion	5.7
Paraoxon	4.8
Physostigmine	2.4
Neostigmine	0.96
OMPA	0.96

	Mortality (%)	
	Aldrin pretreat.	Controls
TEPP, 10 mg./kg.	0	95
DFP, 50 mg./kg.	10	66
EPN, 75 mg./kg.	0	50
Guthion, 15 mg./kg.	15	85

[a] Data adapted from TRIOLO and COON (1966 a and b).

brain barrier, aldrin might also act to alter the penetration of the insecticides into the brain or decrease the susceptibility of the brain cholinesterase to inhibition. WELCH and COON (1964) reported that mice pretreated for four days with the drugs chlorcyclizine and phenobarbital, were less susceptible to parathion, malathion and EPN. They studied the effects of chlorcyclizine pretreatment in detail (Table II). Chlorcyclizine pretreatment had little effect on liver paraoxonase

Table II. *Effects of chlorcyclizine pretreatment in mice* [a]

Measurement	Ratio, chlorcyclizine-pretreat./controls
Oral LD_{50} of parathion	5.0
Liver slice activation of parathion	2.0
Liver paraoxonase activity	1.1
Plasma paraoxonase activity	0.5
Liver triacetin esterase activity	1.5
Plasma triacetin esterase activity	1.6
Plasma TAME esterase activity	2.0

[a] Data adapted from WELCH and COON (1964), mice pretreated with 50 mg./kg. of chlorcyclizine 24 hours previously, except for liver slice activation where pretreatment was 25 mg./kg. twice a day for four days.

(A-esterase, reaction V) and decreased plasma paraoxonase, while it increased the activation of parathion by liver slices. These effects are inconsistent with the protective action of chlorcyclizine against parathion poisoning. However, chlorcyclizine increased the activity of liver and plasma triacetin and TAME esterases. The authors concluded from these findings that the protective action of the drugs against parathion was at least partly due to the capacity of these drugs to increase the activity of certain esterases in liver and plasma. Since paraoxon acts as an inhibitor (but not a substrate) for aliesterases, reaction with these enzymes could spare the more vital target enzyme acetyl cholinesterase. This mechanism might, then, be described as one in which drugs increase the number of non-critical binding sites for the organophosphates.

Recently, NEAL and DU BOIS (1965) working mostly with EPN and its oxygen analog measured the formation of p-nitrophenol by liver homogenates and cell fractions from male and female adult and male weanling rats. When EPN oxygen analog was the substrate, they found no marked age or sex differences and fortification of the enzyme system with an NADPH generating system was not required. When EPN itself was incubated with liver no p-nitrophenol was formed unless an NADPH generating system was present, and then large age and sex differences in the production of p-nitrophenol were observed with rat livers. These differences agreed with the previously reported (MURPHY and DU BOIS 1958 a) age and sex differences in the activity of the microsomal enzyme that catalyzes reaction I. The rate of formation of p-nitrophenol from EPN was only about one-tenth as great as when EPN oxygen analog was used as the substrate. The authors suggested that the microsome enzyme-catalyzed reaction I was the rate-limiting step in the detoxication of EPN in the liver and that this was coupled to a non-limited hydrolysis of the oxygen analog (reaction V). Stimulation of the rate-limiting oxidation step appeared to offer an explanation for the decreased susceptibility of rats to poisoning by DMP and EPN that was produced by pretreatment with microsomal enzyme-inducing compounds: nikethamide, phenobarbital, 3,4-benzpyrene, and several steroids (DU BOIS and KINOSHITA 1965 and 1966). It might provide an additional or alternative explanation for the protective action of organochlorine insecticides and chlorcyclizine against parathion since these compounds also induce the synthesis of microsomal enzymes (CONNEY 1965). However, this coupled reaction mechanism cannot explain why aldrin and chlorcyclizine protected mice against paraoxon, TEPP and physostigmine. It is also difficult to reconcile this mechanism with the generally accepted view that formation of the oxygen analogs (reaction I) is a necessary prerequisite to cholinesterase inhibition, and hence poisoning, by the parent phosphorothioates *in vivo*; unless an extrahepatic tissue, low in A-esterase activity, is the important site of activation of phosphorothioates.

The mechanisms described above were proposed before NEAL (1967) and NAKATSUGAWA and DAHM (1967) discovered that parathion could be enzymatically cleaved to p-nitrophenol and diethyl hydrogen phosphorothionate (reaction III). This reaction is catalyzed by an oxidative, NADPH-dependent microsomal enzyme system. Since p-nitrophenol is one of the products of both this reaction and the hydrolysis of paraoxon (reaction V), it is difficult to differentiate the two reactions in *in vitro* studies in which p-nitrophenol is the only measured product. NEAL (1967) overcame this problem by using ^{32}P-labelled parathion and isolating the products by thin-layer chromatography. He found that when parathion was incubated with rat liver microsomes, fortified with NADPH, that diethyl hydrogen phosphorothionate (reaction III) as well as paraoxon (reaction I) and diethyl hydrogen phosphate (reaction V) were formed. Microsomes from kidney, lung, and brain also produced these metabolites, but in much smaller quantities than liver, and in different proportions. Pretreatment of rats with the microsomal enzyme stimulators, phenobarbital, and 3,4-benzopyrene increased the production of all three compounds by liver microsomes (Table III), but not by microsomes of kidney, lung,

Table III. *Stimulation of formation of metabolites of parathion by rat liver microsomes* [a]

Pretreatment	Ratio, pretreat./control		
	$(C_2H_5O)_2P(S)OH$	$(C_2H_5O)_2P(O)OH$	Paraoxon
Phenobarbital	1.72	1.94	1.08
3,4-Benzopyrene	1.47	2.62	1.64

[a] Data adapted from NEAL (1967). Phenobarbital, 80 mg./kg./day for four days; 3,4-benzopyrene, 20 mg./kg./day for three days; animals sacrificed 24 hours after the last dose.

or brain (NEAL 1967). These findings suggest that stimulation of the direct oxidative cleavage of phosphorothioates is also an important mechanism of protection against parathion by drugs and other chemicals which stimulate liver microsome enzymes.

Thus, at least four plausible metabolic mechanisms have evolved from recent *in vitro* enzyme studies which may explain the protective action of pre-exposure to various drugs and chemicals against poisoning by phosphorothioates. Which of these mechanisms: a) increased liver A-esterase, b) increased non-critical binding sites, c) increased rate of formation of oxygen analogs to provide a substrate for A-esterase, or d) increased rate of direct cleavage of the parent (P = S) compounds is most important, probably differs depending both upon the pretreatment chemical and the insecticide in question. It seems apparent, however, that in future studies on the metabolic mechanisms of toxicologic interactions all of these should be considered.

IV. Interactions that increase organophosphorus toxicity

As discussed above, most recent studies indicate that stimulation of liver microsomal enzymes reduces the toxicity of phosphorothioates. Some exceptions have been reported, however. MURPHY and DU BOIS (1958 a) found that 48 hours after pretreatment with 3-methylcholanthrene, which stimulated the formation of Gutoxon from Guthion (azinphosmethyl) by rat liver microsomes *in vitro*, female rats were more susceptible to poisoning by Guthion. Forty minutes after pretreatment with SKF 525A (which inhibits microsomal enzymes at this time interval), rats were less susceptible to Guthion. SKF 525A did not decrease the toxicity of parathion or EPN, however (MURPHY 1958). O'BRIEN and DAVISON (1958) confirmed the protective action of acute SKF 525A pretreatment against Guthion in mice and found that SKF 525A protected against dimethoate's toxicity, while no protection, in fact potentiation, was obtained for most other phosphorothioates (O'BRIEN 1961). Recently, O'BRIEN (1967) reported that repeated pretreatment with pentobarbital which can induce microsomal enzymes, increased the susceptibility of mice to poisoning by dimethoate and, to a small degree, by Guthion.

In view of the recent reports of the protective action of microsomal enzyme stimulators that were discussed in the preceding section, the effect of 3-methylcholanthrene on Guthion toxicity was recently reinvestigated in this laboratory. Methylcholanthrene pretreatment increased the susceptibility of female rats to Guthion, confirming the previous observation of MURPHY and DU BOIS (1958 a). Phenobarbital pretreatment, however, protected against Guthion toxicity, and both 3-methylcholanthrene and phenobarbital protected against parathion toxicity (MURPHY unpublished data). The precise mechanisms to explain the differing types of interactions with different microsome enzyme inducers and different organophosphorus insecticides have not been elucidated. These findings, however, indicate the need for caution in extrapolating a mechanism of interaction for one inducer and one organophosphorus insecticide to other inducers and other insecticides.

Shortly after FRAWLEY et al. (1957) demonstrated that EPN and malathion were synergistic in rats and dogs, a plausible explanation for the mechanism was developed. MURPHY and DU BOIS (1957 b) demonstrated that single doses or feeding of EPN in concentrations below those which inhibited acetylcholinesterase markedly reduced the capacity of rats' livers to detoxify malaoxon (the oxygen analog of malathion). This suggested that EPN acted as a synergist for malathion by inhibiting enzymes responsible for its detoxication. COOK et al. (1958) developed an *in vitro* test system for measuring the activity of a rat liver esterase which hydrolyzed the carboxyesters of malathion. The oxygen analog of EPN and several other organophosphates inhibited the hydrolysis of malathion *in vitro*. Compared to several other compounds, EPN was outstanding in that it was relatively more potent as an inhibitor of the hydrolysis of malathion and other di- and tri-carboxyesters than as a cholinesterase inhibitor (COOK *et al.* 1958, MURPHY and DU BOIS 1957 b and 1958 b). It appeared that any compound which selectively inhibited carboxyesterases might potentiate malathion. With this in mind, MURPHY *et al.* (1959) tested the potent non-insecticidal aliesterase inhibitor triorthotolyl phosphate (TOTP) and found that under certain conditions it reduced the LD_{50} of malathion to rats by nearly 100-fold. In 1961, DU BOIS reported acute toxicity tests in which about 50 possible pairs of insecticides then in use were given simultaneously to rats. Only four pairs potentiated: malathion with EPN, Dipterex or Co-Ral, and Guthion with Dipterex. All other combinations had either additive or less than additive toxicity. CASIDA (1961), however, found that 26 out of 43 triaryl phosphorus compounds reduced the LD_{50} of malathion to mice by a factor of two-fold or more when the potentiators were given 24 hours before malathion. The degree of potentiation of malathion correlated roughly with the degree of inhibition of its hydrolysis by livers and plasma of the pretreated mice. SEUME and O'BRIEN (1960) found that EPN and TOTP potentiated the toxicity to mice of several other organophosphorus compounds that contained carboxyester groups as well as the insecticide dimethoate which contains a carboxyamide group (see Fig. 2). Subsequently, it was shown that EPN did not potentiate dimethoate's toxicity in guinea pigs which, unlike mice, detoxify dimethoate primarily by phosphatase action rather than carboxyamidase action, and the dimethoate phosphatase was less susceptible to inhibition by EPN than the carboxyamidase (UCHIDA *et al.* 1966).

On the basis of these studies, it appears that organophosphorus esters potentiate malathion and dimethoate if they inhibit carboxyesterases and carboxyamidases at low doses in species in which these enzymes catalyze reactions that are critical and rate-limiting pathways of detoxication. However, this mechanism does not immediately explain the synergism that has been reported for a few pairs of organophosphorus insecticides which contain neither carboxyester nor carboxyamide groups, such as EPN and TOTP with Dowco 109 (SEUME and O'BRIEN 1960), Dipterex with Guthion (DU BOIS 1961), and RONNEL with EPN, Parathion, Guthion, and Systox (McCOLLISTER *et al.* 1959). The mechanism(s) of synergism between these compounds has apparently not been investigated in detail. Possibly it is closely related to the mechanism developed for malathion and dimethoate synergism. That is, if these compounds have a strong affinity for aliesterases (or other proteins), under acute conditions they might compete for these binding sites with a net result that more molecules of one or both compounds are available to inhibit the vital target enzyme acetylcholinesterase in nerve tissue. FLEISHER *et al.* (1963) observed that EPN pretreatment reduced the binding of the nerve gas Sarin to lung proteins, shifted the distribution of Sarin to other tissues (including

brain), and potentiated its toxicity. Their report suggests that competition for binding sites (other than or in addition to degradative enzymes) may play a role in the synergism between organophosphorus
insecticides. In recent research in our laboratory (LAUWERYS and
MURPHY, 1968), we found that the livers from rats pretreated with
TOTP intraperitoneally had a reduced capacity to bind paraoxon *in
vitro* while the hydrolysis of paroxon by A-esterase in these livers was
unaltered. The TOTP pretreated rats were about twice as susceptible
to paraoxon poisoning as controls.

It is probable that various esterases, which for lack of a more
specific term, have been referred to as aliesterases, play an important
role in several types of toxicological interactions of organophosphorus
compounds in mammals. In some cases, such as after pretreatment
with chlorcyclizine and aldrin, their concentration in tissues is increased (WELCH and COON 1964, TRIOLO and COON 1966 a). This,
then, provides an increase in the number of non-critical proteins
capable of binding (more or less irreversibly) organophosphates, thus
sparing the acetylcholinesterase with a net reduction in organophosphate toxicity. In some cases, the insecticides may serve as substrates
for the aliesterases (*e.g.*, malathion and dimethoate), in which case
the toxicity of these compounds would be reduced after pretreatment
with chemicals that increase aliesterase concentration and their toxicity
would be increased by compounds (*e.g.*, EPN and TOTP) which have
a stronger affinity to bind with and inhibit the aliesterases. In other
situations, both compounds of a pair of organophosphorus insecticides
may have a strong affinity to bind with aliesterase without themselves
serving as substrates. In such cases, potentiation would be expected
when one or both of the compounds have a stronger affinity or greater
opportunity to react with aliesterases than with acetylcholinesterases,
and when the total number of reactive (oxygen analogs) molecules
exceeds the number of nonspecific (as well as specific) "sites of loss"
(VELDSTRA 1956). It is even possible to envision a form of self-
potentiation with these mechanisms such as appears to be the case
for recent findings with malathion (MURPHY 1967). In that study, it
was found that malathion itself, in less than cholinesterase-inhibiting
doses, inhibited the further hydrolysis of malathion and increased the
susceptibility of guinea pigs and rats to subsequent doses of the insecticide. The mechanism for this self-potentiation appears to be that
the oxygen analog (malaoxon) formed *in vivo* from malathion can act
both as a substrate and as an inhibitor of carboxyesterases (DAUTER-
MAN and MAIN 1966, MURPHY 1967).

V. Interactions in vertebrate species other than mammals

Desirable vertebrate species in the environment, in addition to man
and other mammals, are likely to be exposed to mixtures of insecti-

cides. Relatively little is known concerning interactions or even the enzymatic metabolism of insecticides in non-mammalian vertebrates. Liver slices and homogenates from avian, piscine and amphibian, as well as mammalian species are capable of forming the oxygen analogs from phosphorothioate insecticides (reaction I, Fig. 1) *in vitro* (POTTER and O'BRIEN 1964, MURPHY 1966, HITCHCOCK and MURPHY 1967). However, liver homogenates from fish and avian species, unlike mammalian livers, produced little or no *p*-nitrophenol from paraoxon (MURPHY 1966). From this information and referring to the hypothesis of NEAL and DU BOIS (1965), we hypothesized that if microsomal enzyme inducers stimulate parathion oxidation in species which lack paraoxonase, the net effect of these inducers might be to increase rather than decrease parathion's toxicity to those species. The results of a preliminary experiment (Table IV) suggest that this hypothesis

Table IV. *Effect of phenobarbital* (PB) *pretreatment on parathion action in vertebrates* [a]

Species	No.	Parathion (mg./kg.)	Inhibition of brain cholinesterase (%) [b]	
			PB pretreated	Not pretreated
Mice	4	20. mg/kg	24 ± 7.8	71 ± 1.3
Chickens	6	2.5 mg/kg	31 ± 8.7	89 ± 4.0
Bullheads	3	200. mg/kg	50 ± 6.2	21 ± 6.3

[a] The animals were given 25 mg./kg. of phenobarbital i.p. twice daily for three days; 24 hours later parathion was injected i.p. and the animals sacrificed after two hours.

[b] Values are means ± S.E.

may hold for fish, but not for chickens. As yet, convincing biochemical evidence to support the hypothesis has not been obtained, but the increased cholinesterase inhibition due to parathion in fish pretreated with phenobarbital has been confirmed in several experiments. The differences in the susceptibilities of different classes of vertebrates to the anticholinesterase action of parathion *in vivo* can be explained, in part, on the basis of differing sensitivities of the target enzyme (brain cholinesterase) to inhibition by paraoxon (MURPHY et al. 1967).

Other comparative studies (MURPHY 1966) showed that livers of birds and fish had much less capacity to hydrolyze the carboxyesters of malathion than mammalian livers, and fish and chickens were more susceptible than mice to poisoning by this insecticide. Table V compares the relative capacities of the livers of a variety of species to hydrolyze malathion. In view of the very low activity in fish and frog livers, it seemed reasonable to question whether the carboxyesterase

Table V. *Comparative hydrolysis of malathion by vertebrate liver homogenates in vitro*

Species	Relative hydrolysis [a]	Species	Relative hydrolysis [a]
Mammals			
Rats	1.00	Sparrow	0.33
Mice	0.70	Quail	0.12
Guinea pigs	1.73	*Fishes*	
Rabbits	3.12	Sunfish	0.05
Dogs	2.04	Bullhead	0.09
Monkey	0.39	Flounder	0.13
Birds		Sculpin	0.05
Chicken	0.25	*Amphibian*	
		Frog	0.09

[a] Activity relative to the activity of rat livers (1.00) measured by previously described methods (MURPHY 1966 and 1967).

enzyme could provide a protection against malathion in these species. However, recent experiments (MURPHY, unpublished data) have shown that pretreatment of frogs with TOTP potentiated malaoxon about two-fold (compared to 15-fold in mice) and TOTP pretreatment of sunfish potentiated malathion's toxicity by about 10-fold. These experiments, incomplete as they are, suggest the many possibilities for basic and applied research on the toxicological interactions of insecticides in eco-systems.

VI. Implications of interaction mechanisms to hazard and safety evaluations

Most studies on the mechanisms of interactions have been done using single or a few relatively high doses of the insecticides or other chemicals in question. What application does the information gained from these mechanism studies have to evaluation of the hazards associated with the handling and use of organophosphorus insecticides? Feeding studies in dogs conducted with several pairs of organophosphorus insecticides at the permitted residue tolerance levels gave no evidence of potentiation as judged by blood cholinesterase assays (WILLIAMS *et al.* 1958). This, however, does not assure safety from interactions that might occur at the level of occupational exposures.

KINOSHITA *et al.* (1966) found that feeding as little as five p.p.m. of DDT in the diet of rats increased the activity of liver enzyme systems that detoxify EPN. Feeding substituted-urea herbicides also increased EPN detoxication by rat livers *in vitro* (KINOSHITA and DU BOIS 1967). The toxicologic significance of these findings, that is how the enzyme changes affected the susceptibilities of the animals to acute or subacute poisoning by challenge doses of EPN (or other organophosphorus compounds) have not, to my knowledge, been reported.

Several years ago, MURPHY and DU BOIS (1957 b) demonstrated that tissues of rats fed diets which contained EPN at concentrations below those which produced cholinesterase inhibition (the criteria usually used for measuring the "no effect" level of organophosphorus insecticides) had reduced capacity to detoxify malaoxon *in vitro*. Tissues from rats fed parathion, Guthion, and Dipterex at less than cholinesterase-inhibiting concentrations in the diet had normal capacities to detoxify malaoxon (MURPHY and DU BOIS 1958 b). Recently, KINOSHITA et al. (1967) showed that feeding Delnav or EPN to rats in concentrations below the levels that caused cholinesterase inhibition resulted in reduced activities of liver and plasma aliesterases which hydrolyze diethylsuccinate and tributyrin and of the acylamidase which hydrolyzed acetanalid. Recent studies in our laboratory have resulted in similar findings for Delnav and also for Ronnel and malathion itself (see Table VI). Additionally, we have found that rats

Table VI. *Summary of comparative esterase inhibition from feeding insecticides*

Insecticide [a]	Dietary concentration (p.p.m.) resulting in 40-60% inhibition of: [b]		
	Red cell cholinesterase	Liver carboxyesterase [c]	Plasma carboxyesterase
Parathion	3	5	5
Delnav	40	10	20
Ronnel	500	30	30
Malathion	>500	100	>500

[a] Insecticides were fed to male rats for seven days except malathion which was fed for 30 days.

[b] The concentrations were derived from results of experiments on three or more concentrations of each insecticide.

[c] Both malathion and diethylsuccinate were used as substrates (0.0067M). The inhibitory effect of the insecticides was approximately equal irrespective of the test substrate.

fed four p.p.m. of Delnav or 30 p.p.m. of Ronnel were more susceptible to poisoning by a single challenge dose of 200 mg./kg. of malathion than rats which received control diets (Table VII). These recent experiments indicate that inhibition of carboxyesterases is a more sensitive test than cholinesterase inhibition for detecting effects of low dietary concentrations of several organophosphorus insecticides. This inhibition of carboxyesterase has toxicologic implications in terms of increased susceptibility to poisoning by relatively high challenge doses

Table VII. *Effect of feeding Delnav and Ronnel on in vivo anticholinesterase action of malathion*

In the Diet [a]		Challenge dose [b] of malathion (mg./kg.)	Inhibition of brain cholinesterase (%)
Compound	p.p.m.		
None	—	200	13
Delnav	4	0	4
	10	0	4
	4	200	37
	10	200	63
Ronnel	30	0	1
	30	200	61

[a] Adult male rats (five/group) were fed Delnav for 28 days and Ronnel for seven days.

[b] Malathion was given intraperitoneally one hour before sacrifice.

of malathion. What is less certain is its physiologic significance. Solution of this problem must await more definitive information on the physiologic role (if any) of the aliesterases in the maintenance of normal health and adaptability of the organism.

Summary

This review summarizes recent research that has contributed to an understanding of mechanisms by which the susceptibility of animals to poisoning by organophosphorus insecticides is altered by exposure to other insecticidal and noninsecticidal chemicals.

Organic phosphorothionate or phosphorodithioate insecticides are toxic to vertebrates by virtue of their capacity to inhibit acetylcholinesterase The actual inhibitors are the oxygen analogs that are metabolically formed from the parent insecticides *in vivo*. Therefore the susceptibility of an animal to poisoning by a phosphorothiate insecticide will depend upon the rate at which its oxygen analog accumulates at critical sites in nerve tissue, and other chemicals which alter this rate can be expected to either increase or decrease the phosphorothioate's toxicity.

Recent research suggests that several biochemical effects of organochlorine insecticides and a number of other drugs and chemicals may provide explanations for their protection of mammals against poisoning by parathion and several related phosphorothioates. These include: a) stimulation of the oxidative cleavage of the phosphorus-*p*-nitrophenol linkage of parathion by liver microsome enzymes, b) stimulation of the microsome enzyme-catalyzed formation of the oxygen

Table VIII. *Common or trademark and chemical names of pesticides and some other chemicals mentioned in text*

Common or trademark name	Chemical name
Aldrin	1,2,3,4,10,10-hexachloro-1,4,4a,5,8,8a-hexahydro-1,4-*endo*,*exo*-5,8-dimethanonaphthalene
Chlordane	1,2,4,5,6,7,8,8-octachloro-3a-4,7,7a-tetrahydro-4,7-methanoindane
Co-Ral	*O*-(3-chloro-4-methyl-2-oxo-2H-1-benzopyran-7-yl) *O*,*O*-diethyl phosphorothioate
DDT	1,1,1-trichloro-2,2,-bis(*p*-chlorophenyl)ethane
DFP	Diisopropyl fluorophosphate
Dieldrin	1,2,3,4,10,10-hexachloro-6,7-epoxy-1,4,4a,5,6,7,8,8a-octahydro-1,4-*endo*,*exo*-5,8-dimethanonaphthalene
Dimethoate	*O*,*O*-dimethyl S-(*N*-methylcarbamoylmethyl) phosphorodithioate
Dipterex	2,2,2-trichloro-1-hydroxyethyl phosphonate
DMP	*O*,*O*-diethyl *O*-(4-methylthio-*m*-tolyl) phosphorothioate
Dowco 109	*O*-(4-tert-butyl-2-chlorophenyl) *O*-methyl phosphoroamidothionate
EPN	*O*-ethyl *O*-*p*-nitrophenyl phenylphosphonothioate
EPN oxygen analog	*O*-ethyl *O*-*p*-nitrophenyl phenylphosphonate
Guthion	*O*,*O*-dimethyl S-4-oxo-1,2,3-benzotriazin-3(4H)-ylmethyl phosphorodithioate
Gutoxon	*O*,*O*-dimethyl S-4-oxo-1,2,3-benzotriazin-3(4H)-ylmethyl phosphorothiolate
Malaoxon	S-(1,2-dicarbethoxyethyl) *O*,*O*-dimethylphosphorothiolate
Malathion	S-(1,2-dicarbethoxyethyl) *O*,*O*-dimethyldithiophosphate
OMPA	octamethylpyrophosphoramide
Paraoxon	*O*,*O*-diethyl *O*-*p*-nitrophenyl phosphate
Parathion	*O*,*O*-diethyl *O*-*p*-nitrophenyl phosphorothioate
Ronnel	*O*,*O*-dimethyl *O*-(2,4,5-trichlorophenyl) phosphorothiate
Sarin	isopropoxymethylphosphoryl fluoride
SKF 525-A	*beta*-diethylaminoethyl diphenyl propylacetate
Systox	mixture of *O*,*O*-diethyl S-(and *O*)-2-((ethylthio)-ethyl) phosphorothioates
TEPP	tetraethyl pyrophosphate
TOTP	tri-*o*-tolyl phosphate

analog which can then be hydrolyzed by A-esterase before the inhibitor reaches nerve tissue, c) stimulation of the activity of A-esterase, and d) increasing the concentration of binding sites (such as aliesterases) which react with organophosphates and spare acetylcholinesterase. The importance of any one of these mechanisms appears to differ with different protective compounds and different organophosphorus insecticides.

Potentiation of phosphorothioates can occur if another chemical stimulates the enzymatic formation of the oxygen analog without an equivalent stimulation of rate-limiting detoxication pathways. This may explain why Guthion (unlike parathion) is potentiated by 3-methylcholantherene. Additional research is needed for a thorough understanding of the quantitative and dynamic effects of chemicals on enzymatic activation and detoxication of phosphorothioates in order to predict whether the toxicity of specific compounds will be reduced or increased.

The tissues of mammalian, avian and piscine species differ greatly with respect to concentrations and activities of the various enzymes that catalyze phosphorothioate activation and detoxication. Therefore extrapolation of biochemical mechanisms of toxicologic interactions in mammals may not be valid for all vertebrate species.

Feeding low dietary concentrations of several insecticides result in changes in the activities of enzymes which metabolize phosphorothioate insecticides as well as several other chemicals and drugs. Measurements of these enzyme changes may be useful criteria for determining a "no effect" dose in subacute and chronic toxicity tests. The physiologic and toxicologic importance of these changes need to be further evaluated, however.

Résumé *

Mécanismes des intéractions entre pesticides chez les vertébrés

Cette mise au point résume les recherches récentes qui ont contribué à la compréhension des mécanismes par lesquels la susceptibilité des animaux à l'empoisonnement par les insecticides organophosphorés est modifiée par l'exposition à d'autres produits chemiques, insecticides ou non.

Les insecticides organiques thionophosphates ou dithiophosphates sont toxiques pour les vertébrés en vertu de leur capacité d'inhiber l'acétylcholinestérase. Les véritables inhibiteurs sont les analogues oxygénés, métabolites formés in vivo des insecticides primitifs. Ainsi, la susceptibilité d'un animal à l'empoisonnement par un insecticide anticholinestérasique dépendra de la vitesse à laquelle son analogue oxygéné s'accumule à des points critiques du tissu nerveux. D'autres produits chimiques qui modifient cette vitesse peuvent donc augmenter ou diminuer la toxicité du thiophosphate.

Un travail récent suggère que plusieurs effets biochimiques des insecticides organo-chlorés et de plusieurs autres drogues et produits chimiques peuvent expliquer leur protection des mammifères contre l'empoisonnement par le parathion et plusieurs thiophosphates voisins. Ils comprennent: a) la stimulation du clivage oxydant dans la liaison phosphore—para nitro phénol du parathion par les enzymes du microsome foie, b) la stimulation de la formation par catalyse enzymatique

---

* Traduit par R. MESTRES.

de l'analogue oxygéné qui peut être hydrolysé par une A-estérase avant que l'inhibiteur n'atteigne le tissu nerveux, c) la stimulation de l'activité de l'A-estérase et d) l'augmentation de la concentration des sites de liaisons (tels que les aliestérases) qui réagissent avec les phosphates organiques et épargnent l'acétylcholinestérase. L'importance de chacun de ces mécanismes paraît varier avec les différents composés protecteurs et les divers insecticides organophosphorés.

La potentialisation des thiophosphates peut survenir si un autre produit chimique stimule la formation enzymatique de l'analogue oxygéné sans une stimulation équivalente des catabolismes. Ceci peut expliquer pourquoi l'azinphos éthyl (contrairement au parathion) est potentialisé par le 3-méthyl cholanthrene. Des recherches complémentaires sont nécessaires pour une bonne compréhension des effets quantitatifs et dynamiques des produits chimiques sur l'activation et la détoxication enzymatique des thiophosphates afin de prédire si la toxicité de composés particuliers sera réduite ou augmentée.

Les tissus des mammifères, des oiseaux et des poissons diffèrent grandement, quant aux concentrations et aux activités des divers enzymes qui catalysent l'activation et la suppression de la toxicité des thiophosphates. Ainsi l'extrapolation des mécanismes biochimiques des interactions toxicologiques chez les mammifères peuvent ne pas être valables pour toutes les espèces de vertébrés.

L'absorption dans la diète de faibles concentrations de plusieurs insecticides conduit à des changements dans les activités des enzymes qui métabolisent les thiophosphates insecticides aussi bien que plusieurs autres produits chimiques et drogues. Les mesures de ces changements peuvent être un critère utile pour déterminer une dose d'innocuité dans les tests de toxicité aigüe et chronique. L'importance physiologique et toxicologique de ces changements a cependant besoin d'être mieux évaluée.

Zusammenfassung *

Mechanismen der Pestizidwechselwirkung bei Wirbeltieren

Dieser Ueberblick fasst neuere Forschung zusammen, welche zum Verständnis der Mechanismen beigetragen hat, durch welche die Empfänglichkeit von Tieren für Vergiftungen mit Organophosphorinsektiziden geändert wird, wenn sie andern insektiziden- und nichtinsektiziden Chemikalien ausgesetzt werden.

Organische Thiophosphat- oder Dithiophosphatinsektizide sind toxisch für Wirbeltiere wegen ihrer Fähigkeit, die Acetylcholinesterase zu hemmen. Die wirklichen Hemmer sind die Sauerstoffanalogen, welche als Metaoliten vom Ausgangsinsektizid *in vivo* gebildet werden.

---

* Übersetzt von A. Schumann.

Darum hängt die Empfänglichkeit eines Tieres für Vergiftung durch ein Thiophosphatinsektizid von der Rate ab, mit welcher sein Sauerstoffanalog sich an den kritischen Punkten im Nervengewebe ansammelt, und man kann erwarten, dass andere Chemikalien, welche diese Rate ändern, die Toxizität des Thiophosphats entweder erhöhen oder vermindern.

Neue Forschungen deuten darauf hin, dass einige biochemische Effekte von Organochlorinsektiziden und einer Anzahl anderer Drogen und Chemikalien Erklärungen für ihren Schutz von Säugetieren gegen Vergiftung durch Parathion und mehrere verwandte Thiophosphate geben können. Diese Effekte schliessen ein: a) Anregung der oxidativen Spaltung der Phosphor-*Para*nitrophenolbindung des Parathion durch Lebermikrosomenenzyme, b) Anregung der Mikrosomenenzymkatalysierten Bildung des Sauerstoffanalogen, welcher dann durch A-Esterase hydrolysiert werden kann, bevor der Hemmer das Nervengewebe erreicht, c) Anregung der A-Esteraseaktivität und d) Zunahme der Konzentration der Bindungspunkte (z. B. Aliesterase), welche mit Organophosphaten reagieren und so die Acetylcholinesterase verschonen. Die Wichtigkeit jedes einzelnen dieser Mechanismen erscheint verschieden mit verschiedenen schützenden Verbindungen und verschiedenen Organophosphorinsektiziden.

Potenzierung von Thiophosphaten kann auftreten, wenn ein anderes Chemikal die enzymatische Bildung des Sauerstoffanalogen ohne eine gleichwertige Anregung von Rate-begrenzenden Detoxifizierungswegen anreizt. Dieses mag erklären, warum Gusathion (anders als Parathion) durch 3-Methylcholantheren potenziert wird. Weitere Forschung ist notwendig für ein gründliches Verständnis der quantitativen und dynamischen Wirkungen von Chemikalien auf enzymatische Aktivierung und Detoxifizierung von Thiophosphaten, um voraussagen zu können, ob die Toxizität von spezifischen Verbindungen reduziert oder erhöht wird.

Die Gewebe von Säugetier-, Vogel- und Fischarten unterscheiden sich sehr in Bezug auf die Konzentration und Aktivität der verschiedenen Enzyme, welche Thiophosphataktivierung und -Detoxifizierung katalysieren. Daher mögen Extrapolationen von biochemischen Mechanismen der toxikologischen Wechselwirkung in Säugetieren nicht für alle Wirbeltierarten gültig sein.

Wenn man niedrige Diätkonzentrationen von einigen Insektiziden verfüttert, ergeben sich Veränderungen in den Aktivitäten der Enzyme, welche Thiophosphatinsektizide als auch mehrere andere Chemikalien und Drogen metabolisieren. Messungen dieser Enzymveränderungen könnten nützliche Kriteria sein, um eine "keine Wirkung"—Dosis in subakuten und chronischen Toxizitätstesten zu bestimmen. Die physiologische und toxikologische Wichtigkeit dieser Veränderungen erfordert jedoch weitere Auswertung.

References

ALDRIDGE, W. N.: Serum esterases. 2. An enzyme hydrolysing diethyl p-nitrophenyl phosphate (E600) and its identity with the A-esterase of mammalian sera. Biochem. J. 53, 117 (1953).

BALL, W. L., J. W. SINCLAIR, M. CREVIER, and K. KAY: Modification of parathion's toxicity for rats by pretreatment with chlorinated hydrocarbon insecticides. Can. J. Biochem. Physiol. 32, 440 (1954).

CASIDA, J. E.: Specificity of substituted phenyl phosphorus compounds for esterase inhibition in mice. Biochem. Pharmacol. 5, 332 (1961).

CONNEY, A. H.: Enzyme induction and drug toxicity. In: Proc. 2nd Internat. Pharmacol. Meeting, Prague, Vol. 4, p. 277. New York: Pergamon Press, (1965).

—— Pharmacological implications of microsomal enzyme induction. Pharmacol. Rev. 19, 317 (1967).

——, R. M. WELCH, R. KUNTZMAN, and J. J. BURNS: Pesticide effects on drug and steroid metabolism: A commentary. Clin. Pharmacol. Therap. 8, 2 (1967).

COOK, J. W., J. R. BLAKE, G. YIP, and M. WILLIAMS: Malathionase. I. Activity and inhibition. J. Assoc. Official Agr. Chemists 41, 399 (1958).

——, and G. YIP: Malathionase. II. Identity of a malathion metabolite. J. Assoc. Official Agr. Chemists 41, 407 (1958).

DAUTERMAN, W. C., and A. R. MAIN: Relationship between acute toxicity and in vitro inhibition and hydrolysis of a series of carbalkoxy homologs of malathion. Toxicol. Applied Pharmacol. 9, 408 (1966).

——, J. E. CASIDA, J. B. KNAAK, J. B. KOWALCZYK, and J. TADEUSZ: Bovine metabolism of organophosphorus insecticides. Metabolism and residues associated with oral administration of dimethoate to rats and three lactating cows. J. Agr. Food Chem. 7, 188 (1959).

DAVISON, A. N.: The conversion of schradan (OMPA) and parathion into inhibitors of cholinesterase by mammalian liver. Biochem. J. 61, 203 (1955).

DU BOIS, K. P.: Potentiation of the toxicity of organo-phosphorus compounds. Adv. Pest Control Research 4, 117 (1961).

——, and F. KINOSHITA: Modification of the anticholinesterase action of O,O-diethyl O-(4-methylthio-m-tolyl) phosphorothioate (DMP) by drugs affecting hepatic microsomal enzymes. Arch. Internat. Pharmacodyn. 156, 418 (1965).

—— —— Stimulation of detoxification of O-ethyl O-(4-nitrophenyl) phenylphosphonothioate (EPN) by nikethamide and phenobarbital. Proc. Soc. Expt. Biol. Med. 121, 59 (1966).

DURHAM, W. F.: The interaction of pesticides with other factors. Residue Reviews 18, 21 (1967).

FLEISHER, J. H., L. W. HARRIS, C. PRUDHOMME, and J. BURSEL: Effects of ethyl p-nitrophenyl thionobenzene phosphonate (EPN) on the toxicity of isopropyl methyl phosphonofluoridate (GB). J. Pharmacol Expt. Therap. 139, 390 (1963).

FRAWLEY, J. P., H. N. FUYAT, E. C. HAGAN, J. R. BLAKE, and O. G. FITZHUGH: Marked potentiation in mammalian toxicity from simultaneous administration of two anticholinesterase compounds. J. Pharmacol. Expt. Therap. 121, 96 (1957).

HITCHCOCK, M., and S. D. MURPHY: Activation of parathion and Guthion by tissues of mammalian, avian and piscine species. Federation Proc. 26, 427 (1967).

KINOSHITA, F. K., and K. P. DU BOIS: Effects of urea-substituted herbicides on activity of hepatic microsomal enzymes. Toxicol. Applied Pharmacol. 10, 410 (1967).

—— ——, and J. P. FRAWLEY: Aliesterase and cholinesterase inhibition by Delnav and EPN. The Pharmacologist 9, 206 (1967).

——, J. P. FRAWLEY, and K. P. DU BOIS: Quantitative measurement of induction of hepatic microsomal enzymes by various dietary levels of DDT and toxaphene in rats. Toxicol. Applied Pharmacol. 9, 505 (1966).

KRUEGER, H. R., and R. D. O'BRIEN: Relation between metabolism and differential toxicity of malathion in insects and mice. J. Econ. Entomol. 52, 1063 (1959).

—— ——, and W. C. DAUTERMAN: Relationship between metabolism and differential toxicity in insects and mice of diazinon, dimethoate, parathion and acethion. J. Econ. Entomol. 53, 25 (1960).

KUBISTOVA, J.: Parathion metabolism in female rat. Arch. Internat. Pharmacodyn. 18, 308 (1959).

LAUWERYS, R. R., and S. D. MURPHY: Comparison of in vitro assay methods for studying metabolic mechanisms that affect paraoxon toxicity. Toxicol. Applied Pharmacol. 12, 306 (1968).

MAIN, A. R.: The role of A-esterase in the acute toxicity of paraoxon, TEPP, and parathion. Can. J. Biochem. Physiol. 34, 197 (1956).

MARCH, R. B., T. R. FUKUTO, R. L. METCALF, and M. G. MOXON: Fate of P 32-labeled malathion in the laying hen, white mouse and American cockroach. J. Econ. Entomol. 49, 185 (1956).

McCOLLISTER, D. D., F. OYEN, and V. K. ROWE: Toxicological studies of O,O-dimethyl-O-(2,4,5-trichlorophenyl) phosphorothioate (Ronnel) in laboratory animals. J. Agr. Food Chem. 7, 689 (1959).

MURPHY, S. D.: The enzymatic conversion of organic thiophosphates to anticholinesterase agents and factors which influence this reaction. Ph.D Thesis, Univ. Chicago (1958).

—— Liver metabolism and toxicity of thiophosphate insecticides in mammalian, avian and piscine species. Proc. Soc. Expt. Biol. Med. 123, 392 (1966).

—— Malathion inhibition of esterases as a determinant of malathion toxicity. J. Pharmacol Expt. Therap. 156, 352 (1967).

——, R. L. ANDERSON, and K. P. DU BOIS: Potentiation of the toxicity of malathion by triorthotolyl phosphate. Proc. Soc. Expt. Biol. Med. 100, 483 (1959).

——, and K. P. DU BOIS: Enzymatic conversion of the dimethoxy ester of benzotriazine dithiophosphoric acid to an anticholinesterase agent. J. Pharmacol. Expt. Therap. 119, 208 (1957 a).

—— —— Quantitative measurement of inhibition of the enzymatic detoxification of malathion by EPN (ethyl p-nitrophenyl thionobenzene phosphonate). Proc. Soc. Expt. Biol. Med. 96, 813 (1957 b).

—— —— The influence of various factors on the enzymatic conversion of organic thiophosphates to anticholinesterase agents. J. Pharmacol. Expt. Therap. 124, 194 (1958 a).

—— —— Inhibitory effect of dipterex and other organic phosphates on detoxification of malathion. Federation Proc. 17, 396 (1958 b).

——, R. R. LAUWERYS, and K. L. CHEEVER: Comparative anticholinesterase action of organophosphate insecticides in mammals, birds and fishes. Toxicol. Applied Pharmacol. 10, 391 (1967).

NAKATSUGAWA, T., and P. A. DAHM: Microsomal metabolism of parathion. Biochem. Pharmacol. 16, 25 (1967).

NEAL, R. A.: Studies on the metabolism of diethyl 4-nitrophenyl phosphorothionate (parathion) in vitro. Biochem. J. 103, 183 (1967).

——, and K. P. DU BOIS: Studies on the mechanism of detoxification of cholinergic phosphorothioates. J. Pharmacol. Expt. Therap. 148, 185 (1965).

O'BRIEN, R. D.: The effect of SKF 525A (2-diethylaminoethyl 2:2-diphenylvalerate hydrochloride) on organophosphate metabolism in insects and mammals. Biochem. J. 79, 229 (1961).

—— Effects of induction by pentobarbital upon susceptibility of mice to insecticides. Bull. Environ. Contamination Toxicol. 2, 163 (1967).

——, and A. N. Davison: Antagonists to schradan poisoning in mice. Canad. J. Biochem. Physiol. 36, 1203 (1958).

Plapp, F. W., and J. E. Casida: Hydrolysis of alkyl-phosphate bond in certain dialkyl aryl phosphorothioate insecticides by rats, cockroaches, and alkali. J. Econ. Entomol. 51, 800 (1958).

Potter, J. L., and R. D. O'Brien: Parathion activation by livers of aquatic and terrestrial vertebrates. Science 144, 55 (1964).

Seume, F. W., and R. D. O'Brien: Potentiation of the toxicity to insects and mice of phosphorothionates containing carboxyester and carboxyamide groups. Toxicol. Applied Pharmacol. 2, 495 (1960).

Street, J. C., and R. W. Chadwick: Stimulation of dieldrin metabolism by DDT. Toxicol. Applied Pharmacol. 11, 68 (1967).

—— ——, M. Wang, and R. L. Phillips: Insecticide interactions affecting residue accumulation in animal tissues. J. Agr. Food Chem. 14, 545 (1966 a).

——, M. Wang, and A. D. Blau: Drug effects on dieldrin storage in rat tissue. Bull. Environ. Contamination Toxicol. 1, 6 (1966 b).

Triolo, A. J., and J. M. Coon: Toxicologic interactions of chlorinated hydrocarbon and organophosphate insecticides. J. Agr. Food Chem. 14, 549 (1966 a).

—— —— The protective effect of aldrin against the toxicity of organophosphate anticholinesterases. J. Pharmacol. Expt. Therap. 154, 613 (1966 b).

Uchida, T., W. C. Dauterman, and R. D. O'Brien: The metabolism of dimethoate by vertebrate tissue. J. Agr. Food Chem. 12, 48 (1964).

Uchida, T., J. Zschintzsch, and R. D. O'Brien: Relation between synergism and metabolism of dimethoate in mammals and insects. Toxicol. Applied Pharmacol. 8, 259 (1966).

Veldstra, H.: Synergism and potentiation with special reference to the combination of structural analogues. Pharmacol. Rev. 8, 339 (1956).

Welch, R. M., and J. M. Coon: Studies on the effect of chlorcyclizine and other drugs on the toxicity of several organophosphate anticholinesterases. J. Pharmacol. Expt. Therap. 144, 192 (1964).

Williams, M. W., H. N. Fuyat, J. P. Frawley, and O. G. Fitzhugh: In vivo effects of paired combinations of five organic phosphate insecticides. J. Agr. Food Chem. 6, 514 (1958).

The in vitro metabolism

of organophosphorus insecticides by

tissue homogenates from mammal and insect

by

Kazuo Fukunaga,* Jun-ichi Fukami,* and Takashi Shishido **

Contents

I. Introduction

In Japan, dialkylaryl phosphorothioate insecticides such as ethyl parathion (O,O-diethyl O-p-nitrophenyl phosphorothioate), methyl parathion (O,O-dimethyl O-p-nitrophenyl phosphorothioate) and Sumithion [O,O-dimethyl O-(4-nitro-m-tolyl)-phosphorothioate] are widely used to control the larvae of the rice stem borer, *Chilo suppressalis*. Sumithion has very low toxicity against mammals (Nishizawa et al. 1961), although its chemical structure is very similar to that of methyl parathion which is highly toxic to mammals. Several years ago, we found the development of resistance for ethyl parathion to the

---

* Laboratory of Insect Toxicology, The Institute of Physical and Chemical Research, Bunkyoku, Tokyo, Japan.
** Division of Agricultural Chemicals, National Institute of Agricultural Sciences, Nishigahara, Kitaku, Tokyo, Japan.

larvae of rice stem borer. Therefore, it is worth while to study differences of metabolism of these insecticides in mammal, insect, and plant.

It has been assumed that hydrolysis of dialkylaryl phosphate insecticides occurs only at the aryl phosphate bond (ALDRIDGE 1953 a). An important paper by PLAPP and CASIDA (1958 b) presented data on the exact site of cleavage of alkylaryl phosphates and phosphorothioates by rat and American cockroach, *Periplaneta Americana* (L.), *in vivo*. This cleavage is of considerably greater importance in mammal (PLAPP and CASIDA 1958 c). The rate of metabolic attack on different parts of the insecticide molecule varies with different organisms. To summarize the differences between mammals and insects in the *in vivo* metabolism of organophosphorus from several reports (PLAPP and CASIDA 1958 b and c, DAUTERMAN *et al.* 1959, KNAAK and O'BRIEN 1960, BULL *et al.* 1963), the P-O-methyl bond is easier to destroy than the P-O-ethyl bond with mammals, and also a P-O-methyl or an ethyl P-O-bond is nearly not destroyed with insects. These conclusions are based on *in vivo* experiments in which degradation products are isolated. However, the nature and distribution of the responsible enzyme system have not been clarified.

In the present report, the metabolism and nature of the products formed, and also the investigation to clarify the nature and distribution of the responsible enzyme on the *in vitro* degradation of ethyl parathion, methyl parathion, methyl paraoxon (*O,O*-dimethyl *O-p*-nitrophenyl phosphate) and Sumithion in mammals, insects, and plants were studied using homogenate preparations from several tissues.

II. Results and discussion

a) Degradation of dialkylaryl phosphorothioate in the subcellular fraction by tissue homogenate of mammal and insects

The insecticides and their derivatives used in the experiment were prepared according to the description of PLAPP and CASIDA (1958 a) and KOSALAPOFF (1950). The fractionations of homogenates of the rat liver, the rice stem borer larvae, and the adult cockroach were carried out by the author's preceding report (FUKAMI and SHISHIDO 1963 b). The liver was chosen because it caused highest degradation of phosphorus insecticides among several organs of the rat (FUKAMI and SHISHIDO 1963 b). Because of the small sizes of the rice stem borer larvae and the adult cockroach, whole insect bodies were used. Subcellular fractionation of the homogenate of cauliflower was prepared by the method of WEDDING and BLACK (1962). Incubation was carried out in a Warburg apparatus. The incubation mixture consisted of 2.4 ml. of homogenate, 4×10^{-3} M of cofactor, and 0.3 ml. of P^{32}-labeled insecticide in a final volume of 3.0 ml., pH 7.4. The mixture was incubated for two hours at 37° C., with shaking in the presence of air. At the end of the incubation period, the reaction was stopped by the addition of one ml. of a 10 percent TCA solution (W/V). The mixture was then macerated in a Waring Blendor. The macerates were held for 90 minutes in crushed ice, and four ml. of chloroform were then added. After macerating this mixture in the Blendor again, the chloroform and water extractables were separated by centrifugation, and the total radioactivity of the water and chloroform was estimated by counting an aliquot. The results on the degradation of methyl parathion, methyl paraoxon, Sumithion, and ethyl parathion in the subcellular fractions of rat, rice stem borer larvae, adult cockroach, and cauliflower are shown in Table I.

The order of activity for the degradation of methyl parathion, methyl paraoxon, and Sumithion by rat liver homogenate was supernatant > mitochondria $\rightleftharpoons$ washed microsome. Methyl paraoxon was almost completely degradated in the supernatant. In insects, there is little difference in the amounts of degradation products between supernatant, washed microsome, and mitochondria. There are no differences in the water-extractable metabolite of methyl parathion between the homogenate of the rice stem borer larvae and of the adult of the American cockroach. No degradation of methyl parathion and methyl paraoxon occurred in cauliflower.

HODGSON and CASIDA (1962) reported that in rat liver homogenate, DDVP was hydrolyzed by the soluble and mitochondrial fractions but not by the microsomes, and desmethyl DDVP was hydrolyzed by the soluble fraction but not by the microsome or mitochondria. This agrees with the present result in that degradation occurs in the soluble fraction of rat liver homogenate. On the other hand, MATUMURA and BROWN (1961) reported that in the hydrolysis of malathion and malaoxon in the homogenate of *Culex tarsales* larvae the order of the activities of carboxyesterase and phosphatase was mitochondria > microsome > supernatant. Their result is rather opposite to the present one with the homogenates of rice stem borer larvae and adult cockroaches.

As the specific activity of P^{32}-labeled ethyl parathion was low, its degradation in the homogenates of rat liver and rice stem borer larvae was investigated by incubating with one mg./three ml. of system. For comparison, the degradations of Sumithion and methyl parathion were examined at the same concentration as that of ethyl parathion. As shown in Table I, although Sumithion and methyl parathion were effectively degradated at this concentration, the degradation of ethyl parathion was very small. In the case of the rice stem borer larvae, the degradation of ethyl parathion was also very small.

b) Identification for the products of reaction by degradation of alkylaryl phosphorothioate

The metabolites in the water-extractable portion were identified by means of ion exchange and paper chromatography according to the

Table I. *Amounts of degradation of methyl parathion, methyl paraoxon, and Sumithion by the subcellular fraction of several tissues.*
(SHISHIDO and FUKAMI 1963)

$\mu g./120\ min./N$ mg. as metabolites on water extractable (calc. as ethyl parathion, parathion, methyl paraoxon, or Sumithion) (dosage: 360 $\mu g.$)

Insecticide	$\begin{array}{c}CH_3O\ \ S\\ \ \ \ \ \ \diagdown P-O-\!\!\langle\ \rangle\!-NO_2\\ CH_3O\end{array}$	$\begin{array}{c}CH_3O\ \ S\ \ \ \ \ \ \ CH_3\\ \ \ \ \ \ \ \diagdown P-O-\!\!\langle\ \rangle\!-NO_2\\ CH_3O\end{array}$	$\begin{array}{c}CH_3O\ \ O\\ \ \ \ \ \ \ \diagdown P-O-\!\!\langle\ \rangle\!-NO_2\\ CH_3O\end{array}$	$\begin{array}{c}C_2H_5O\ \ S\\ \ \ \ \ \ \ \ \diagdown P-O-\!\!\langle\ \rangle\!-NO_2\ {}^{*a1}\\ C_2H_5O\end{array}$
	Rat			
Mitochondria	2.1	19.8	1.9	0.54
Microsome	3.6	43.1	3.6	1.9
Supernatant	23.4 (90.3) ᵃ	119.4	19.9 (107.3) ᵃ	2.4
	Rice stem borer			
Mitochondria	1.4	—	2.1	0.39
Microsome	2.0	—	2.8	0.96
Supernatant	2.8	—	4.9	0.31
	Cockroach			
Mitochondria	0.23	0.82	—	—
Microsome	0.38	1.3	—	—
Supernatant	1.0	2.6	—	—
	Cauliflower			
Mitochondria	Nil	Nil	—	—
Microsome	Nil	Nil	—	—
Supernatant	0.1	0.1	—	—

ᵃ This dosage is one mg.

Table II. *Water extractable metabolites produced by the subcellular fraction to radioactive insecticides: Amounts as percentage of total recovered (dosage: 360 μg.)* (SHISHIDO and FUKAMI 1963)

Insecticide	Peak	Rat[a]			Rice stem borer[a]			Cockroach[a]
		Mit.	Mic.	Sup.	Mit.	Mic.	Sup.	Sup.
A. CH₃O S=P–O–⟨NO₂⟩ / CH₃O	1	2.0	2.0	0.5	2.4	3.7	5.7	—
unknown H₃PO₄ or H₃PO₃S	2	35.6	67.5	9.2	52.1	50.3	38.9	—
CH₃O S=P–OH / CH₃O	3	14.2	16.7	5.3	19.2	18.6	23.6	—
CH₃O O=P–O–⟨NO₂⟩ / OH	4	8.5	7.8	1.5	5.5	5.3	4.2	—
CH₃O S=P–O–⟨NO₂⟩ / OH	5	33.3	9.2	81.8	17.3	17.1	23.6	—
B. CH₃O O=P–O–⟨NO₂⟩ / CH₃O	1	2.0	2.0	2.0	—	—	—	—
unknown H₃PO₄ or H₃PO₃S	2	78.0	90.0	8.0	—	—	—	80.0
CH₃O O=P–O–⟨NO₂⟩ / OH	3	20.0	8.0	90.0	—	—	—	20.0

Table II. (continued)

Insecticide	Peak	Rat [a]			Rice stem borer [a]			Cockroach [a]
		Mit.	Mic.	Sup.	Mit.	Mic.	Sup.	Sup.
C. CH_3O \ $S=P-O-$ (ring with CH_3, NO_3) / CH_3O	1	1.5	1.6	0.7	1.5	2.3	4.3	—
unknown								
H_3PO_4 or H_3PO_3S	2	35.6	66.5	11.2	69.8	63.4	53.9	—
CH_3O \ $S=P-OH$ / CH_3O	3	26.1	21.8	5.3	16.1	20.6	20.2	—
CH_3O \ $O=P-O-$ (ring with CH_3, NO_2) / OH	4	4.1	4.4	2.3	3.4	3.0	2.8	—
CH_3O \ $S=P-O-$ (ring with CH_3, NO_2) / OH	5	39.7	3.7	78.6	6.3	7.4	15.6	—

Table II. (continued)

Insecticide	Peak	Rat [a]			Rice stem borer [a]			Cockroach [a]
		Mit.	Mic.	Sup.	Mit.	Mic.	Sup.	Sup.
D. $(C_2H_5O)_2P(=S)-O-\bigcirc-NO_2$ *[b]	1	—	—	—	—	—	—	—
unknown H_3PO_4 or H_3PO_3S	2	44.5	73.8	18.7	—	—	—	—
$(C_2H_5O)_2P(=S)-OH$	3	42.6	20.8	56.3	—	—	—	—
$(C_2H_5O)_2P(=O)-\bigcirc-NO_2$ (OH)	4	4.8	1.6	2.0	—	—	—	—
$(C_2H_5O)_2P(=S)-\bigcirc-NO_2$ (OH)	5	5.6	0.7	17.1	—	—	—	—

[a] Mit. = mitochondria, Mic. = microsome, Sup. = supernatant.
[b] Dosage one mg.

method of Plapp and Casida (1958 a). Identification of eluted metabolites was based on comparison with the result of an ion-exchange chromatogram carried out with known compounds. In the case of radioactive metabolites, paper chromatography was carried out with concentrates of the water-soluble and chloroform-soluble fractions, and the spots were detected by autoradiography.

As shown in Table II, five peaks of metabolites were found for methyl parathion and Sumithion, four peaks for ethyl parathion, and three peaks for methyl paraoxon. With the supernatant of rat liver homogenate, the principal metabolites of methyl parathion, methyl paraoxon, and Sumithion were O-methyl O-p-nitrophenyl phosphorothioate (desmethyl parathion), O-methyl O-p-nitrophenyl phosphate (desmethyl paraoxon), and O-methyl 3-methyl 4-nitrophenyl phosphorothioate (desmethyl Sumithion) (81.8, 90.0, and 78.6 percent, respectively) with the remaining radioactivity appearing in phosphorothioate or phosphate, dimethyl phosphorothioate, desmethyl paraoxon, or desmethyl Sumioxon.

However, with the supernatant from the rice stem borer larvae and adult cockroach, these principal metabolites were not found in such large amounts as in the case of the rat liver; desmethyl parathion was 23.6 percent, desmethyl paraoxon was 20 percent, and desmethyl Sumithion was 15.6 percent. Amounts of their degradation products were considerably different from those in the rat liver. In the experiment of Tables I and II, NAD and $NADH_2$ were used as cofactors for the rat liver homogenate, whereas in the insect NADP and $NADPH_2$ were added. It may be that the difference of degradation products in the supernatant between insect and rat liver is due to the difference in the cofactor. This possibility is, however, excluded by the finding that no change in the amounts of degradation products was found when $NADP + NADPH_2$ was used in place of $NAD + NADH_2$ in the rat liver homogenate (Fukami and Shishido 1963 b).

The toxicity of ethyl parathion to the rat is higher than that of methyl parathion and Sumithion. The production of desethyl parathion in the supernatant was very slight, 17.1 percent (Table II). Hodgson and Casida (1962) examined the nature of initial hydrolysis of DDVP by the mitochondrial and soluble enzyme preparations. The enzyme hydrolyzing the P-O-methyl precipitated predominately between 60 and 80 percent saturation of ammonium sulfate fractionation of the soluble liver fraction, while P-O-vinyl hydrolyzing esterase precipitated between 40 and 60 percent saturation. In the present investigation, it was shown that the reaction systems cleaving P(S)O-methyl and P(O)O-methyl of methyl compounds existed in the supernatant of rat liver homogenate. Therefore, it is of interest to see whether these reactions are due to the same enzyme as that of DDVP. It is also necessary to show the difference of degrading reactions between metabolism of the P(S)O-methyl bond and that of the P(O)O-methyl bond.

c) *The characters of enzyme in cleavage of methyl parathion to des-methyl parathion in the supernatant of several species of homogenate*

It has been shown that an enzyme present in the rabbit serum hydrolyzes diisopropyl fluorophosphate (DFP) to diisopropyl phosphate and hydrofluoric acid (MAZUR 1946) and diethyl *p*-nitrophenyl phosphate (paraoxon) to diethyl phosphate and *p*-nitrophenol (ALD-RIDGE 1953 b). ALDRIDGE (1953 a and b) concluded that the paraoxon-hydrolyzing enzyme in serum was identical with A-esterase which was known to hydrolyze *p*-nitrophenyl acetate but not to be inhibited by paraoxon. Later, several investigators (MOUNTER 1953 and 1954) have confirmed the observation of ALDRIDGE. MAIN (1960 a and b) purified an enzyme that hydrolyzed paraoxon in sheep serum, and showed that the paraoxon-hydrolyzing enzyme was different from A-esterase. Thus, he called the enzyme paraoxonase to distinguish from the other A-type esterase. HODGSON and CASIDA (1962) examined the hydrolysis of P^{32}-DDVP by mitochondria and soluble enzymes. In the mitochondria the only significant hydrolysis occurred at the P-O-vinyl bond, whereas in the soluble fraction both the P-O-vinyl and the P-O-methyl bonds were attacked.

From the above-mentioned results, it has been demonstrated that the reaction system to hydrolyze methyl parathion, Sumithion, and methyl paraoxon to desmethyl parathion, desmethyl Sumithion, and desmethyl paraoxon, respectively, is located in the supernatant fraction of rat liver homogenate. It was not yet known whether the reaction to degrade methyl parathion to desmethyl parathion in the supernatant of rat liver homogenate is enzymatic or not.

The present experiment was undertaken to study systematically the reaction system from methyl parathion to desmethyl parathion in the tissues of rat, guinea pig, rabbit, and several insects.

1. Optimum pH.—It is reasonable to assume that the radioactivity remaining in the water after extraction of the incubation mixtures with chloroform is indicative of desmethyl compounds produced by degradation of methyl parathion and Sumithion (SHISHIDO and FUKAMI 1963, FUKAMI and SHISHIDO 1963 a). For this reason, the radioactivity in the water phase was counted to estimate degradation.

The effects of varying the hydrogen ion concentration on the degradation of methyl parathion in the supernatant of rat liver homogenate were studied using citric-hydrochloric acid buffer, citric-sodium hydroxide buffer, phosphate buffer, Tris buffer, and glycine-sodium hydroxide buffer. It is evident from Figure 1 that the optimum pH lies between 8.5 and 9.5.

2. Effects of various inhibitors and ions.—The results are given in Tables III and IV. The percentage of inhibition was obtained after a

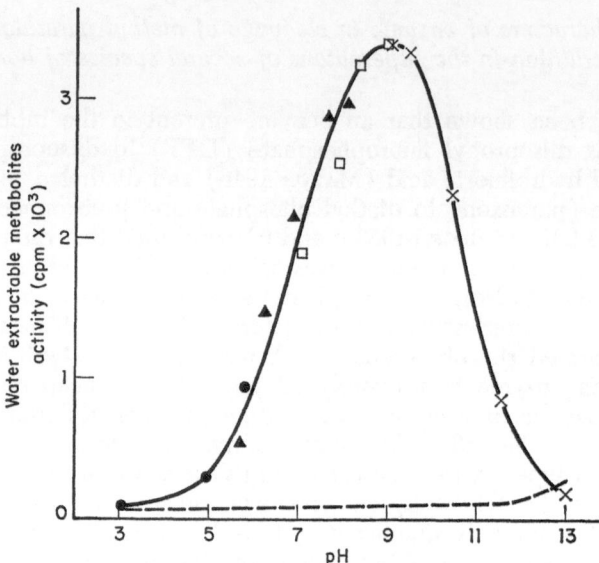

Fig. 1. The activities of water extractable metabolites of methyl parathion by supernatant of rat liver homogenate at different pH values; 360 μg. of insecticides: —●—●— citric-HCl (or NaOH) buffer, —▲—▲— phosphate buffer, —□—□— Tris buffer, —×—×— glycine-NaOH buffer, and .... no enzymatic activity (Fukami and Shishido 1963 a)

Table III. *Effects of several inhibitors on water extractable metabolites of methyl parathion by supernatant of rat liver homogenate* (360 μg. of insecticides) (Fukami and Shishido 1963 a)

Inhibitor added	Final conc. (M)	Inhibition of production of water-extractable metabolite (%)
Antimycin A	10^{-5}	0
Rotenone	5×10^{-5}	0
Potassium cyanide	10^{-2}	0
Sodium fluoride	10^{-1}	12
Sodium monofluoroacetate	10^{-2}	0
Phenyl mercuric acetate (PMA)	10^{-3}	97.4
($10^{-3}M$ PMA + $6 \times 10^{-2}M$ GSH)	—	91.0
PMA	10^{-4}	49.6
($10^{-4}M$ PMA + $6 \times 10^{-2}M$ GSH)	—	0
p-Chloromercuribenzoate (PCMB)	10^{-2}	95.3
PCMB	10^{-3}	53.3
($10^{-3}M$ PCMB + $6 \times 10^{-2}M$ GSH)	—	2.5

Table IV. *Effects of cations on water extractable metabolites of methyl parathion by supernatant of rat liver homogenate (360 μg. of insecticides)*
(FUKAMI and SHISHIDO 1963 a)

Cations added	Final conc. (M)	Inhibition of production of water extractable metabolite (%)
Ca++	10^{-4}	0.2
Mg++	10^{-4}	2.8
Fe++	10^{-4}	10.2
Fe+++	10^{-4}	9.2
Mn++	10^{-4}	10.6
Cu++	10^{-4}	50.4
Cu++	10^{-3}	91.3
Co++	10^{-4}	5.5
Ni++	10^{-4}	3.7
Al+++	10^{-4}	11.3
Zn++	10^{-4}	6.8
Ba++	10^{-4}	−7.2
EDTA	10^{-4}	−2.2

120-minute incubation period. The most effective inhibitors were phenyl mercuric acetate and p-chloromercuribenzoate. The addition of Antimycin A, rotenone, potassium cyanide, sodium fluoride, or sodium monofluoroacetate had no effect on the activity. The addition of reduced glutathione as an activator to the homogenate which was inhibited by phenyl mercuric acetate or p-chloromercuribenzoate restored the activity. This suggest that SH groups are an essential part of this reaction system. The activity was not affected by Ca^{++}, Mg^{++}, Fe^{++}, Fe^{+++}, Mn^{++}, Co^{++}, Ni^{++}, Al^{+++}, Zn^{++}, Ba^{++}, and EDTA; however, the activity was inhibited by Cu^{++}.

3. **Anaerobic condition.**—The experiments so far described were carried out in the presence of air. In order to determine whether or not the reaction from methyl parathion to desmethyl parathion in the supernatant of rat liver is an oxidation, incubation was made in the presence of 95 percent nitrogen and five percent carbon dioxide. The water-extractable metabolites were identified by ion-exchange column chromatography (Table V). Most of the degradation products in the water extractables were desmethyl parathion (80 percent) as in the presence of air. This result leads us to the assumption that the reaction from methyl parathion to desmethyl parathion in the supernatant of rat liver is not due to oxidation but due to hydrolysis.

4. **Difference for A-esterase or paraoxonase.**—Based on the foregoing results, it is reasonable to assume that the reaction from methyl parathion to desmethyl parathion in the supernatant of rat liver homogenate is due to hydrolytic enzyme systems. It is known that paraoxon

Table V. *Activities of the water-extractable metabolites of methyl parathion and methyl paraoxon by the subcellular fraction of rat liver homogenate under anaerobic conditions* (Fukami and Shishido 1963 a)

Insecticides	Water-extractable metabolites (μg./120 min./mg. N, calc. as methyl parathion and methyl paraoxon) [a]
Methyl parathion	
Mitochondria	3.7
Microsome	2.5
Supernatant	30.0
Methyl paraoxon	
Mitochondria	26.7
Microsome	43.4
Supernatant	183.3

[a] Each three ml. of incubation mixture contained 2.5 ml. of homogenate and the indicated concentration: 360 μg. of insecticides, $2.3 \times 10^{-2} M$ $NaHCO_3$; incubation: 120 min. at 37° C., $95 N_2$: 5 CO_2, pH 7.4.

is destroyed by A-esterase and paraoxonase of rabbit serum, and is hydrolyzed to diethyl phosphoric acid and p-nitrophenol (Aldridge 1953 a and b, Main 1960 a and b, Mounter and Whittaker 1953, Mounter 1954). In order to see whether the enzyme system is the same as A-esterase or paraoxonase, an experiment was made by incubating the rabbit serum with P^{32}-labeled methyl parathion and methyl paraoxon. The incubation mixture of this experiment was the same as described by Aldridge (1953 b). As shown in Table VI, the

Table VI. *Activities of the water-extractable metabolites of methyl parathion and methyl paraoxon by rabbit serum under anaerobic conditions* (Fukami and Shishido 1963 a)

Substrates	$\mu l CO_2$/60 min.	Water-extractable metabolites (μg./60 min., calc. as methyl parathion and methyl paraoxon) [a]
Methyl parathion	Nil	Nil
Methyl paraoxon	120	910

[a] Each four ml. of incubation mixture contained 0.5 ml. of the rabbit serum and the indicated concentration: $3.1 \times 10^{-2} M$ $NaHCO_3$, $1.6 \times 10^{-1} M$ NaCl, 1000 μg. of insecticides, gelatine, 0.1% (w/v); incubation: 60 min. at 37° C., 95 N_2: 5 CO_2, pH 7.7.

value of the radioactivity of the water-extractable metabolites of methyl paraoxon was very large, whereas it was negligible in methyl

Table VII. *Effect of various cofactors on methyl parathion-degrading activity by gel filtration supernatant of rat liver* [a] (FUKAMI and SHISHIDO 1966)

Cofactor added to gel filtration supernatant	μg. methyl parathion destroyed/reaction mixture/120 min.
	6.1
ATP	5.7
L-ascorbic acid	9.8
Cocarboxylase	5.4
Coenzyme A	8.2
Cytochrome c	3.9
GSH	129.3
GSSG	20.1
NAD	15.4
NADH$_2$	24.8
NADP	15.4
NADPH$_2$	18.6
Nicotinamide	6.6
Heated fresh supernatant [b]	13.0
Supernatant [c] alone	52.7
Supernatant [d] alone	2.5

[a] Each reaction mixture contained 2.4 ml. of enzyme solution (1.21 mg. N/ml.), $4 \times 10^{-3}M$ of cofactor, and 300 μg. of methyl parathion, in a final volume of 3.0 ml., pH 7.4. The mixture was incubated for 120 minutes at 37° C. The supernatant was prepared in isotonic buffer: 0.25M sucrose, 0.05M phosphate buffer (pH 7.4), 0.01M EDTA, pH 7.4. The gel filtration supernatant was prepared through Sephadex G-50 column previously equilibrated with 0.25M sucrose, 0.01M EDTA, and 0.05 M phosphate buffer, pH 7.4.

[b] 0.3 ml. of supernatant of fresh homogenate heated at 70° C. for three minutes.

[c] Fresh supernatant.

[d] Standing at 4° C. for 24 hours in the air.

parathion. It can therefore be assumed that the reaction system from methyl parathion to desmethyl parathion in the supernatant of rat liver homogenate is not the same as A-esterase or paraoxonase.

5. Cofactor for methyl parathion and Sumithion degradation enzyme system.—*Rat liver.*—Homogenates of the several tissues of rat and insect were fractionated as described previously (FUKAMI and SHISHIDO 1963 b). Soluble and particulate fractions were prepared by centrifugation of the homogenate at 105,000 g for 60 minutes. The particulate and supernatant portions were designated as insoluble and soluble fractions, respectively. The insoluble fraction consists of mitochondria, microsomes, nucleus, and debris. To examine the effects of several cofactors on the enzyme system, the supernatant was applied to a Sephadex G-50, 1.2 x 25 cm., previously equilibrated with 0.25 M sucrose, 0.01 M EDTA, and 0.05 M phosphate buffer (pH 7.4) at 5° C. Various cofactors were then added to the eluate fractions prior to assay. The ability of the rat liver homogenate supernatant fraction to degrade methyl parathion and Sumithion was lost after passing through Sephadex G-50. As shown in Table VII, the addition of GSH as a cofactor to the Sephadex-treated supernatant restored the activity. Activity was not restored by addition of an extract when it had been heated at 70° C. for three minutes, or by the addition of GSSG or many other potential cofactors. After standing at 4° C. for 24 hours in the air, enzyme activity decreased but could be restored by adding GSH, and again the addition of other cofactors had no effect on the activity.

Insect mid-gut.—The effect of several cofactors on the degradation of methyl parathion and Sumithion by the Sephadex-treated supernatant of homogenizing the mid-gut from silkworms and horn beetle larvae was investigated. The results are presented in Table VIII. For

Table VIII. *Effect of various cofactors on methyl parathion-degrading activity by the gel filtration supernatant of mid-gut of horn beetle larvae* [a]
(Fukami and Shishido 1966)

Cofactor added to gel filtration supernatant	μg. methyl parathion destroyed/reaction mixture/120 min.
	1.5
ATP	2.5
L-ascorbic acid	1.7
Cocarboxylase	1.5
Coenzyme A	2.4
Cytochrome c	2.6
GSH	27.1
GSSG	5.3
NAD	1.0
NADH$_2$	3.3
NADP	4.6
NADPH$_2$	5.1
Nicotinamide	3.4
Heated fresh supernatant [b]	4.5
Supernatant [c] alone	7.3
Supernatant [d] alone	2.1

[a] Incubation conditions were as described in Table VII; nitrogen mg. of enzyme preparation/ml. = 1.05 mg./ml.

[b] 0.3 ml. of supernatant of fresh homogenate heated at 70° C. for three minutes.

[c] Fresh supernatant.

[d] Standing at 4° C. for 24 hours in the air.

one of these experiments, using methyl parathion and the horn beetle larvae mid-gut, essentially the same results had offered with other source enzyme combinations, as follows: methyl parathion-silkworm larvae mid-gut, Sumithion-horn beetle larvae mid-gut, and Sumithion-silkworm larvae mid-gut. GSH was the most effective cofactor, as demonstrated for rat-liver. As described previously (FUKAMI and SHISHIDO 1963 a), the total amount of degradation resulting from the supernatant of the whole insect homogenate was less than that of the rat liver homogenate. Although the ability of the insect mid-gut to degrade methyl parathion and Sumithion was also low in the present experiment, it was effectively activated by the addition of GSH, but not by the addition of other cofactors.

Degradation products.—The water-extractable metabolites of methyl parathion as formed by the supernatant from rat liver and insect mid-gut homogenate were identified by ion-exchange column chromatography and paper chromatography. Conditions for preparation of these products were as follows: (a) with and without GSH addition to the supernatant and (b) with and without GSH addition to the Sephadex-treated supernatant. Metabolites were detected from the rat liver supernatant with or without added GSH. The principal metabolite was desmethyl parathion (85 percent). The same two metabolites were produced in the Sephadex-treated supernatant, but only when GSH was added as a cofactor. The supernatant fraction of the insect mid-gut homogenate produced only small amounts of water-extractable metabolites. However, when GSH was added either to the supernatant or to the Sephadex-treated supernatant, three different metabolites were detected, the principal metabolites being desmethyl parathion (65 percent) and phosphate (30 percent). From these results, it is apparent that the degradation products are different as formed by the supernatant of insect mid-gut and rat liver when both are fortified with GSH.

6. **Distribution of degrading enzyme in several tissues.**—As described previously (FUKAMI and SHISHIDO 1963 a), the radioactivity of water-extractable metabolites of methyl parathion in the supernatant fraction of homogenates of several rat tissues was estimated. The liver showed the highest activity, which was even greater than the total of the remaining tissues studied. In the present study the degradation of methyl parathion and ethyl parathion in several tissues of rat and insects was measured with and without GSH (Tables IX and X). In both the insoluble and soluble fractions, GSH was effective in increasing degradation of methyl parathion. The liver showed the highest degradation activity among the several rat tissues studied, and in the presence of GSH the degradation was more active in the soluble fraction than in the insoluble fractions. In insects, in the absence of GSH, the degradation was very small in the soluble and insoluble fractions of all tissues except the mid-gut. The addition of GSH improved the degradation in the soluble fraction of the mid-gut and fat body of silkworm and horn beetle larvae, whereas, in the insoluble fraction,

Table IX. *Methyl parathion and ethyl parathion degrading activity of soluble and insoluble fractions of rat tissue homogenates with and without GSH* [a]
(Fukami and Shishido 1966)

| Tissue | μg. insecticide destroyed/g. of equiv. fresh wt./120 min. | | | |
| | Soluble fraction | | Insoluble fraction | |
	Methyl para-thion	Ethyl para-thion	Methyl para-thion	Ethyl para-thion
Liver	165.3	3.92	23.5	0.35
Liver + GSH	243.8	4.73	70.6	.87
Brain	5.3	0.27	5.7	.11
Brain + GSH	24.9	1.88	13.6	.37
Spleen	2.8	—	2.5	—
Spleen + GSH	20.2	—	7.4	—
Lung	2.2	—	2.9	—
Lung + GSH	13.7	—	4.8	—
Kidney	4.5	.14	3.2	.05
Kidney + GSH	10.0	1.60	5.7	.06
Heart	3.2	—	4.3	—
Heart + GSH	9.3	—	8.9	—
Muscle	1.9	—	3.5	—
Muscle + GSH	6.2	—	7.8	—
Blood	3.9	—	3.8	—
Blood + GSH	4.0	—	3.5	—

[a] Each reaction mixture contained 2.4 ml. of enzyme solution (0.25g. of equiv fresh wt./ml.), $4 \times 10^{-3}M$ of GSH, and 300μg. of insecticide in a final volume of 3.0 ml., pH 7.4. The soluble and insoluble fractions of rat tissues were prepared in a medium of 0.25M sucrose 0.05M phosphate buffer (pH 7.4), 0.01M EDTA, pH 7.4. Results were converted into μg. of insecticide decomposed/g. of equiv. fresh wt./120 min. Other details are as described in Table VII.

addition of GSH only slightly increased degradation. The degradative activity of the soluble fraction of the rat liver homogenate was high, but the effect of the addition of GSH on the degradation was preferably rather small. In contrast to this, the degradation in the soluble fractions of the mid-gut of horn bettles and silkworms was very small, but the effect of GSH was preferably very large. Degradation of ethyl parathion was very slight in the enzyme preparation for insect and rat tissue homogenates with or without added GSH.

7. **Proposed the metabolic pathway for demethylation from alkylaryl phosphorothioate.**—The metabolic pathways for the degradation of methyl parathion, methyl paraoxon and Sumithion to desmethyl parathion, desmethyl paraoxon, and desmethyl Sumithion by rat-liver homogenate were investigated. At first, the effect of three SH compounds is shown in Table XI: when these compounds were added to the deionized enzyme solution, only GSH had the ability to restore

Table X. *Methyl parathion and ethyl parathion degrading activity by soluble and insoluble fractions of insect tissue homogenates with and without GSH* [a]
(FUKAMI and SHISHIDO 1966)

Tissue	μg. insecticide destroyed/g. of equiv. fresh wt./120 min.		
	Methyl parathion		Ethyl parathion
	Horn beetle larvae	Silk worm larvae	Horn beetle larvae
Soluble fraction			
Mid-gut	11.9	4.2	0.12
Mid-gut + GSH	44.8	10.8	.17
Fat body	2.8	5.5	—
Fat body + GSH	10.8	15.4	—
Cuticle and muscle	0.3	1.3	—
Cuticle and muscle + GSH	7.2	3.6	—
Haemolymph	1.4	1.6	—
Haemolymph + GSH	2.3	6.4	—
Insoluble fraction			
Mid-gut	.5	3.4	.11
Mid-gut + GSH	4.4	3.9	.14
Fat body	.4	2.0	—
Fat body + GSH	4.1	2.2	—
Cuticle and muscle	2.8	2.9	—

[a] Enzyme preparation and incubation conditions were as described in Table VII.

activity; the other SH compounds had no such restorative activity.

It remains to be decided whether GSH participates directly in the reaction with methyl parathion, *i.e.*, as a direct methyl group acceptor on the enzyme. In this connection, it should be pointed out that L-cysteine and thioglycolic acid had no activation effect. This fact favors the view that GSH functions as a methyl group acceptor in the same manner as reported on the metabolism of iodomethane and on the glutathione-S-alkyl-transferase (JOHNSON 1966 a and b), and also the differential glutathione-dependent detoxication of two geometrical vinyl organophosphorus (mevinphos) isomers by rat-liver solubles (MORELLO *et al.* 1967). These results using methyl parathion and Sumithion may show reduced glutathione-depended form, but it is unknown whether or not "the product of reaction" is a GSH-conjugated one. This product shows a ninhydrin-positive spot. This ninhydrin-positive substance was separated by TLC with silica gel G developed by isopropyl alcohol: water: ammonia (25:12:2). Further purification of this substance was by TLC with a different developing

Table XI. *Effect of various SH compounds on degradation of methyl parathion in gel filtration supernatant of rat liver homogenate* [a]
(Fukami and Shishido 1966)

SH compound added	Molarity	μg. methyl parathion destroyed/reaction mixture/120 min.
GSH	4×10^{-3}	189.7
	4×10^{-4}	45.2
	4×10^{-5}	18.7
	4×10^{-6}	6.3
	4×10^{-7}	5.4
L-Cysteine	1×10^{-2}	25.0
	1×10^{-3}	11.3
	1×10^{-4}	8.0
Thioglycolic acid	1×10^{-2}	15.5
	1×10^{-3}	8.8
	1×10^{-4}	5.3
		5.1

[a] Each reaction mixture contained 2.4 ml. of enzyme solution (1.85 mg. N/ml.), 300 μg. of methyl parathion, and SH compound as shown, in a final volume of 3.0 ml., pH 7.4. Other details are as described in Table VII.

system (n-butyl alcohol: acetic acid: water, 4:1:5) successfully. By these chromatographic operations, GSH and the "product of reaction" which was ninhydrin-positive were separated. This metabolite was also hydrolyzed by 6N-hydrochloric acid for 12 hours at 105° C., and was spotted on paper with phenol: isopropyl alcohol: water: ammonia (9:5:2:1) or n-butyl alcohol: acetic acid: water (4:1:5). The chromatographic pattern of this hydrolyzed material had R_f values corresponding to glutamic acid, glycine, and S-methyl cysteine, respectively.

Table XII. *Identification of the glutathione metabolite by paper chromatography*
(Shishido *et al.* 1968)

Compounds	R_f value [a]		
	Solvent A	Solvent B	Solvent C
S-Methyl glutathione	0.09	0.47	0.35
S-Methyl cysteine	0.46	0.75	0.40
Glutamic acid	0.04	0.23	0.29
Glycine	0.20	0.49	0.24
Reaction product	0.09	0.47	0.35
Reaction product hydrolyzed			
(a)	0.46	0.75	0.40
(b)	0.04	0.23	0.29
(c)	0.20	0.49	0.24

[a] Solvent A = phenol:isopropyl alcohol:aq. ammonia:water (9:5:1:2).
  Solvent B = phenol:water:aq. ammonia (8:1:1).
  Solvent C = n-Butyl alcohol:acetic acid:water (4:1:5).

Furthermore, "the product of reaction" (described above) is identical with a standard S-methyl glutathione on the same paper chromatograph (Table XII). From these results, it has been concluded that "the product of reaction" is a GSH-conjugated one. The present finding shows the same results as cited in the report on the character of the glutathione dependent system and the reaction product on the detoxication of cis-phosdrin by rat-liver homogenate (MORELLO et al. 1967).

A metabolic pathway for the degradation of methyl parathion by the tissue homogenate of rat liver and cockroach fat body was proposed as follows:

d) Selective toxicity

1. **The relationship between the selective toxicity and chemical structure of alkylaryl phosphorothioate.**—The relationship between the selective toxicity and the chemical structure of alkylaryl phosphorothiate to explain the nature of Sumithion's low toxicity has been reported (VARDANIS and CRAWFORD 1964). This report was on the basis that (a) mouse liver slice degraded Sumioxon about four times faster than an equal weight of American cockroach fat body and (b) the slices degraded Sumithion 1.4 times faster than methyl parathion (but this difference may be negligible).

HOLLINGWORTH et al. (1967) published that the high selective level of Sumithion depended on the ability of the system cleaving the P-O-alkyl bond enhancing detoxication as the dosage is increased in an in vivo experiment, whereas MIYAMOTO (1964) described that methyl paraoxon penetrated into the brain considerably faster than Sumioxon. No attempt was made to show whether the in vitro difference between methyl parathion and Sumithion was associated with a difference in the intact animal, resulting in significantly different levels of degradation. The interesting mystery, therefore, remains unsolved. From the conclusion described above, a comparison was made of the ability of degradation of methyl parathion and Sumithion by the Sephadex-treated supernatant of rat liver homogenate in the presence of different concentration of GSH (FUKAMI and SHISHIDO 1966). No significant difference was found in the effect of GSH on the metabolism of two compounds (Table XIII).

2. **The difference in the metabolism of alkylaryl phosphorothioate between mammal and insect.**—As described previously, degradation of methyl parathion is rapid in the soluble fraction of rat liver but very slow in some fractions of the whole insect homogenate (Tables VII and VIII). Following these findings, measurements were made of the time course of degradation of methyl parathion in the supernatant of the homogenates of rat liver and the mid-gut of horn beetle larvae

Table XIII. *Effect of GSH on degradation of methyl parathion and Sumithion by gel filtration supernatant of rat liver homogenate* [a]
(Fukami and Shishido 1966)

Molarity of GSH	μg. insecticide destroyed/ reaction mixture/120 min.	
	Sumithion	Methyl parathion
4×10^{-3}	133.0	146.4
2×10^{-3}	121.8	135.4
1×10^{-3}	118.6	116.8
8×10^{-4}	109.4	102.1
4×10^{-4}	64.8	61.0
2×10^{-4}	30.4	34.2
1×10^{-4}	12.1	8.5
0	1.0	1.7

[a] Each reaction mixture contained 2.4ml. of enzyme solution (1.05mg.N/ml.), GSH as indicated, and 300μg. of insecticide, in a final volume of 3.0ml., pH 7.4. Other details are as described in Table VII.

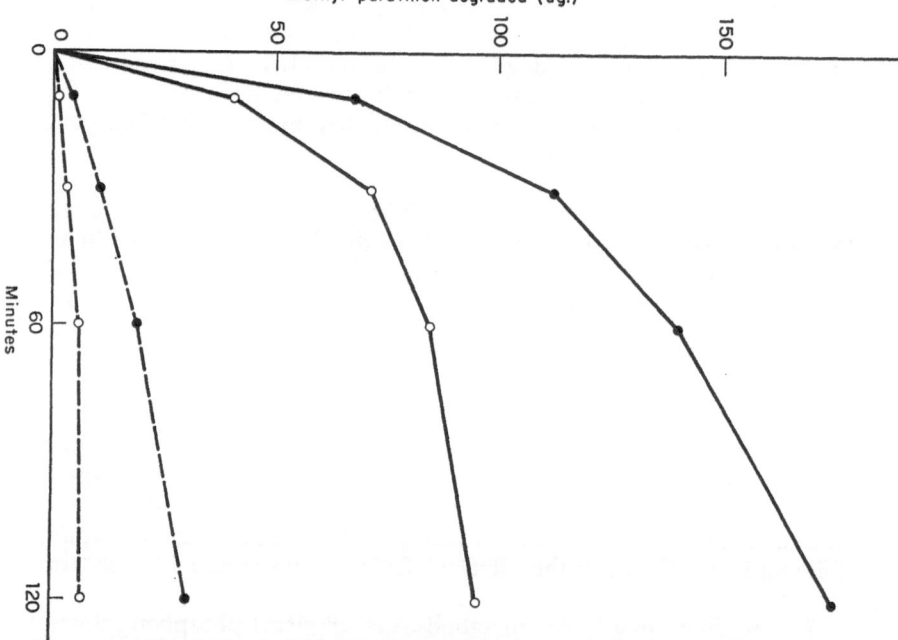

Fig. 2. Time course of degradation of methyl parathion in supernatant of rat liver and of mid-gut of horn beetle larvae with and without GSH. Each reaction mixture contained 2.4 ml. of supernatant (0.25 g. of equiv. fresh wt./ml.), $4 \times 10^{-3}M$ GSH, and 300μg. of methyl parathion, in a final volume of 3.0 ml., pH 7.4: O—O liver, ●—● liver with GSH, O—O mid-gut, ●—● mid-gut with GSH (Fukami and Shishido 1966)

with and without added GSH. Degradation proceeded rapidly in the rat liver, increasing further with the addition of GSH. In the insect mid-gut, degradation was small, and although the rate of degradation was effectively increased in the addition of GSH, it was still less than the rate of degradation in the rat preparation without addition of GSH (Fig. 2).

It appears that when methyl parathion and Sumithion are enzymically degraded, the amount of GSH available for demethylation in the tissue is important. Table XIV indicates the content of GSH

Table XIV. *Levels of reduced glutathione in total rat liver and mid-gut of horn beetle larvae* (FUKAMI and SHISHIDO 1966)

Tissue	GSH (mg./g. wet tissue wt.)
Liver	2.88
Mid-gut	0.52

determined manometrically using the glyoxylase method (PATTERSON and LAZAROW 1955) in the rat liver and the mid-gut of horn beetle larvae. The rat liver contains 5.5 times more GSH than the mid-gut of the insect, on a wet-weight basis.

In the presence of GSH the principal metabolite in the rat liver is desmethyl parathion. In the insect mid-gut, two principal metabolites are found: desmethyl parathion and phosphate (described above). The metabolic pathway for the degradation of methyl parathion in the rat liver and in the insect mid-gut is proposed as follows:

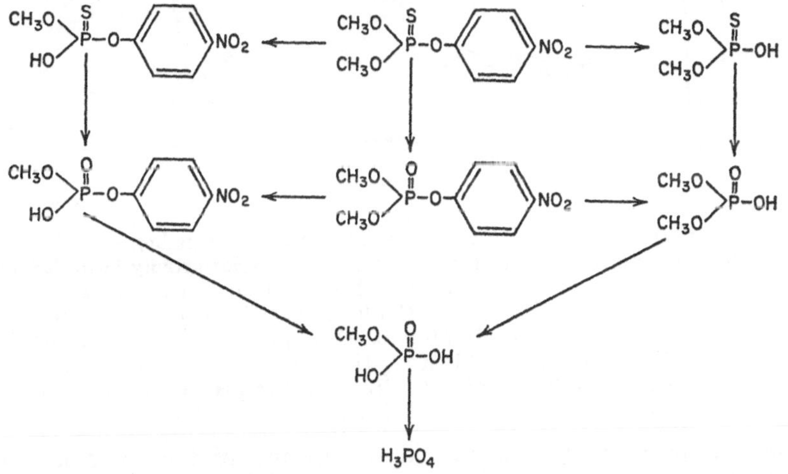

In the rat and in the insect, the initial demethylation reaction depends entirely on the presence of GSH. However, in the insect it must further be studied if the reactions leading to phosphate are due to GSH other than in the formation of the first product, desmethyl parathion.

As a subsequent experiment, the enzyme preparations from dialyzed supernatant of the rat liver and the homogenate of the insect mid-gut were fractionated by passage through CM and DEAE cellulose columns for the purification of a soluble, glutathione-dependent enzyme system which cleaves methyl parathion to desmethyl parathion. Chromatographic columns of DEAE and CM cellulose, 1.2 x 20 cm., were prepared according to the method of Moore and Lee (1960) without using air pressure.

The supernatant was dialyzed through cellophane overnight at 4° C. against 0.005 M tris-hydrochloric acid buffer, pH 7.0. The column was developed using this buffer gradient. In the rat liver, the enzyme activity was eluted in the 57th to 67th tubes on passage through the CM cellulose column (Fig. 3). The enzyme was activated by GSH

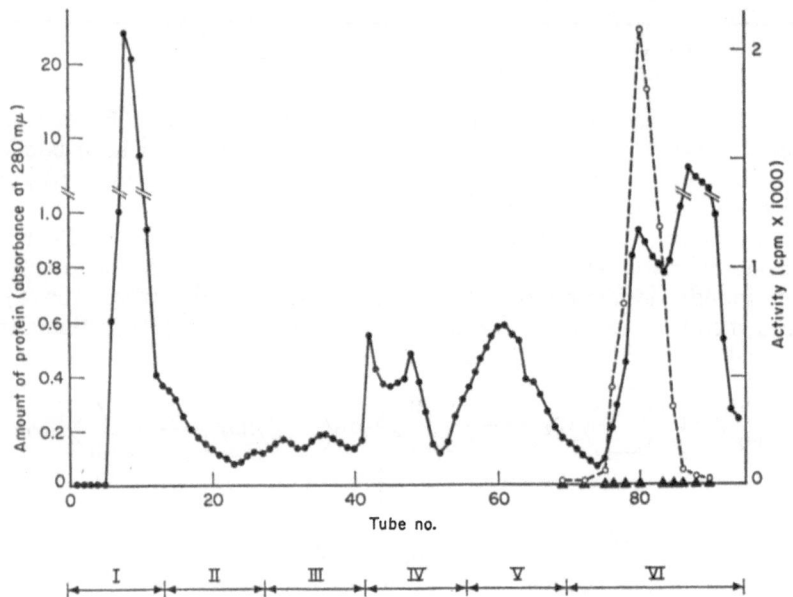

Fig. 3. CM-cellulose column chromatography of enzyme degrading methyl parathion in rat liver supernatant: ● protein, ○ enzyme activity (with GSH), and ▲ enzyme activity (without GSH). The elution system is as follows: I = 0.005M tris-HCl buffer at pH 7.0, II = 0.005M at pH 7.0 to 0.02M at pH 7.0, III = 0.02M at pH 7.0 to 0.09M at pH 7.0, IV = 0.09M at pH 7.0 to 0.12M at pH 7.0, V = 0.12M at pH 7.0 to 0.33M at pH 7.0, VI = 0.33M at pH 7.0 to 0.33M at pH 8.0, VII = 0.33M at pH 8.0 to 1.0M NaCl in 0.33M tris-HCl buffer at pH 8.0 (Fukami and Shishido 1966)

and was purified at least 15-fold by the use of the column. The preparation from rat liver, however, was not adsorbed on the DEAE cellulose column. On the contrary, the enzyme of the dialyzed supernatant from the insect mid-gut was adsorbed by the DEAE cellulose column, and the activity was eluted in the 75th to 86th tubes (Fig. 4); the enzyme was activated by GSH and was purified at least

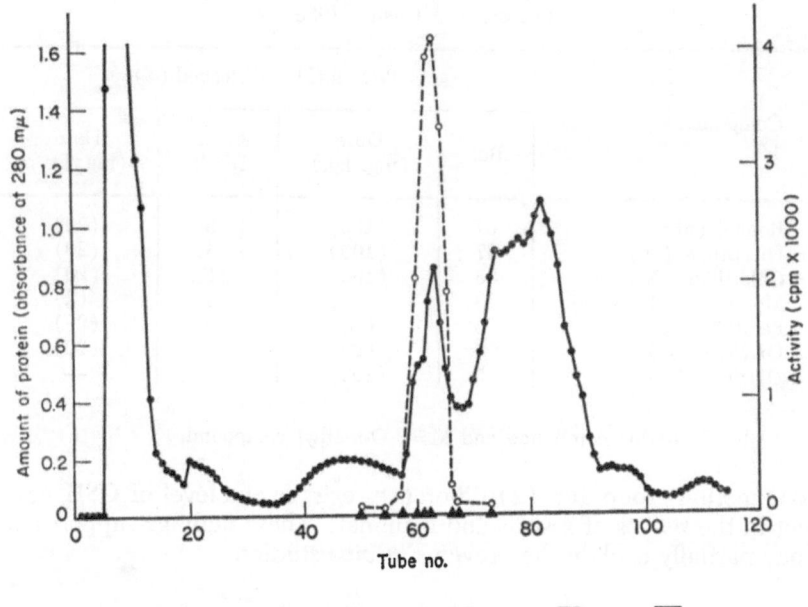

Fig. 4. DEAE-cellulose column chromatography of enzyme degrading methyl parathion in supernatant of mid-gut of horn beetle larvae: ● protein, ○ enzyme activity (with GSH), and ▲ enzyme activity (without GSH). The eluting system is as described in Figure 3 (FUKAMI and SHISHIDO 1966)

10-fold. The water-extractable metabolites produced by the purified rat liver enzyme in the presence of GSH were identified using ion-exchange column chromatography. One of the metabolites was des-methyl parathion, which accounts for 95 percent of the total metabo-lites. The same result was obtained with insect mid-gut.

The degrading enzyme in the rat liver preparation was adsorbed on CM cellulose, but not on DEAE cellulose, whereas the enzyme in the insect mid-gut preparation was adsorbed on DEAE and not on CM cellulose; this fact indicates that the nature of these two enzyme pro-teins is different.

As described previously, PLAPP and CASIDA (1958 b) presented the data on the exact site of cleavage of alkylaryl phosphates and phos-phorothioates by the rat and the cockroach, in vivo. These results are cited in Table XV.

As is shown in Table XV, much less cleavage of the P-O-methyl group occurs in insects than in the mammal. In mammals, the P-O-methyl group is destroyed more easily than the P-O-ethyl group. From these results, the present in vitro study indicates that (1) qualitative and quantitative differences between mammals and insects exist in the enzymes which degrade organophosphorus insecticides containing the

Table XV. *Variations in place of cleavage of dialkylaryl phosphorothioates*
(Plapp and Casida 1958 a)

Compound [a]	P-O-alkyl bond cleaved (%)			
	Rat	Dose (mg./kg.)	Amer. roach	Dose (mg./kg.)
Ronnel (M)	67	(100)	8	(20)
Dicapthon (M)	37	(100)	4	(20)
Chlorthion (M)	38	(100)	1	(20)
Methyl parathion (M)	13	(2)	26	(2)
Parathion (E)	6	(15)	5	(20)
Diazinon (E)	5	(20)	0	(20)
Diazinon (E)	5	(100)	—	—

[a] E = P-O-ethyl compound and M = P-O-methyl compound.

P-O-methyl group and (2) differences exist in the level of GSH present in the tissues of insects and mammals. These findings support and may partially explain the previous *in vivo* studies.

Summary

The nature and distribution of enzyme systems responsible for the selective toxicities of dialkylaryl phosphorothioates were investigated. A system hydrolyzing P-O-methyl compounds to their respective desmethyl derivatives has been demonstrated to be located in the soluble fractions of rat liver homogenates.

The GSH content available for demethylation of P-O-methyl compounds is much higher in the rat liver than in the mid-gut of the insect larvae. The differential ratio of the degrading action on P-O-methyl compounds in the liver of rat and in insects is due, in part, to the difference in degrading activity of the enzyme. No significant difference is, however, found in the effect of GSH on the metabolism of methyl parathion and Sumithion.

Résumé *

Le métabolisme in vitro des insecticides organo-phosphorés dans les homogénats de tissus de mammifères et d'insectes

La nature et la distribution des systèmes enzymatiques responsables de la toxicité sélective des thiophosphates dialkylaryliques ont été examinées. On a démontré qu'un système hydrolysant les composés

---

* Traduit par S. Dormal-van den Bruel.

P-O-méthylés en dérivés déméthylés correspondants se trouve dans les fractions solubles des homogénats de foie de rat. La teneur en GSH disponible pour la déméthylation des composés P-O-méthylés est plus élevée dans le foie du rat que dans l'intestin moyen des larves d'insectes. Le rapport différentiel de la dégradation des composés P-O-méthylés dans le foie du rat et chez les insectes est dû, en partie, à la différence de l'activité dégradante de l'enzyme. Aucune différence significative n'a cependant été trouvée dans l'effet du GSH sur le métabolisme du parathion-méthyl et du Sumithion.

Zusammenfassung *

Der in vitro Metabolismus von Organophosphorinsektiziden durch Gewebehomogenisate von Säugetieren und Insekten

Die Beschaffenheit und Verteilung von Enzymsystemen, welche für die selektiven Toxizitäten von Dialkylaryl-thiophosphaten verantwortlich sind, wurde untersucht. Es wurde gezeigt, dass ein System, welches P-O-Methylverbindungen zu ihren entsprechenden Demethylderivaten hydrolisiert, sich in den löslichen Fraktionen von Rattenleberhomogenisaten befindet. Der GSH-Gehalt, welcher für die Demethylierung von P-O-Methylverbindungen verfügbar ist, ist viel höher in Rattenleber als im Mitteldarm von Insektenlarven. Das Differentialverhältnis der abbauenden Wirkung bei P-O-Methylverbindungen in Rattenleber und in Insekten beruht teilweise auf dem Unterschied in der Abbauaktivität des Enzyms. Jedoch wurde kein bedeutsamer Unterschied in der Wirkung des GSH auf den Metabolismus von Methylparathion und Sumithion gefunden.

References

ALDRIDGE, W. N.: Serum esterases. 1. Two types of esterase (A and B) hydrolysing p-nitrophenyl acetate, propionate and butyrate, and a method for their determination. Biochem. J. 53, 110 (1953 a).
—— Serum esterases. 2. An enzyme hydrolysing diethyl p-nitrophenyl phosphate (E 600) and its identity with the A-esterase of mammalian sera. Biochem. J. 53, 117 (1953 b).
BULL, D. L., D. A. LINDQUIST, and J. HACSKAYLO: Absorption and metabolism of dimethoate in the ballworm and ball weevil. J. Econ. Entomol. 56, 129 (1963).
DAUTERMAN, W. C., J. E. CASIDA, J. B. KNAAK, and T. KOWALCZYK: Bovine metabolism of organophosphorus insecticides. Metabolism and residues associated with oral administration of dimethoate to rats and three lactating cows. J. Agr. Food Chem. 7, 188 (1959).
FUKAMI, J., and T. SHISHIDO: Studies on the selective toxicities of organic phosphorus insecticides (III). The characters of the enzyme system in cleavage

* Übersetzt von A. SCHUMANN.

of methyl parathion to desmethyl parathion in the supernatant of several species of homogenates (Part 1). Botyu-Kagaku 28, 77 (1963 a).

—— —— Studies on the selective toxicities of organic phosphorus insecticides (I). Activation of ethyl parathion in mammal and insect (Part 1). Botyu-kagaku 28, 63 (1963 b).

—— —— Nature of soluble, glutathione-dependent enzyme system active in cleavage of methyl parathion to desmethyl parathion. J. Econ. Entomol. 59, 1338 (1966).

Hodgson, F., and J. E. Casida: Mammalian enzymes involved in the degradation of 2, 2-dichlorovinyl dimethyl phosphate. J. Agr. Food Chem. 10, 208 (1962).

Hollingworth, R. M., R. L. Metcalf, and T. R. Fukuto: The selectivity of Sumithion compared with methyl parathion. Metabolism in the white mouse. J. Agr. Food Chem. 15, 242 (1967).

Johnson, M. K.: Metabolism of iodomethane in the rat. Biochem. J. 98, 38 (1966 a).

—— Studies on glutathione S-alkytransferase of the rat. Biochem. J. 98, 44 (1966 b).

Knaak, J. B., and R. D. O'Brien: Effect of EPN on in vivo metabolism of malathion by the rat and dog. J. Agr. Food Chem. 8, 198 (1960).

Kosolapoff, G. M.: Organophosphorus compounds. New York: Wiley (1950).

Main, A. R.: The purification of the enzyme hydrolysing diethyl p-nitrophenyl phosphate (paraoxon) in sheep serum. Biochem. J. 75, 10 (1960 a).

—— The differentiation of the A-type esterases in sheep serum. Biochem. J. 75, 188 (1960 b).

Matumura, F., and A. W. A. Brown: Biochemistry of malathion resistance in Culex tarsalis. J. Econ. Entomol. 54, 1176 (1961).

Mazur, A.: An enzyme in animal tissues capable of hydrolyzing the phosphorus-fluorine bond of alkyl fluorophosphates. J. Biol. Chem. 164, 271 (1946).

Miyamoto, J.: Studies on the mode of action of organophosphorus compounds. Part IV. Penetration of Sumithion, methyl parathion and their oxygen analogs into guinea pig brain and inhibition of cholinesterase in vivo. Agr. Biol. Chem. 28, 422 (1964).

Moore, B. W., and R. H. Lee: Chromatography of rat liver soluble proteins and localization of enzyme activities. J. Biol. Chem. 235, 1359 (1960).

Morello, A., A Vardan's, and E. Y. Spencer: Differential glutathione-dependent detoxication. Biochem. Biophys. Research Comm. 29, 241 (1967).

Mounter, L. A., and V. P. Whittaker: The effect of thiol and other group-specific reagents on erythrocyte and plasma cholinesterases. Biochem. J. 54, 551 (1953).

—— —— Some studies of enzymatic effects of rabbit serum. J. Biol. Chem. 209, 813 (1954).

Nishizawa, Y., Fujii, T., Kadota, T., T. Miyamoto, and H. Sakamoto: Studies on organophosphorus insecticides. Part VII. Chemical and biological properties of new low toxic organophosphorus insecticide O,O-dimethyl-O-(3-methyl-4-nitrophenyl) phosphorothioate. Agr. Biol. Chem. 25, 605 (1961).

Patterson, J. W., and A. Lazarow: Determination of glutathione. In: D. Glick (ed.), Methods of biochemical analysis, vol. 2, pp. 259-78. New York: Interscience (1955).

Plapp, F. W., and J. E. Casida: Hydrolysis of the alkyl-phosphorothioate insecticides by rat, cockroaches, and alkali. J. Econ. Entomol. 51, 800 (1958 a).

—— —— Bovine metabolism of organophosphorus insecticides. Metabolic fate of O,O-dimethyl O-(2,4,5-trichlorophenyl) phosphorothioate in rats and a cow. J. Agr. Food Chem. 6, 662 (1958 b).

—— —— Ion-exchange chromatography for hydrolysis products of organophosphate insecticides. Anal. Chem. 30, 1622 (1958 c).

SHISHIDO, T., and J. FUKAMI: Studies on the selective toxicities of organic phosphorus insecticides (II). The degradation of ethyl parathion, methyl, parathion, methyl paraoxon and Sumithion in mammal, insect, and plant. Botyu-kagaku **28**, 69 (1963).

—— ——, and K. FUKUNAGA: The degradation of dimethyl arylphosphates and phosphorothioates by glutathione S-methyltransferase of rat liver and American cockroach. Unpublished data (1968).

VARDANIS, A., and L. G. CRAWFORD: Comparative metabolism of *O,O*-dimethyl *O-p*-nitrophenyl phosphorothioate (Methyl Parathion) and *O,O*-dimethyl *O*-(3-methyl-4-nitrophenyl) phosphorothioate (Sumithion). J. Econ. Entomol. **57**, 136 (1964).

WEDDING, R. T., and M. K. BLACK: Response of oxidation and coupled phosphorylation in plant mitochondria to 2,4-dichlorophenoxyacetic acid. Plant Physiol. **37**, 364 (1962).

Mechanism of low toxicity of Sumithion toward mammals

by
JUNSHI MIYAMOTO *

Contents

I. Introduction

Although Sumithion [fenitrothion, O,O-dimethyl O-(3-methyl-4-nitrophenyl) phosphorothioate] has a chemical structure very similar to that of methyl parathion [O,O-dimethyl O-(4-nitrophenyl) phosphorothioate], it has strikingly low-toxicity to warm-blooded animals, as is evident in Table I. The work described in this paper is a part of biochemical studies carried out in the author's laboratory on the mode of action of the organophosphorus insecticides, especially in hopes of elucidating the mechanism of their selective toxicity to mammals.

II. Cholinesterase inhibition in vivo

Sumithion or methyl parathion, when administered orally to guinea pigs or white rats, inhibited brain and blood cholinesterases. As Figure 1 shows, methyl parathion inhibited these enzymes more powerfully than Sumithion at the same dosage. On increasing the dosage, methyl parathion hindered both brain and plasma enzyme activity with equal potency, while Sumithion inhibited the brain enzyme far less

---

* Agricultural Chemicals Research Department, Osaka Works, Sumitomo Chemical Co., Ltd. Konohana-ku, Osaka, Japan.

Table I. *Toxicity of Sumithion, methyl parathion and their oxygen analogs* LD$_{50}$ (mg./kg.)

Mammal	Route	Sumithion	Methyl parathion	Sumioxon	Methyl paraoxon
White rat	Oral	200	24.5	24	4.5
	Intravenous	33	4.1	3.3	0.5
Mouse	Oral	870	17	90	10.8
	Intravenous	220	13	—	—
Guinea pig	Oral	1850	417	221	83
	Intravenous	112[a]	50	32	2.2

[a] Death occurred immediately after or even during injection. Symptoms different from acetylcholine poisoning such as paralysis and head drop were observed. Such symptoms were not observed by methyl parathion or by Sumioxon. Injection of 30 mg./kg. of Sumithion every 30 minutes gave an LD$_{50}$ value of approximately 250 mg./kg.

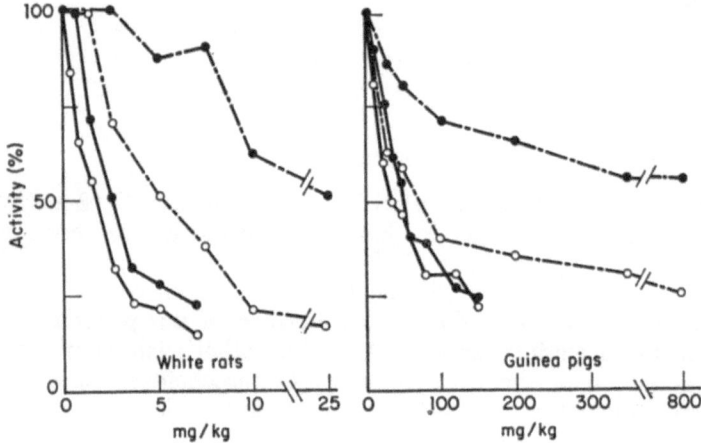

Fig. 1. Inhibition of cholinesterase by orally administered Sumithion and methyl parathion. Assay was made two hours after treatment: ○ plasma, ● brain, —.— Sumithion, and —— methyl parathion

than the plasma enzyme. Symptoms of intoxication such as dispnoea, twitch, clonic convulsion, salivation, bloody tears, and exophthalmos which are considered typical of acetylcholine poisoning were observed when brain cholinesterase activity was inhibited fairly heavily and the enzyme activity of the animals dead of intoxication was decreased to less than 20 percent of the control, often to five to seven percent (MIYAMOTO *et al.* 1963 b). And with regard to these points there was found hardly any clear-cut distinction between Sumithion and methyl parathion.

The above-mentioned weaker inhibitory activity of Sumithion was also observed in intravenous administration (Fig. 2) (MIYAMOTO *et al.*

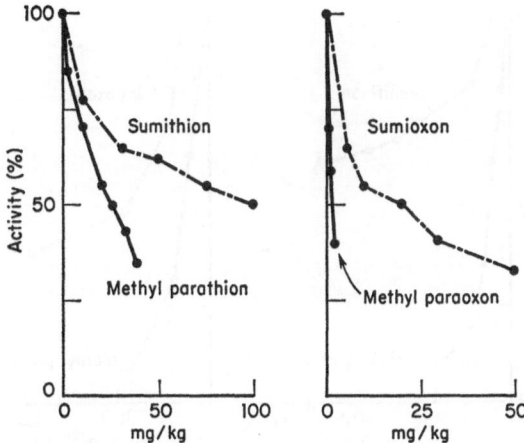

Fig. 2. Inhibition of guinea pig brain cholinesterase by intravenously administered Sumithion, methyl parathion, and their oxygen analogs. Assay was made one hour after treatment

1963 b); brain cholinesterase activity of guinea pigs and white rats declined rather linearly as the dosage of methyl parathion increased, whereas the enzyme was found more resistant to Sumithion and its increased dosage was not so effective in causing expected inhibition. Similar results were obtained, too, in intravenous administration of Sumioxon and methyl paraoxon, the presumed physiologically active form of Sumithion and methyl parathion, respectively. Sumioxon was a far weaker inhibitor of cholinesterase than methyl paraoxon, although these oxygen analogs hindered the enzyme activity more powerfully than their parent compounds. These relationships held true in the case of inhalation experiments (KADOTA 1967), that is, as reproduced in Figure 3, Sumithion inhibited brain cortex cholinesterase of mice far less than methyl parathion, and Sumioxon far less than methyl paraoxon.

Thus, in the light of these results, lower toxicity of Sumithion toward mammals might come from the lesser inhibitory activity of this compound on cholinesterases, especially on the brain enzyme *in vivo*. The weaker inhibitory activity of Sumioxon seemingly participates, too.

III. Metabolic factors

As is widely accepted, phosphorothioates such as Sumithion and methyl parathion, once they are absorbed into the animal body, are converted by microsomal oxidation enzymes in tissues into their oxygen analogs, which really exert harmful effects upon the whole

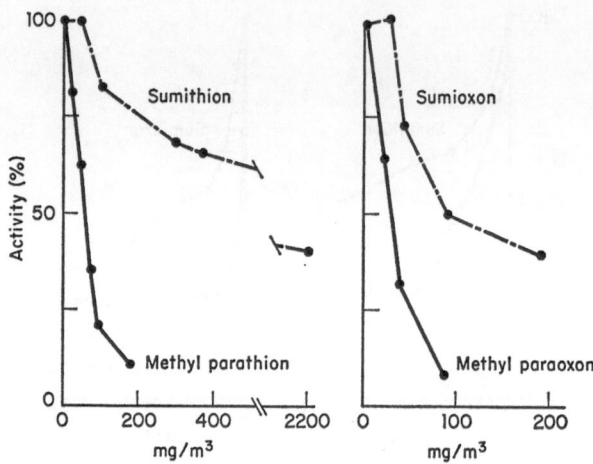

Fig. 3. Inhibition of brain cortex cholinesterase of mouse by inhalation of Sumithion, methyl parathion, and their oxygen analogs. The compounds, dissolved in a mixture of toluene and kerosene, were inhaled for four hours
at a pressure of 0.5 kg./cm.[2]

animal. On the other hand, these phosphorothioates and phosphorates
are subject to decomposition and detoxication. Roughly speaking,
therefore, toxicity of a phosphorothioate is to some extent dependent
upon the ratio of these activation and degradation reaction rates.
Sumithion and methyl parathion orally administered to guinea pigs
and white rats were readily absorbed from the gastro-intestinal tracts
of the animals and similar degradation products were excreted into
the urine (MIYAMOTO et al. 1963 a). Concerning these points there
was little difference between those phosphorothioates. Further experiments were made to elucidate if Sumithion is actually degraded
more rapidly in vivo than is methyl parathion, and/or if the former
is converted more slowly into the toxic oxygen analog than the latter
(MIYAMOTO 1964 a). Therefore, toxicity ratio of these phosphorothioates was neglected and the same dosage was administered intravenously. Various assays were made at comparatively short intervals after
treatment, because they were acute toxicants and symptoms of poisoning began appearing several minutes afterward.

Phosphorus-32 labeled Sumithion and methyl parathion administered intravenously to guinea pigs disappeared equally rapidly from
blood and were found to be distributed into various tissues. The
presence of the oxygen analog was demonstrated in every tissue
tested (brain, lung, heart, liver, kidney, spleen, and muscle) and in
blood even at such a short period after treatment as one minute.
These tissues contained a little larger amount of Sumioxon than methyl
paraoxon in most cases, as shown in Table II. A similar tendency was

Table II. *Oxygen analog content in several tissues of guinea pig after intravenous treatment of 40 mg./kg. of methyl parathion or Sumithion*

Minutes after treatment	Oxygen analog content (μg. equiv./g. tissue)			
	Brain	Liver	Lung	Kidney
Methyl paraoxon				
1	0.01	0.71	2.67	0.01
2.5	0.11	1.77	2.97	0.51
5	0.30	2.81	1.63	0.66
15	0.26	0.27	0.42	0.30
30	0.01	0.21	0.13	0.18
Sumioxon				
1	0.27	1.46	6.18	0.14
2.5	0.45	2.33	7.69	1.17
5	0.42	2.84	2.79	1.02
15	0.55	0.97	1.58	1.29
30	0.29	0.20	0.58	0.36

observed for white rats treated intravenously with Sumithion or methyl parathion (Table III). Furthermore, microsomal fractions from the

Table III. *Oxygen analog content in rat liver*

Conditions	Oxygen analog content (μg. equiv./g. tissue)	
	Methyl paraoxon	Sumioxon
Dosage [a] *15 mg./kg.* *Minutes after treatment*		
0.5	2.40	4.07
1	2.30	3.48
1.5	2.49	4.00
2.5	1.80	2.60
Assayed one minute after treatment *Dosage* [a] *in mg./kg.*		
2.5	0.11	—
5	0.33	0.48
10	1.36	1.43
15	2.30	3.48
25	—	3.11

[a] Of methyl parathion or Sumithion.

liver of guinea pig, white rat, or mouse fortified with NADPH$_2$ oxidized *in vitro* tritium-labeled Sumithion slightly more rapidly than
methyl parathion (Table IV) (MIYAMOTO *et al.* 1968 a), which was in

Table IV. *Oxygen analog formation by liver microsomes*

Phosphorothioate added	Phosphorothioate decomposed (%) [a]	Oxygen analog formed (%) [a]
	Guinea pig	
Methyl parathion	20.7	7.7
Sumithion	24.3	9.8
	Rat	
Methyl parathion	9.0	5.4
Sumithion	12.5	6.0
	Mouse	
Methyl parathion	12.6	5.0
Sumithion	15.3	6.4

[a] Phosphorothioate initially added was taken as 100 percent.

accordance with the above results *in vivo*. The formation rate of oxygen
analog from the respective phosphorothioate would not, of course, be
accurately determined *in vivo*, because some portions of the oxygen
analog would be either decomposed or bound with the susceptible
proteins and thus be exempted from assay as such. However, the
larger content of Sumioxon in most tissues might rule out the possibility that the lower toxicity of Sumithion to mammals results from
its slower oxidation in the animal body.

Liver and kidney of guinea pigs were found principally active
in decomposing these phosphorothioates, and phosphorates and water-
soluble metabolites such as the desmethyl compound (desmethyl Sumithion and desmethyl parathion), dimethylphosphorothioic acid, and
dimethylphosphoric acid were identified (Table V). The most abundant product in the liver was dimethylphosphoric acid, presumably
derived from Sumioxon or methyl paraoxon, while in the kidney dimethyl phosphorothioic acid and the desmethyl compound predomi-
nated. In both tissues as well as in blood, water-soluble metabolites
were formed more from methyl parathion than from Sumithion, which
may imply that the former is really subject to easier decomposition.
Similar results were obtained with white rats treated intravenously with
Sumithion or methyl parathion; the presence of lesser amounts of degradation products from Sumithion was observed. In searching for an
explanation of the selective toxicity of Sumithion, VARDANIS and CRAW-

Table V. *Water-soluble metabolites of Sumithion and methyl parathion in guinea pig liver and kidney* [a]

Minutes after treatment	Metabolite [b] content [c]					
	Total	I	II	III	IV	V
Liver						
Sumithion						
1	9.9	1.1	5.0	3.2	0.6	—
2.5	26.2	3.4	14.4	5.2	3.2	—
5	44.7	5.0	30.6	8.5	0.7	—
15	52.9	4.1	41.8	5.4	1.6	—
Methyl parathion						
1	20.9	1.8	13.8	4.0	1.3	—
2.5	36.9	4.9	19.1	7.4	5.5	—
5	58.5	9.4	38.3	10.8	—	—
15	68.5	2.3	54.9	11.3	—	—
Kidney						
Sumithion						
1	14.9	5.2	—	3.0	6.7	—
2.5	15.7	4.3	—	2.7	8.6	—
5	22.9	6.6	3.7	8.3	4.2	—
15	51.9	12.3	9.1	7.3	23.3	—
Methyl parathion						
1	25.1	10.7	—	7.5	6.9	—
2.5	30.5	15.8	—	2.8	5.4	6.6
5	37.6	18.1	—	3.3	9.9	6.3
15	92.7	50.3	6.4	8.6	15.1	12.3

[a] Dosage, 40 mg./kg.
[b] I = phosphoric and/or phosphorothioic acid.
  II = dimethylphosphoric acid.
  III = dimethylphosphorothioic acid.
  IV = desmethyl compound and unidentified.
  V = presumably monomethylphosphoric and/or monomethyl-phosphorothioic acid.
[c] Content was expressed as μg. equivalent of treated compound/g. of tissue.

FORD (1964) tested the transformation of Sumithion and methyl parathion by mouse liver *in vitro* and surmised that faster degradation of the former compound than the latter causes the selective toxicity, but their assay methods were indirect and the results were rather ambiguous. As to the degradation of these phosphorothioates and their oxygen analogs, HOLLINGWORTH *et al.* (1967) stated in their publication dealing with the *in vivo* metabolism of the above two compounds in mice that the high selectivity level of Sumithion depends on the ability of cleaving the P-O-alkyl bond thus forming the desmethyl compound to play an enhanced role in detoxication as the dosage is increased. Moreover, they examined the toxicity and metabolism of O-methyl phosphonate analogs of Sumithion and methyl parathion to verify the above hypothesis. As results, they found that the toxicity ratio of these Sumiphonothion and methyl paraphonothion was less than that of Sumithion and methylparathion, also that desmethylation of these phosphono analogs occurred only slightly. However, their results might not always be so conclusive, for the enhanced desmethylation of Sumithion was presumably the result of lower toxicity of Sumithion, rather than the cause, as suggested, for example, by PLAPP and CASIDA (1958), and that their conclusion apparently lacks the evidences that the reduction of toxicity ratio observed in the phosphono analogs comes solely from the inability of these compounds to be desmethylated. Although indeed mouse liver homogenate decomposed tritium labeled Sumithion and Sumioxon a little more easily *in vitro* than their counterparts, as shown in Table VI, owing princi-

Table VI. *Identification of decomposition products from phosphorothioates and their oxygen analogs by mouse liver homogenate*

Product	Compound added [a] and decomposition (%)			
	Sumithion	Methyl parathion	Sumioxon	Methyl paraoxon
Oxygen analog	1.96	2.43	—	—
Desmethyl phosphorothioate	21.58	18.25	—	—
Desmethyl phosphorate	3.74	2.40	40.59	34.36
3-Methyl-4-nitrophenol (4-nitrophenol)	3.54	3.55	3.62	6.24
Unidentified [b]	0.07	0.26	2.30	0.30
Total decomposed	30.89	26.89	46.51	40.90

[a] Compound initially added was taken as 100 percent.
[b] Chloroform soluble.

pally to the enzyme in the supernatant fraction desmethylating the former compounds faster than the latter compounds, the difference in the enzyme activity was found to be not large (MIYAMOTO et al. 1968 a). Thus, the results obtained hitherto seemingly suggest that the difference in respect of the degradation rate in mammals plays a role of rather minor importance in lowering the toxicity of Sumithion.

IV. Penetration into mammalian brain

As stated above, Sumithion and Sumioxon inhibited cholinesterase of mammals far less potently *in vivo* than methyl parathion and methyl paraoxon, respectively. Among many possible reasons presumed, the slower oxidizability of Sumithion, and the easier change of those organophosphorus compounds into non-toxic substances in the animal body were tentatively excluded. In the next experiment phosphorus content and inhibition of cholinesterase in brain were investigated *in vivo* after Sumioxon and methyl paraoxon had been administered intravenously to guinea pigs. The purpose was to know if there be such relationships between penetrability of these phosphorates into brain and concomitant inhibition of the enzyme activity that might be helpful for elucidating the selective toxicity of Sumithion.

At intervals after administration of phosphorus-32 labeled Sumioxon or methyl paraoxon, guinea pig brain was perfused with Lock-Ringer solution through the common carotid artery (MIYAMOTO *et al.* 1963 a), the brain was dissected out, and the phosphorus-32 as well as cholinesterase activity were determined. As shown in Figure 4, the

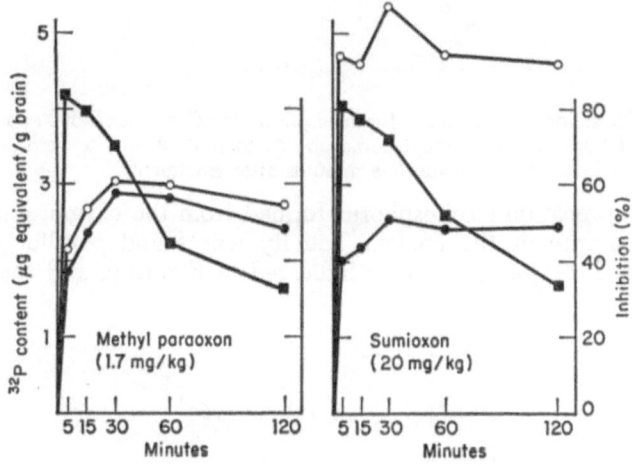

Fig. 4. Phosphorus-32 content and inhibition of cholinesterase in guinea pig brain after intravenous administration of Sumioxon or methyl paraoxon: ○ total P-32, ● acid-precipitated P-32, and ■ inhibition

inhibition of cholinesterase activity reached its maximum within five minutes and the activity recovered gradually thereafter, although the phosphorus content increased during 30 minutes and then remained nearly unchanged. Therefore, phosphorus-32 which penetrated the brain after the initial five minutes seemed to have no effects upon the cholinesterase. If more dosage of either oxygen analog were given, the content of total and acid-precipitated phosphorus (phosphorus co-precipitated with protein by perchloric acid) in brain increased and cholinesterase activity decreased. Total and acid-precipitated phosphorus concentrations in the brain at five minutes were always higher than that of administered methyl paraoxon, whereas those concentrations were much lower than that of administered Sumioxon (Fig. 5). When inhibition caused was plotted against

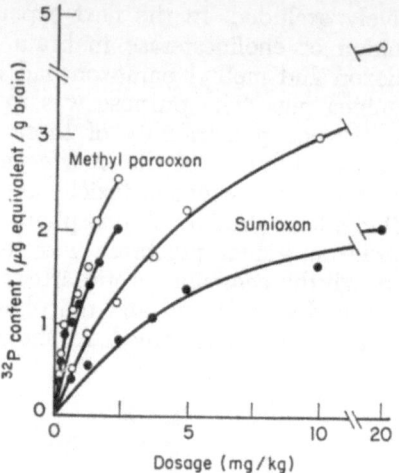

Fig. 5. Phosphorus-32 content of guinea pig brain after intravenous administration of Sumioxon or methyl paraoxon: ○ total P-32 and ● acid-precipitated P-32. Assay was made five minutes after treatment

the acid-precipitable phosphorus formed from the oxygen analog, the reduction rate of the enzyme activity was found parallel with the amount of the phosphorus, as indicated in Figure 6, and the activity

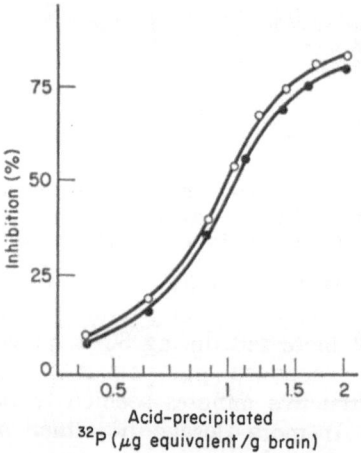

Fig. 6. Dependence of guinea pig brain cholinesterase inhibition on acid-precipitated phosphorus from oxygen analogs: ● Sumioxon and ○ methyl paraoxon

was inhibited to approximately the same degree by such dosages of Sumioxon or methyl paraoxon to produce the same amount of the acid-precipitated phosphorus. Thus, the weaker inhibitory potency of Sumioxon on the brain enzyme *in vivo* may be a consequence of poorer penetration of the compound into the brain and the formation of lesser amounts of the acid-precipitable phosphorus. As already mentioned above, the cholinesterase activity was decreased to less than 20 percent, often five to seven percent of the initial level in the brain of animals dead of poisoning. From these results the amount of the acid-precipitable phosphorus for reducing the enzyme activity to such a level was calculated to be two μg. or more equivalents of either oxygen analog/g. of brain under the present experimental conditions. This means approximately ten times or more of Sumioxon than methyl paraoxon must be given to guinea pigs for the formation of such amounts of phosphorus. This ratio is roughly equal to the toxicity ratio of Sumioxon and methyl paraoxon (Sumioxon/methyl paraoxon, LD$_{50}$ mg./kg.; 32/2.2). As the activity in mammalian brain of the enzyme systems capable of converting Sumithion or methyl parathion into the oxygen analog has been found negligible (FUKAMI and SHISH-IDO 1963, NEAL 1967, MIYAMOTO 1968 b), also that these phosphorothioates are intrinsically but weak inhibitors of cholinesterases (MIYA-MOTO *et al.* 1963 b), the inhibition of the cholinesterase caused after treatment with the phosphorothioates would be ascribed to the exogenous oxygen analog which penetrated.

When Sumithion and methyl parathion were administered to guinea pigs intravenously, considerable quantities of them transferred to the brain rapidly, as already shown (MIYAMOTO 1964 b). Effects of their presence upon the penetration of the oxygen analog were therefore investigated. Radioactive Sumioxon or methyl paraoxon and the non-active corresponding phosphorothioate were given simultaneously to guinea pigs, and radioactive phosphorus and inhibition of cholinesterase in the brain were determined *in vivo*. As indicated in Table VII, the content of radioactive phosphorus was not affected significantly, at least during the initial period after treatment, by the corresponding phosphorothioate which was co-existent. With larger dosages of phosphorothioate, inhibition was heavier due to the fact that the non-active oxygen analog was newly formed from the phosphorothioate and this also penetrated the brain. This preliminary test seems to imply that the poorer penetrability of Sumioxon-phosphorus into the brain is also observed in the presence of Sumithion.

In summary, therefore, it is possible to conclude that the above-described behavior of Sumioxon in the animal body may reflect the low toxicity of Sumithion toward mammals.

Summary

The low mammalian toxicity of Sumithion [O,O-dimethyl O-(3-

Table VII. *Phosphorus-32 content and inhibition of cholinesterase of guinea pig brain after intravenous treatment of radioactive oxygen analog and non-active phosphorothioate simultaneously* [a]

Dosage of phosphorothioate (mg./kg.)	Inhibition (%)	Phosphorus-32	
		Total (μg./g.)	Acid-precipitated (μg./g.)
P-32 Sumioxon (5 mg./kg.) Sumithion			
0	68	2.21	1.38
20	69	2.09	1.33
30	73	2.25	1.35
40	76	2.49	1.44
P-32 Methyl paraoxon (0.9 mg./kg.) Methylparathion			
0	68	1.32	1.18
10	76	1.30	1.20
20	78	1.13	1.11
30	81	1.28	1.21

[a] Assay was made five minutes after treatment.

methyl-4-nitrophenyl) phosphorothioate] was presumed to come from the poorer penetration of its toxic oxygen analog, Sumioxon, into the brain and formation therein of lesser amounts of phosphorus coprecipitated with proteins, upon which inhibition of cholinesterase activity was dependent.

Metabolic factors such as conversion of Sumithion into Sumioxon and degradation of these compounds into non-toxic substances in the animal body were found to play a role of rather minor importance in the selective toxicity of this organophosphorus insecticide.

Résumé *

Mécanisme de la faible toxicité du Sumithion à l'égard des mammifères

On a présumé que la faible toxicité du Sumithion [thiophosphate de O,O-diméthyle et de O-(3-méthyl-4-nitrophényle)] à l'égard des mammifères provient du plus faible pouvoir de pénétration de son homologue toxique oxygéné, le Sumioxon, dans le cerveau et de la formation dans cet organe de plus petites quantités de phosphore coprécipité avec les protéines, dont dépend l'inhibition de l'activité

* Traduit par S. DORMAL-VAN DEN BRUEL.

cholinestérasique. On a trouvé que les facteurs du métabolisme, tels que la conversion du Sumithion en Sumioxon et la dégradation de ces composés en substances non toxiques dans l'organisme animal, jouent un rôle assez peu important dans la toxicité sélective de cet insecticide organo-phosphoré.

Zusammenfassung *

Der Mechanismus der niedrigen Toxizität von Sumithion gegenüber Säugetieren

Es wurde angenommen, dass die niedrige Toxizität von Sumithion [O,O-Dimethyl-O-(3-methyl-4-nitrophenyl)-thiophosphat] gegenüber Säugetieren von der geringeren Durchdringung seines Sauerstoffanalogen, Sumioxon, ins Gehirn kommt und der darin erfolgenden Bildung von geringeren Mengen von Phosphor, welcher mit Proteinen zusammen ausfällt; die Hemmung der Cholinesteraseaktivität war hiervon abhängig. Es wurde gefunden, dass Metabolitfaktoren wie die Umwandlung von Sumithion zu Sumioxon und der Abbau dieser Verbindungen im Tierkörper zu nicht toxischen Substanzen eine ziemlich unbedeutende Rolle in der selektiven Toxizität dieses Organophosphorinsektizids spielen.

References

FUKAMI, J., and T. SHISHIDO: Studies on the selective toxicities of organophosphorus insecticides. Part I. Activation of ethyl parathion in mammal and insect. Botyu-Kagaku 28, 63 (1963).

HOLLINGWORTH, R. M., R. L. METCALF, and T. R. FUKUTO: The selectivity of Sumithion compared with methylparathion. Metabolism in the white mouse. J. Agr. Food Chem. 15, 242 (1967).

KADOTA, T.: Private communication (1967).

MIYAMOTO, J., Y. SATO, T. KADOTA, A. FUJINAMI, and M. ENDO: Studies on the mode of action of organophosphorus compounds. Part I. Metabolic fate of P32-labeled Sumithion and methylparathion in guinea pig and white rat. Agr. Biol. Chem. (Tokyo) 27, 381 (1963 a).

—— —— —— —— Studies on the mode of action of organophosphorus compounds. Part II. Inhibition of mammalian cholinesterase in vivo following the administration of Sumithion and methylparathion. Agr. Biol. Chem. (Tokyo) 27, 669 (1963 b).

—— Studies on the mode of action of organophosphorus compounds. Part III. Activation and degradation of Sumithion and methylparathion in mammals in vivo. Agr. Biol. Chem. (Tokyo) 28, 411 (1964 a).

—— Studies on the mode of action of organophosphorus compounds. Part IV. Penetration of Sumithion, methylparathion and their oxygen analogs into guinea pig brain and inhibition of cholinesterase in vivo. Agr. Biol. Chem. (Tokyo) 28, 422 (1964 b).

——, Y. SATO, K. YAMAMOTO, and S. SUZUKI: Activation and degradation of

* Übersetzt von A. SCHUMANN.

Sumithion, methylparathion and their oxygen analogs by mammalian enzymes in vitro. Botyu-Kagaku 33, (1968 a).

—— Unpublished observation (1968 b).

NEAL, R. A.: Studies on the metabolism of diethyl 4-nitrophenyl phosphorothioate (parathion) in vitro. Biochem. J. 103, 183 (1967).

PLAPP, F. W., and J. E. CASIDA: Hydrolysis of the alkyl-phosphate bond in certain dialkyl aryl phosphorothioate insecticides by rats, cockroaches, and alkali. J. Econ. Entomol. 51, 800 (1958).

VARDANIS, A., and L. G. CRAWFORD: Comparative metabolism of O,O-dimethyl O-p-nitrophenyl phosphorothioate (methylparathion) and O,O-dimethyl O-(3-methyl-4-nitrophenyl) phosphorothioate (Sumithion). J. Econ. Entomol. 57, 136 (1964).

Comparative mechanisms of insecticide binding
with nerve components of insects and mammals

by

Fumio Matsumura * and M. Hayashi *

Contents

I. Introduction

It is well established now that many modern insecticides owe their toxicities to their ability to attack the central nervous system of animals, yet it is surprising that the actual mechanisms through which the insecticides disrupt the normal function of the nervous system have not been fully understood with the exception of various inhibitors of important nerve enzymes such as the cholinesterases. This situation is particularly true with the insecticides that belong to the group of organochlorine compounds which are regarded as being almost unreactive from a chemical standpoint. There has never been a confirmed report that this type of compound can block any particular enzyme system which is vitally related to the cause of poisoning.

In studying the binding phenomena of insecticides with the nerve components we have selected three representative organochlorine insecticides, namely, DDT, gamma BHC, and dieldrin. In addition two insecticides, phthalthrin, a synthetic pyrethrin-like insecticide, and nicotine have been selected because of the fact that these compounds

* Department of Entomology, University of Wisconsin, Madison, Wisconsin. Supported by a PHS grant CC00252 from the National Communicable Disease Center, Atlanta, Georgia. Approved for publication by the Director of the Wisconsin Agricultural Experimental Station.

(*e.g.* pyrethrin and nicotine) have already been extensively studied in insects and mammals from pharmacological standpoints.

For an insecticide to be an active nerve poison, first it is necessary to reach the nerve tissue, and second it should be able physically to come in contact with the nerve components which play a vital role in the nerve function (*i.e.*, the target substance). Competition should exist among various neural substances for the insecticides, since most insecticidal materials are apolar and the nervous system is rich with lipid-containing substances. To circumvent the problems of spacial differences among various components, first it was decided to study the binding phenomenon with nerve homogenates at a low insecticide concentration under which condition the insecticide is expected to bind with the substances of high affinity.

II. Methods

The rat brain homogenate was made in 0.32M sucrose solution as described by Whittaker (1959) by using a 25-ml. Teflon-glass homogenizer (rotating at 1,000 rev./min.) at 0° C. To eight-ml. portions of this homogenate each radioactive insecticide was added with 10 μl. of acetone to make the final concentration of insecticide 1×10^{-5} M, and the system was maintained at 24° C. for ten minutes. The centrifugation procedures adopted were essentially those of Marchbanks *et al.* (1964) who modified the method of Whittaker (1959). Figure 1 schematically illustrates the centrifugal procedures followed in this

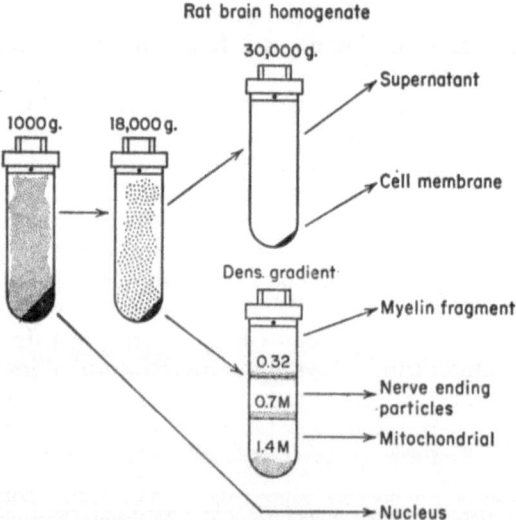

Fig. 1. Schematic illustration of the centrifugal method employed to separate various nerve components of the rat brain

study. The fractions were carefully separated and particulate portions were washed once. In the case of the fractions from the density gradient technique, this was done by diluting the sucrose solution to 0.32 M and by centrifuging and resuspending the sediment in fresh sucrose solution at 0.32 M.

The protein content of each fraction was measured by using Folin-Ciocalteu's method with crystalline bovine albumin in the same incubation medium as a standard. The amounts of insecticides distributed in each fraction were measured through radio-assay with a liquid scintillation counter.

III. Distribution of insecticides among mammalian nerve components

The Table I is a brief summary of the results from such binding experiments. As a whole all organochlorine insecticides exhibited a

Table I. Binding [a] of various insecticides with various nerve components of rat brain

Nerve component	Insecticide				
	Dieldrin	DDT	BHC	Phthalthrin	Nicotine
Supernatant [b]	1.00	1.12	4.98	3.74	18.26
Cell membrane	1.92	2.49	1.02	1.64	1.55
Myelin fragment	6.05	6.17	5.45	10.25	2.58
Nerve ending particles	4.60	5.73	1.81	0.16	3.62
Mitiochondrial	1.23	1.16	0.73	0.71	1.90
Nucleus	8.32	8.12	8.00	7.52	1.97

[a] Data expressed in mμmole of insecticide bound/mg. of protein.
[b] Including free insecticides.

similar binding pattern that could be distinguished from others, although the pattern shown by gamma BHC somewhat differed from those shown by DDT and by dieldrin, i.e., the former binding pattern was characterized by a high affinity to the supernatant and a low affinity to the nerve-ending particles, while the latter compounds showed a reverse tendency. The patterns were different with nicotine and phthalthrin. As compared with other insecticides, nicotine showed a relatively low binding capacity toward all particulate fractions as expected by the high water-solubility of this insecticide. It had, however, a moderately high affinity toward the nerve-ending particles. The binding characteristics of phthalthrin were rather similar to gamma BHC except that the former showed much higher affinity to the myelin sheath, and lower affinity to the nerve-ending particles than did the latter.

IV. Distribution of insecticides among insect nerve components

Studies with the insect nervous system were more difficult because there was no established method of fractionating various insect nerve components. Figure 2 illustrates the experimental approach we have adopted to establish such a fractionation method. Thus, obtained fractions were tentatively designated as shown in Figure 2 as a result

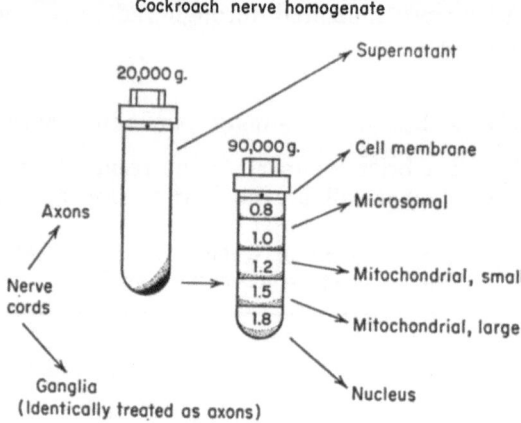

Fig. 2. Schematic illustration of the centrifugal method employed to separate various nerve components of the American cockroach

of lipid phosphorus and cholinesterase assay as well as microscopic observations. Each fraction was incubated with the corresponding insecticide as before. Table II indicates the results of the binding study

Table II. *Binding* [a] *of insecticides with various nerve components of axonic portions of roach nerve cords*

Axonic component	Insecticide				
	Dieldrin	DDT	BHC	Phthalthrin	Nicotine
Supernatant [b]	246.8	231.7	324.2	502.0	560.0
Cell membrane	55.8	72.8	47.4	66.7	25.9
Microsomal	59.2	21.6	6.4	15.4	3.2
Mitochondrial, small	37.2	25.9	1.5	12.5	2.5
Mitochondrial, large	53.4	66.5	1.1	6.3	1.7
Nucleus	34.2	66.0	0.6	3.3	3.2

[a] Data expressed in mμmole of insecticide bound/mg. of protein.
[b] Including free insecticides.

Table III. *Binding [a] of insecticides with various nerve components of ganglionic portions of roach nerve cords*

Ganglionic component	Insecticide				
	Dieldrin	DDT	BHC	Phthalthrin	Nicotine
Supernatant [b]	144.6	209.0	382.3	512.5	568.5
Cell membrane	151.2	104.8	59.6	62.1	10.4
Microsomal	99.6	40.2	6.1	9.7	4.0
Mitochondrial, small	58.3	40.2	1.6	4.6	0.1
Mitochondrial, large	27.5	45.6	0.6	4.9	0.5
Nucleus	21.0	42.2	0.5	3.9	1.4

[a] Data expressed in mμmole of insecticide bound/mg. of protein.
[b] Including free insecticides.

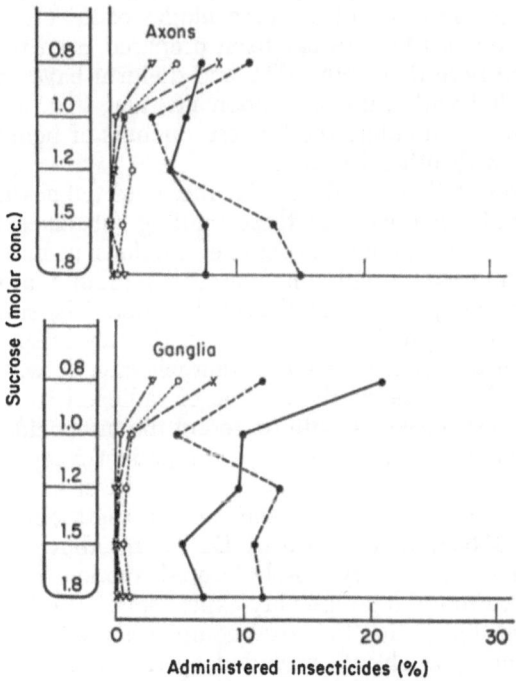

Fig. 3. Total distribution patterns of bound insecticides. The data are expressed in terms of percent of each compound recovered in each centrifugal fraction from the originally administered insecticide: ......o..... phthalthrin, —•—•∇•—•— nicotine, ——•—— dieldrin, ---- • ---- DDT, and — —×— — BHC

with components from the axonic portion of the roach nerve cord. It must be mentioned here that in the experiments with roach nerve cords the protein content of each fraction was much less than that of the rat fractions resulting in much higher binding values for the insect components than for the mammalian counterparts (10 mg. of neural material/ml. vs. 100 mg./ml). The total pattern of distribution of the insecticides was, however, similar to the case with the rat brain. For instance, gamma BHC and phthalthrin showed a high affinity toward the supernatant and the cell membrane fractions. Table III indicates the result of identical binding experiment with the ganglionic portion of the roach nerve cord. Dieldrin appeared to have a high affinity toward relatively small particles whereas no such preference pattern was observed with DDT. Contrary to the mammalian case, nicotine did not show a particularly high affinity toward particulate nerve components, i.e., only a modest binding capacity was observed toward small particles. All the above data have been presented in terms of the specific binding capacity/unit protein. To show the total distribution pattern in terms of the percentages of insecticides recovered from each fraction, Figure 3 has been prepared from the above data. It can be seen here that both DDT and dieldrin have an outstanding capacity to bind with various components, and that, as a whole, the cell membrane fraction had the highest amounts of bound insecticides as compared with other fractions.

The natures of these binding substances are unknown at this time. To study the characteristics of these binding substances in the insect nervous system, the above particulate fractions containing each insecticide were treated with one percent sodium taurocholate, and the system was subjected to brief centrifugation. The supernatant was passed through a Sephadex G-100 column (ANDREWS 1964) previously saturated with one percent sodium taurocholate. It was found (Fig. 4) that for each insecticide the portion of radioactivity appeared as a single peak. These peaks can still represent the insecticide-taurocholate complex rather than insecticide-nerve component complexes. To study this problem further, DDT was chosen as the test material since this was by far the best studied substance among the group (MATSUMURA and O'BRIEN 1966). In the case of DDT, the above two groups of complexes are eluted at very closely located fractions on the Sephadex G-100 column. From the preliminary experiment with Sephadex G-200, however, it was found that the latter complexes could be distinguished from the former (which was excluded by the column), and that the above eluate from the G-100 consisted of the latter complexes alone. The DDT-nerve component complexes are apparently macromolecules as judged by the retention volume of the G-200 column eluate. It was not clear, however, whether the above macromolecular fraction was composed of one or more components. An attempt was made, therefore, to analyze the above DDT-complexes in sodium taurocho-

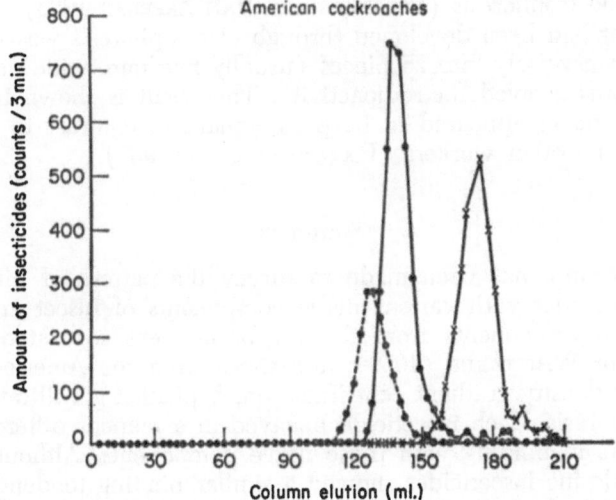

Fig. 4. Sephadex G-100 column elution patterns of bound insecticides. The bound insecticide was obtained by suspending the combined cell membrane fraction for each insecticide test in one percent of sodium taurocholate-buffer solution; it was then developed through a 51 to 52 cm. Sephadex G-100 column which had been saturated with the same one percent of sodium taurocholate-buffer medium: ——•—— dieldrin, ----•---- DDT, and ——×—— BHC

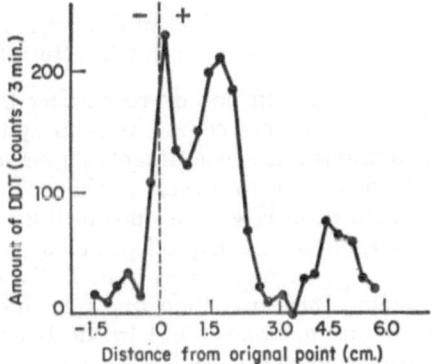

Fig. 5. Separation of DDT-complexes by agar gel electrophoresis. The purified fraction obtained by the Sephadex column treatment (see Fig. 4) in one percent of sodium taurocholate-buffer medium was electrophoretically developed. The agar plate was then cut into 25 pieces (each piece five mm. wide) perpendicular to the direction of development. Each agar piece was assessed for its radioactivity to estimate the amount of insecticide bound to the proteins which had migrated to the corresponding position of the agar plate

late on electrophoresis (VELTHUIS and VAN ASPEREN 1963). The agar
plate that had been developed through electrophoresis was then sec-
tioned transversely into 25 pieces (usually five mm. wide) and each
section was assayed for radioactivity. The result is shown in Figure
5. Three peaks appeared to be predominant in contrast to the case
reported by other workers (HATANAKA et al. 1967).

Summary

An attempt has been made to survey the pattern of binding of
five insecticides with various nerve components of insect and mammals. The components from the rat brain were separated by the
method of WHITTAKER (1959), and those from the American cockroach by density gradient centrifugation, Sephadex gel filtration, and
electrophoresis. Each insecticide behaved in a manner different from
the others in binding with those nerve components. Although three
organochlorine insecticides showed a similar binding tendency, DDT
and dieldrin had a much higher binding capacity with the nerve-
ending particles than did BHC, while BHC had a very high rate
of binding with soluble nerve components. Phthalthrin showed an
outstanding affinity toward the fraction containing myelin sheath materials, and nicotine showed a similar tendency toward the nerve-
ending particles.

Résumé *

Mécanismes comparatifs de la liaison des insecticides avec les composants du système nerveux chez les insectes et les mammifères

Une tentative a été faite en vue de rechercher le mode de liaison
de cinq insecticides avec divers composants du système nerveux des
insectes et des mammifères. Les composants du cerveau du rat furent
séparés par la méthode de WHITTAKER (1959), et ceux de la blatte
américaine par densité selon la vitesse de centrifugation, filtration sur
gel sephadex et électrophorèse. Chaque insecticide se comportait d'une
façon différente par rapport aux autres dans la liaison avec les composants de ces systèmes nerveux. Bien que trois insecticides organo-
chlorés présentaient une tendance similaire de liaison, le DDT et la
dieldrine ont montré un pouvoir de liaison avec les particules nerveuses
terminales beaucoup plus élevé que le HCH; par contre, le HCH a
révélé un pouvoir de liaison très élevé avec les composants solubles
du système nerveux. La phtalthrine présentait une affinité remarquable à l'égard de la fraction contenant des matériaux de gaine de
myéline, la nicotine une tendance analogue à l'égard des particules
nerveuses terminales.

---

* Traduit par S. DORMAL-VAN DEN BRUEL.

Zusammenfassung *

Vergleichende Mechanismen von Insektizidbindungen mit Nervenkomponenten von Insekten und Säugetieren

Es wurde der Versuch unternommen, die Form der Bindung von 5 Insektiziden mit verschiedenen Nervenkomponenten von Insekten und Säugetieren zu prüfen. Die Komponenten des Rattenhirns wurden nach der Methode von WHITTAKER (1959) getrennt, die der Amerikanischen Küchenschabe durch Dichtegradienten-Zentrifugierung, Sephadex Gelfiltration und Elektrophorese. Jedes Insektizid verhielt sich in seiner Art unterschiedlich von den andern im Verbinden mit diesen Nervenkomponenten. Obwohl drei der Chlorkohlenwasserstoffinsektizide eine ähnlicher Bindungstendenz zeigten, hatten DDT und Dieldrin eine viel höhere Bindungskapazität mit den Nervenendenteilen als BHC, während BHC eine sehr hohe Bindungsrate mit löslichen Nervenkomponenten hatte. Phthalthrin zeigte eine aussergewöhnliche Affinität gegenüber der Fraktion, welche Markscheidematerialien enthielt, und Nikotin zeigte eine ähnliche Tendenz gegenüber den Nervenendenteilen.

References

ANDREWS, P.: Estimation of the molecular weights of proteins by Sephadex gel filtration. Biochem. J. 72, 694 (1959).

HATANAKA, A., B. D. HILTON, and R. D. O'BRIEN: The apparent binding of DDT to tissue components. J. Agr. Food Chem. 15, 854 (1967).

MARCHBANKS, R. M., F. ROSENBLATT, and R. D. O'BRIEN: Serotonin binding to nerve-ending particles of the rat brain and its inhibition by lysergic acid diethylamide. Science 144, 1135 (1964).

MATSUMURA, F., and R. D. O'BRIEN: Absorpton and binding of DDT by the central nervous system of the American cockroach. J. Agr. Food Chem. 14, 36 (1966).

VELTHUIS, H. H. W., and K. VAN ASPEREN: Occurrence and inheritance of esterases in Musoa domestica. Entomol. Expt. Appl. 6, 79 (1963).

WHITTAKER, V. P.: The isolation and characterization of acetylcholine-containing particles from brain. Biochem. J. 72, 694 (1959).

* Übersetzt von A. SCHUMANN.

Mode of action of DDT and allethrin on nerve: Cellular and molecular mechanisms

by

Toshio Narahashi [*]

Contents

I. Introduction

In spite of a wealth of knowledge accumulated concerning the biochemical aspects of the mode of action of insecticides, little is known about the mechanism whereby the function of the nervous system, which is the major site of action of most insecticides, is affected and impaired. The only system that has been studied rather extensively in connection with the neurotoxic action of insecticides is cholinesterase. Since the bioelectric signal is the only parameter observable during nerve excitation, and also since the excitation does not directly depend on the metabolism in nerve as will be described later, electrophysiological experiments are most rewarding to elucidate the mechanism of action of insecticides on the nervous tissue.

[*] Department of Physiology and Pharmacology, Duke University Medical Center, Durham, North Carolina.

275

The electrophysiological approach is useful at least in three respects as far as the study of the mode of action of insecticides is concerned: a) The site of action of an insecticide may be located, if it acts on the nerve or muscle tissue. b) Bioelectric potential changes may be observed as indices to compare the sensitivity of nerve to an insecticide. c) The cellular or even molecular mechanism of action on nerve or muscle may be studied by advanced electrophysiological techniques.

There are many examples of the category (a) above. For instance, DDT stimulates the nerve to produce repetitive discharges, and the sensory nervous system is generally most sensitive to this action of DDT (LALONDE and BROWN 1954, ROEDER and WEIANT 1948, YAMASAKI and ISHII 1954 a). BHC stimulates the central nervous system to produce synaptic after-discharges (YAMASAKI and ISHII 1954 d), and many organophosphorus insecticides affect synaptic transmission through the inhibition of cholinesterase (NARAHASHI and YAMASAKI 1960 a, YAMASAKI and NARAHASHI 1960).

The study of category (b) above involves the comparison of the nerve sensitivity to insecticides between susceptible and resistant strains of insects. For example, in some DDT-resistant strains of houseflies, the nerve is less sensitive to DDT than that of normal susceptible strains, and this is controlled by a recessive gene located on the second chromosome (TSUKAMOTO et al. 1965). The sensitivity of the cockroach nerve to the action of DDT was compared at high and low temperatures, and this factor was found to be one of the major contributions to the negative temperature coefficient of the insecticidal action of DDT (YAMASAKI and ISHII 1954 b).

Attempts have also been made to elucidate the mechanism of action of insecticides at the cellular level [category (c) above]. For example, by means of intracellular microelectrode techniques, it was suggested that DDT partially inhibits the mechanism by which the potassium conductance of the nerve membrane is increased upon stimulation and the mechanism by which the increased sodium conductance of the nerve membrane is turned off (NARAHASHI 1960 and 1963, NARAHASHI and YAMASAKI 1960 b and c). The inhibition of the increases in membrane sodium and potassium conductances was suggested to be responsible for the nerve blockage by allethrin (NARAHASHI 1965).

Three major questions arise concerning the cellular or molecular mechanism of action of drug: 1) How does the drug act? 2) Which form of the drug is effective in case it is dissociated into a charged and uncharged form? Or, which group of the drug molecule is responsible for the action? 3) Where is the site of action? Is it on the external or internal surface of the nerve membrane, or on the inside of the membrane?

The present article is primarily concerned with question (1) above. In order to interpret the mechanisms of action of DDT and allethrin in terms of nerve membrane conductances to sodium and potassium, voltage-clamp experiments have been performed. This technique is one of the most powerful and straightforward electrophysiological approaches to the mode of action of drugs on the nerve. In addition, the possible sites of action of DDT and allethrin in the nerve membrane are discussed in an attempt to clear the road to the molecular aspects of the insecticidal action.

The detailed experimental results have been published elsewhere (NARAHASHI and ANDERSON 1967, NARAHASHI and HAAS 1967 and 1968).

II. Principle of approach

a) Mechanism of nerve excitation

Because of concentration gradients of sodium and potassium across the nerve membrane and because of the selective permeability of the membrane to potassium at resting conditions, the resting membrane potential approaches the equilibrium potential for potassium,

$$E_K = \frac{RT}{F} \ln \frac{[K]_o}{[K]_i}$$

where R, T, and F are the gas constant, the absolute temperature, and the Faraday constant, respectively, and $[K]_o$ and $[K]_i$ are the external and internal potassium concentrations. The resting potential is of the order of 50 to 100 mv, the inside being negative with respect to the outside. When the membrane is depolarized, the conductance of the membrane to sodium increases rapidly allowing the membrane potential to approach the sodium equilibrium potential,

$$E_{Na} = \frac{RT}{F} \ln \frac{[Na]_o}{[Na]_i}$$

where $[Na]_o$ and $[Na]_i$ are the external and internal sodium concentrations, respectively. This forms the rising phase of an action potential. The increased sodium conductance soon starts decreasing and the potassium conductance now starts increasing, so that the membrane potential returns to the resting level. Figure 1 shows this process (HODGKIN and HUXLEY 1952).

Owing to these ionic conductance changes, sodium ions enter the nerve fiber down the electrochemical gradient during the rising phase of the action potential, while potassium ions exit during the falling phase. The gained sodium is pumped out and the lost potassium is absorbed by the metabolic energy after the excitation. It should be emphasized that the metabolic energy contributes only indirectly to the excitation phenomenon in the form of the "sodium pump" men-

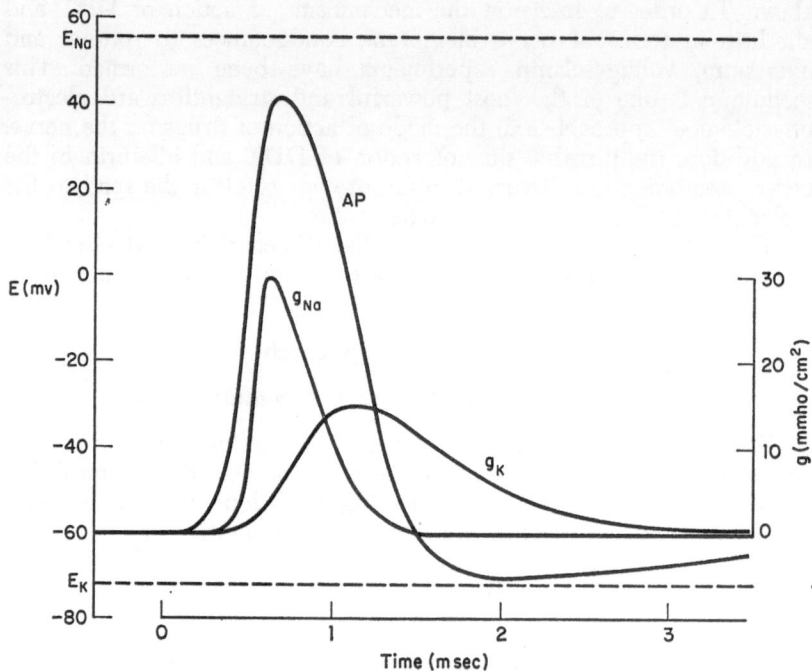

Fig. 1. Schematic representation of the nerve excitation: AP, action potential; E, membrane potential; E_K, potassium equilibrium potential; E_{Na}, sodium equilibrium potential; g, membrane conductance; g_K, membrane potassium conductance; g_{Na}, membrane sodium conductance; and t, time (adapted from HODGKIN and HUXLEY 1952)

tioned above. The production of action potential does not require such energy.

b) Voltage-clamp study

In order to measure the membrane conductances to sodium and potassium, the voltage clamp is most convenient and straightforward. The idea is to observe ionic currents flowing across the nerve membrane when the membrane potential is suddenly shifted from one value to another and held at the latter level by means of an electronically controlled feed-back circuit. The membrane current associated with a moderate depolarization consists of an inward peak transient current followed by an outward steady-state current (see Figs. 3 and 4). The transient current is carried mostly by sodium, while the steady-state current is carried mostly by potassium. These currents are plotted as a function of membrane potential to obtain current-voltage relations, from which sodium and potassium conductances can be calculated.

III. Methods

Giant nerve fibers of the squid, *Loligo pealii*, and of the lobster, *Homarus americanus*, were used as material. The diameter of the axons was about 80 μ in the lobster and about 400 to 500 μ in the squid. The resting and action potentials were recorded by means of capillary microelectrodes filled with $3M$ potassium chloride. The sucrose-gap voltage-clamp technique was essentially the same as that described previously (JULIAN *et al.* 1962 b, MOORE *et al.* 1967 b). The method of internal perfusion with squid giant axons was also described previously (NARAHASHI and ANDERSON 1967, NARAHASHI *et al.* 1967 a).

IV. Results

a) DDT

1. Effect on nerve excitability and action potential.—DDT has long been known to exert a dual action on nerve. It causes the nerve fibers and synapses to produce repetitive discharges, and this action is directly responsible for the symptoms of poisoning in insects, *i.e.*, hyperexcitability and convulsion (GORDON and WELSH 1948, LALONDE and BROWN 1954, NARAHASHI 1964, NARAHASHI and YAMASAKI 1960 b, ROEDER and WEIANT 1948, WELSH and GORDON 1947, YAMASAKI and ISHII 1952 a and 1954 a, b, and c, YAMASAKI and NARAHASHI 1958 and 1962). DDT also increases the negative after-potential of nerve; this effect was first observed with the crab nerve by SHANES (1949) and with the cockroach nerve by YAMASAKI and the present author (1952 b). The increased negative after-potential is at least partly responsible for the repetitive discharges in the DDT-poisoned nerve.

Further analyses by means of conventional external electrode and intracellular microelectrode techniques have led to the hypothesis that the inhibition of the mechanism whereby the sodium conductance is turned off or the so-called "sodium-inactivation" or the inhibition of the mechanism whereby the potassium conductance is turned on or the inhibition of both mechanisms is responsible for the large negative after-potential in the DDT-poisoned nerve fibers (NARAHASHI 1960 and 1963, NARAHASHI and YAMASAKI 1960 b and c, YAMASAKI and NARAHASHI 1957 a and b).

Figure 2 illustrates the action potentials from the normal and DDT-treated lobster giant axons under sucrose-gap conditions. Because of the hyperpolarization caused by sucrose (JULIAN *et al.* 1962 a, BLAUSTEIN and GOLDMAN 1966), the apparent resting potentials were high. In the DDT-poisoned axon, although the action potential rose as quickly as in the normal axon, it decayed very slowly forming a plateau phase. The plateau phase usually lasted for several hundred milliseconds resembling a cardiac action potential.

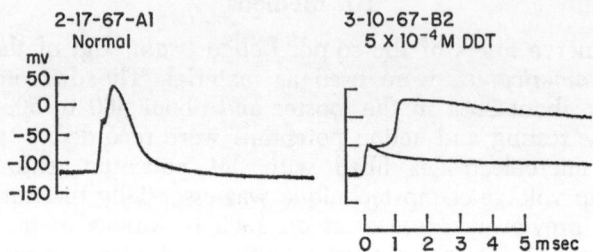

Fig. 2. Action potentials from a normal and a DDT-treated lobster giant axon
under sucrose-gap conditions. A subthreshold response is superimposed
on the action potential in the DDT-treated axon. Because of the hyperpolarization by sucrose, the apparent resting potentials are higher than
the normal. Note the falling phase of the action potential is greatly
slowed by DDT

2. Effect on membrane currents under voltage-clamp conditions.—
The schematic drawings of the membrane currents associated with
step depolarizations in the normal and DDT-poisoned axons are illus-

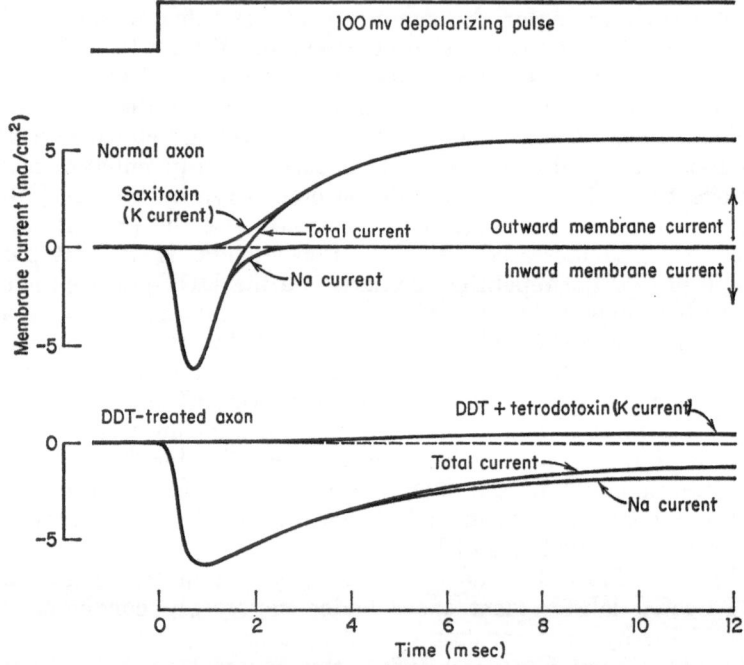

Fig. 3. Tracings of the membrane ionic currents from a normal and a DDT-
treated lobster giant axon under voltage-clamp conditions. The top set
represents the step depolarization applied. Note that the sodium current
is turned off very slowly in the DDT-treated axon. For further explanation, see text

trated in Figure 3. Two kinds of toxins, tetrodotoxin (TTX) and saxitoxin (STX), were used here. Since both toxins selectively block the peak transient sodium component of membrane current without affecting the late steady-state potassium component (NARAHASHI et al. 1964 and 1967 b, TAKATA et al. 1966, MOORE et al. 1967 b, NAKA-MURA et al. 1965), they are very useful to dissociate the membrane current into the two components. It should also be mentioned here that leakage currents have been subtracted in drawing the scheme.

In the normal axon, there appeared an inward current upon step depolarization. This inward current lasted only for about one millisecond and was followed by a steady-state outward current. The latter current kept flowing as long as the nerve membrane was held depolarized. When STX was applied at a concentration of 3 x $10^{-7}M$, the inward transient component was completely abolished, leaving the outward steady-state component intact. Hence the latter represents the potassium current. The portion of the current that has disappeared after application of STX represents the sodium current.

In contrast to the transient time course of the sodium current in the normal axon, the sodium current in the DDT-poisoned axon was turned off very slowly (Fig. 3, bottom set). In the DDT-poisoned axon, the inward current started flowing upon step depolarization, and was followed by a steady-state *inward* current instead of outward current. The steady-state inward current cannot be due to the potassium flowing inwardly across the nerve membrane, because the potassium concentration is high inside and low outside. It is most likely that this represents a residual component of the sodium current which is turned off slowly. The application of TTX has proved that this is actually the case. When $3 \times 10^{-7}M$ TTX was applied to the DDT-poisoned axon, both the initial transient phase and the late steady-state phase of the inward current started decreasing in magnitude. The inward steady-state current was finally converted into a small outward steady-state current, while the initial transient component completely disappeared. The small outward steady-state current represents the potassium current, and the portion of the current that has been abolished by the application of TTX represents the sodium current. Thus it can be concluded that DDT slows the turning-off process of the sodium conductance and inhibits the turning-on process of the potassium conductance.

b) Allethrin

1. **Effect on nerve excitability and action potential.**—At relatively low concentrations, allethrin increases and prolongs the negative after-potential. Repetitive after-discharges are often superimposed on the negative after-potential. At higher concentrations, however, it blocks the nerve conduction (NARAHASHI 1962 a and b). From microelectrode experiments it was suggested that the conduction block is due

to the inhibition of both sodium and potassium conductance changes (NARAHASHI 1965).

Although the squid giant axons were slightly less sensitive to the blocking action of allethrin than the cockroach giant axons, the effects from the outside of the nerve were generally the same for both axons. However, when applied from the inside of the squid axon under the sucrose-gap conditions, allethrin increased the negative after-potential very effectively, eventually producing a plateau phase resembling a cardiac action potential. In order to interpret these effects of allethrin in terms of ionic conductances of the membrane, voltage-clamp experiments have been performed.

2. **Effect on membrane currents under voltage-clamp conditions.—** The middle set of Figure 4 represents the membrane currents before

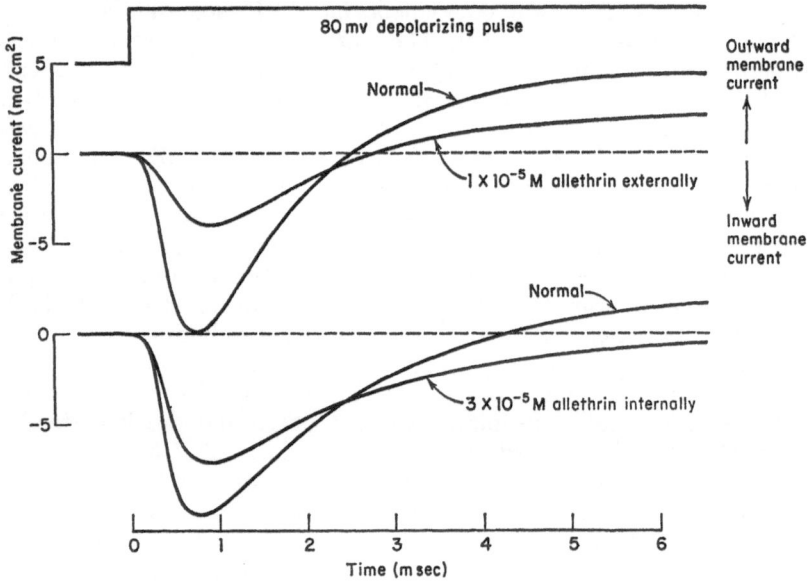

Fig. 4. Tracings of the membrane ionic currents from squid giant axons before and after application of allethrin externally (middle set) or internally (bottom set). The top set represents the step depolarization applied. Note that both the sodium current and the potassium current are inhibited by allethrin and that the outward steady-state current is converted into an inward current by internal application of allethrin. For further information, see text

and after application of $1 \times 10^{-5}M$ allethrin to the outside of a squid giant axon. It is clear that allethrin inhibits both the peak transient sodium current and the late steady-state potassium current. However, when applied inside of the nerve membrane, allethrin exerts an additional effect which is shown in the bottom set of Figure 4. After internal application of 3 x $10^{-5}M$ allethrin, the outward steady-state current was converted into an inward current. Although the record in Figure 4 does not show the time course of the steady-state current beyond 6.5 milliseconds, other experiments in which longer pulses of depolarization are applied show that the steady-state current is still flowing in inward direction some 15 milliseconds after the onset of the pulse. Although no experiment was made with the additional application of TTX or STX as in the DDT experiments, it is reasonable to assume that the inward steady-state current under the influence of internally applied allethrin is due to the residual component of the sodium current which is turned off very slowly.

V. Discussion

a) Ionic mechanisms of the actions of DDT and allethrin

Since the turning-off process of the sodium conductance or the so-called sodium inactivation and the turning-on process of the potassium conductance are responsible for the falling phase of the action potential, the delayed falling phase of the action potential observed in the DDT-poisoned axons can well be interpreted on the basis of the inhibitions of these conductance components observed under voltage-clamp conditions. This conclusion unequivocally supports the previous hypothesis based on the conventional microelectrode experiments (NARAHASHI 1960 and 1963, NARAHASHI and YAMASAKI 1960 b and c, YAMASAKI and NARAHASHI 1957 a and b).

The action potential prolonged by internal application of allethrin can also be interpreted on the same basis. The present voltage-clamp experiments demonstrate that the turning-on process of both sodium and potassium conductances are inhibited by allethrin. This explains the conduction block and also supports the previous hypothesis (NARAHASHI, 1965).

b) Interactions of DDT and allethrin with the nerve membrane channels

The nerve membrane may be assumed to have channels through which ions flow at rest or during activity. However, it should be emphasized that the term channel does not necessarily mean an anatomical structure but simply refers to a conceptual pathway.

There are two examples of studies in which the site of action of drug in the nerve membrane has been located. Since lipid-insoluble TTX has virtually no effect on the action potential or sodium current when applied to the inside of the nerve membrane, its site of action is located on the external surface of the nerve membrane (NARAHASHI et al. 1966 and 1967 a). The sodium channel can be visualized as having a gate on the external surface of the membrane, and one TTX molecule may plug one sodium channel at its gate (Moore *et al.* 1967 a). Evidence has also been obtained that TTX acts in cationic forms rather than in a zwitterion form (Camougis *et al.* 1967).

Contrariwise, the experiments in which local anesthetic derivatives are applied to the outside or inside of the nerve membrane at different pH values suggest that the local anesthetics act on the internal surface of the nerve membrane in charged forms (Narahashi and Frazier 1968). Tetraethylammonium (TEA) ions are also known to act on the internal surface of the squid nerve membrane to block the potassium current selectively (Armstrong and Binstock 1965, Tasaki and Hagiwara 1957).

It seems likely that the mechanism that controls the turning-off process of the sodium channel may be located near the internal surface of the nerve membrane, because this process is highly sensitive to changes in ionic composition of internal solution or to drugs applied internally (Adelman and Senft 1966, Adelman *et al.* 1965 a and b, Chandler and Meves 1966).

However, DDT and allethrin are entirely different from TTX, local anesthetics, and TEA in two respects: a) the insecticides are not in charged forms, whereas TTX and local anesthetics are dissociated into cations depending on pH, and TEA is permanently charged; b) the insecticides are highly soluble in oil or lipid, whereas TTX and TEA are not lipid-soluble and local anesthetics are generally permeable to the lipid membrane only in uncharged forms. Owing to the lipid solubility of these insecticides, it is difficult to locate their sites of action in the nerve membrane. However, it is reasonable to assume that DDT and allethrin interact with the channels of the membrane in the manner quite different from TTX, local anesthetics, and TEA, because the latter three act in charged forms and, hence, the electrostatic binding is most probably the major initial force for the interaction. It seems possible for the insecticides to penetrate the lipid nerve membrane and exert their channel blocking action through the lateral pressure. Alternatively, it is suggested that the target component where the channel mechanisms are located is protein rather than phospholipids in view of the recent finding on the possible role of protein in the excitation phenomenon (Mueller and Rudin 1967).

Acknowledgment

This study was supported by grants from the National Institutes of Health (NB 06855 and NB 03437) and from the Duke Endowment (84-0954).

Summary

The usefulness of electrophysiological approach to the mode of action of insecticides is described and classified. Voltage-clamp experiments have demonstrated that DDT slows the turning-off process of the initial transient sodium conductance increase of the nerve membrane and inhibits the turning-on process of the late steady-state potassium conductance increase. These effects explain the prolonged action potential in the DDT-poisoned axons. Allethrin inhibits the turning-on processes of both the sodium and the potassium conductances. In addition, allethrin slows the turning-off process of the sodium conductance increase when applied from the inside of the nerve membrane.

Résumé *

Mode d'action du zeidane et de l'allethrine sur le nerf: mécanismes cellulaires et moléculaires

L'utilité de l'étude électrophysiologique du mode d'action des insecticides est décrite et discutée. Des expériences de conductibilité ont démontré que le zeidane ralentit le processus d'arrêt de l'augmentation initiale provisoire de la conductibilité du sodium de la membrane nerveuse et inhibe l'ouverture du processus de l'augmentation finale de la conductibilité normale du potassium.

Ces effets expliquent l'effet retard constaté dans les axones intoxiqués au zeidane.

L'allethrine inhibe simultanément les processus d'ouverture des conductibilités du sodium et du potassium. De plus, l'allethrine ralentit le processus de coupure de l'augmentation de la conductibilite du sodium lorsqu'elle est appliquée à l'intérieur de la membrane nerveuse.

Zusammenfassung **

Die Wirkungsweise von DDT und Allethrin auf die Nerven: zellulare und molekulare Mechanismen

Die Brauchbarkeit der elektrophysiologischen Methode zur Wirkungsweise von Insektiziden ist beschrieben und klassifiziert. "Voltage-clamp"—Experimente haben gezeigt, dass DDT den Ausschaltprozess der ursprünglichen momentanen Natriumleitungsfähigkeitszunahme der Nervenmembrane verlangsamt und den Einschaltprozess der verspäteten "steady-state" Kaliumleitungsfähigkeitszunahme hemmt. Diese Wirkungen erklären das verlängerte Wirkungspotential in den mit

---

* Übersetzt von A. Schumann.
** Traduit par R. Mestres.

DDT vergifteten Axonen. Allethrin hemmt den Einschaltprozess von sowohl der Natrium- als auch der Kaliumleitungsfähigkeit. Ausserdem verlangsamt Allethrin den Ausschaltprozess der Natriumleitungsfähigkeitszunahme, wenn es von der Innenseite der Nervenmembrane appliziert wird.

References

ADELMAN, W. J., JR., and J. P. SENFT: Voltage clamp studies on the effect of internal cesium ion on sodium and potassium currents in the squid giant axon. J. Gen. Physiol. 50, 279 (1966).

——, F. M. DYRO, and J. SENFT: Long duration responses obtained from internally perfused axons. J. Gen. Physiol. 48 (No. 5, Pt. 2), 1 (1965).

—— —— —— Prolonged sodium currents from voltage clamped internally perfused squid axons. J. Cell. Comp. Physiol. 66 (No. 3, Suppl. 2), 55 (1965).

ARMSTRONG, C. M., and L. BINSTOCK: Anomalous rectification in the squid giant axon injected with tetraethylammonium chloride. J. Gen. Physiol. 48, 859 (1965).

BLAUSTEIN, M. P., and D. E. GOLDMAN: Origin of axon membrane hyperpolarization under sucrose-gap. Biophys. J. 6, 453 (1966).

CAMOUGIS, G., B. H. TAKMAN, and J. R. P. TASSE: Potency difference between the zwitterion form and the cation forms of tetrodotoxin. Science 156, 1625 (1967).

CHANDLER, W. K., and H. MEVES: Incomplete sodium inactivation in internally perfused giant axons from Loligo forbesi. J. Physiol. 186, 121P (1966).

GORDON, H. T., and J. H. WELSH: The role of ions in axon surface reactions to toxic organic compounds. J. Cell. Comp. Physiol. 31, 395 (1948).

HODGKIN, A. L., and A. F. HUXLEY: A quantitative description of membrane current and its application to conduction and excitation in nerve. J. Physiol. 117, 500 (1952).

JULIAN, F. J., J. W. MOORE, and D. E. GOLDMAN: Membrane potentials of the lobster giant axon obtained by use of the sucrose-gap technique. J. Gen. Physiol. 45, 1195 (1962 a).

—— —— —— Current-voltage relations in the lobster giant axon membrane under voltage clamp conditions. J. Gen. Physiol. 45, 1217 (1962 b).

LALONDE, D. I. V., and A. W. A. BROWN: The effect of insecticides on the action potentials of insect nerve. Canad. J. Zool. 32, 74 (1954).

MOORE, J. W., T. NARAHASHI, and T. I. SHAW: An upper limit to the number of sodium channels in nerve membrane? J. Physiol. 188, 99 (1967 a).

——, M. P. BLAUSTEIN, N. C. ANDERSON, and T. NARAHASHI: Basis of tetrodotoxin's selectivity in blockage of squid axons. J. Gen. Physiol. 50, 1401 (1967 b).

MUELLER, P., and D. O. RUDIN: Action potential phenomena in experimental bimolecular lipid membranes. Nature 213, 603 (1967).

NAKAMURA, Y., S. NAKAJIMA, and H. GRUNDFEST: The action of tetrodotoxin on electrogenic components of squid giant axons. J. Gen. Physiol. 48, 985 (1965).

NARAHASHI, T.: Excitation and electrical properties of giant axon of cockroaches. In: Electrical activity of single cells, ed. Y. Katsuki, p. 119. Tokyo: Igakushoin (1960).

—— Effect of the insecticide allethrin on membrane potentials of cockroach giant axons. J. Cell. Comp. Physiol. 59, 61 (1962 a).

—— Nature of the negative after-potential increased by the insecticide allethrin in cockroach giant axons. J. Cell. Comp. Physiol. 59, 67 (1962 b).

—— The properties of insect axons. In: Advances in insect physiology, ed. J. W. L. Beament et al., Vol. 1, p. 175. London and New York: Academic Press (1963).

—— Insecticide resistance and nerve sensitivity. Japan J. Med. Sci. Biol. 17, 46 (1964).

—— The physiology of insect axons. In: The physiology of the insect central nervous system, ed. J. E. Treherne and J. W. L. Beament, p. 1. London and New York: Academic Press (1965).

——, and N. C. ANDERSON: Mechanism of excitation block by the insecticide allethrin applied externally and internally to squid giant axons. Toxicol. Applied Pharmacol. 10, 529 (1967).

——, and D. T. FRAZIER: Site of action and active form of local anesthetics in nerve fibers. Fed. Proc. 27, 408 (1968).

——, and H. G. HAAS: DDT: Interaction with nerve membrane conductance changes. Science 157, 1438 (1967).

—— —— Interaction of DDT with the components of lobster nerve membrane conductance. J. Gen. Physiol. 50, 177 (1968).

——, and T. YAMASAKI: Nervous and cholinesterase activities in the cockroach as affected by demeton and methyldemeton. Japan. J. Applied Entomol. Zool. 4, 64 (1960 a).

—— —— Mechanism of increase in negative after-potential by dicophanum (DDT) in the giant axons of the cockroach. J. Physiol. 152, 122 (1960 b).

—— —— Behaviors of membrane potential in the cockroach giant axons poisoned by DDT. J. Cell. Comp. Physiol. 55, 131 (1960 c).

——, N. C. ANDERSON, and J. W. MOORE: Tetrodotoxin does not block excitation from inside the nerve membrane. Science 153, 765 (1966).

—— —— —— Comparison of tetrodotoxin and procaine in internally perfused squid giant axons. J. Gen. Physiol. 50, 1413 (1967 a).

——, H. G. HAAS, and E. F. THERRIEN: Saxitoxin and tetrodotoxin: Comparison of nerve blocking mechanism. Science 157, 1441 (1967 b).

——, J. W. MOORE, and W. R. SCOTT: Tetrodotoxin blockage of sodium conductance increase in lobster giant axons. J. Gen. Physiol. 47, 965 (1964).

ROEDER, K. D., and E. A. WEIANT: The effect of DDT on sensory and motor structures in the cockroach leg. J. Cell. Comp. Physiol. 32, 175 (1948).

SHANES, A. M.: Electrical phenomena in nerve. II. Crab nerve. J. Gen. Physiol. 33, 75 (1949).

TAKATA, M., J. W. MOORE, C. Y. KAO, and F. A. FUHRMAN: Blockage of sodium conductance increase in lobster giant axon by tarichatoxin (tetrodotoxin). J. Gen. Physiol. 49, 977 (1966).

TASAKI, I., and S. HAGIWARA: Demonstration of two stable potential states in the squid giant axon under tetraethylammonium chloride. J. Gen. Physiol. 40, 859 (1957).

TSUKAMOTO, M., T. NARAHASHI, and T. YAMASAKI: Genetic control of low nerve sensitivity to DDT in insecticide-resistant houseflies. Botyu-Kagaku (Scientific Pest Control) 30, 128 (1965).

WELSH, J. H., and H. T. GORDON: The mode of action of certain insecticides on the arthropod nerve axon. J. Cell. Comp. Physiol. 30, 147 (1947).

YAMASAKI, T., and T. ISHII [1]: Studies on the mechanism of action of insecticides. IV. The effects of insecticides on the nerve conduction of insect. Oyo-Kontyu (J. Nippon Soc. Applied Entomol.) 7, 157 (1952 a).

—— —— Studies on the mechanism of action of insecticides. V. The effects of DDT on the synaptic transmission in the cockroach. Oyo-Kontyu (J. Nippon Soc. Applied Entomol.) 8, 111 (1952 b).

---

[1] Former name of T. Narahashi.

—— —— Studies on the mechanism of action of insecticides. VII. Activity of neuron soma as a factor of development of DDT symptoms in the cockroach. Botyu-Kagaku (Scientific Insect Control) 19, 1 (1954 a).

—— —— Studies on the mechanism of action of insecticides. VIII. Effects of temperature on the nerve susceptibility to DDT in the cockroach. Botyu-Kagaku (Scientific Insect Control) 19, 39 (1954 b).

—— —— Studies on the mechanism of action of insecticides. IX. Repetitive excitation of the insect neuron soma by direct current stimulation and effects of DDT. Japan. J. Applied Zool. 19, 16 (1954 c).

—— —— Studies on the mechanism of action of insecticides. X. Nervous activity as a factor of development of γ-BHC symptom in the cockroach. Botyu-Kagaku (Scientific Insect Control) 19, 106 (1954 d).

——, and T. Narahashi: Increase in the negative after-potential of insect nerve by DDT. Studies on the mechanism of action of insecticides, XIII. Botyu-Kagaku (Scientific Insect Control) 22, 296 (1957 a).

—— —— Intracellular microelectrode recordings of resting and action potentials from the insect axon and the effects of DDT on the action potential. Studies on the mechanism of action of insecticides. XIV. Botyu-Kagaku (Scientific Insect Control) 22, 305 (1957 b).

—— —— Resistance of houseflies to insecticides and the susceptibility of nerve to insecticides. Studies on the mechanism of action of insecticides. XVII. Botyu-Kagaku (Scientific Insect Control) 23, 146 (1958).

—— —— Synaptic transmission in the last abdominal ganglion of the cockroach. J. Insect Physiol. 4, 1 (1960).

—— —— Nerve sensitivity and resistance to DDT in houseflies. Japan. J. Applied Entomol. Zool. 6, 293 (1962).

Biochemical genetics of insecticide resistance in the housefly

by

MASUHISA TSUKAMOTO [*]

Contents

I. Introduction

The process of evolution is usually so slow that we cannot observe it within a short period. However, development of insecticide resistance in insects after extensive uses of insecticides is an example of rapid microevolution caused by artificial selections. The modern theory of evolution is based on mutation-selection mechanism and changes of gene frequencies in a Mendelian population. Thus the development of resistance to this insecticide and its derivatives but also population during and after selection pressures with an insecticide have to be considered from such a standpoint of the population genetics.

---

[*] Institute for Tropical Medicine, Nagasaki University, Nagasaki, Japan.

Selections of insects by an insecticide result not only in the development of resistance to this insecticide and is derivatives but also often in changes of morphological, ecological, biochemical, or physiological characters, and they may depend on each genetic factor. On the basis of comparison of such differences between insecticide resistant and susceptible strains of an insect, various plausible explanations for mechanism of the resistance have been proposed by many investigators. Some of these characters may be an expression of protective resistance mechanisms. Some of them, however, may be a concomitant or indirect phenomenon accompanied with some resistance mechanism during the course of selection for the resistance. So-called cross resistance may also be considered under the similar situation. Therefore, it is necessary to distinguish and/or isolate the "true" resistance mechanism from others. For example, chemists usually have to purify a single compound from its crude material before measuring the chemical and physical properties or biological activity, because the contamination of even small amounts of impurities often misleads them to a meaningless conclusion. In the case of insecticide resistance problem, such a "purification" of heterogeneous insect materials is also very important. By usual toxicological techniques based on the comparison of resistant and susceptible strains, it is not easy to distinguish which one is truly due to the resistance mechanism or which one is due to an indirect influence of the resistance mechanism.

Among various suspect mechanisms of resistance, metabolism of an insecticide to a less toxic compound or compounds is usually considered to be the most important factor and, indeed, extensive studies have been done along this direction. In most cases, resistant strains have higher detoxication rates than those in susceptible strains under *in vivo* conditions. Such an *in vivo* experiment, however, cannot distinguish whether the higher metabolic rate in resistant strains is a cause or result of resistance. On the contrary, it seems more likely that the higher enzyme activity for metabolism of a given insecticide *in vitro* is a reflection of major causes of the resistance. *In vitro* enzyme activity by insect materials is sometimes too low to account for resistance level or *in vivo* metabolism, and some investigators are doubtful to ascribe the higher metabolic rate of an insecticide to a cause of resistance. Of course, other resistance mechanisms than metabolism should also be considered in such cases. Improvements of biochemical techniques, however, often may give rise to a good result for demonstrating very active and stable enzyme activity with insect enzyme preparations.

In earlier studies on the mode of inheritance of resistance with moderately resistant strains, the reported results were often complicated, but recent investigations indicate that a high level of resistance is controlled by a relatively simpler genetic system than that was assumed before. However, too many important questions still remain

unanswered on over-all aspects of relationships from a resistance gene
(a local coding pattern of the DNA molecule) to an observable terminal phenomenon (death or survival). Figure 1 is a schematic illus-

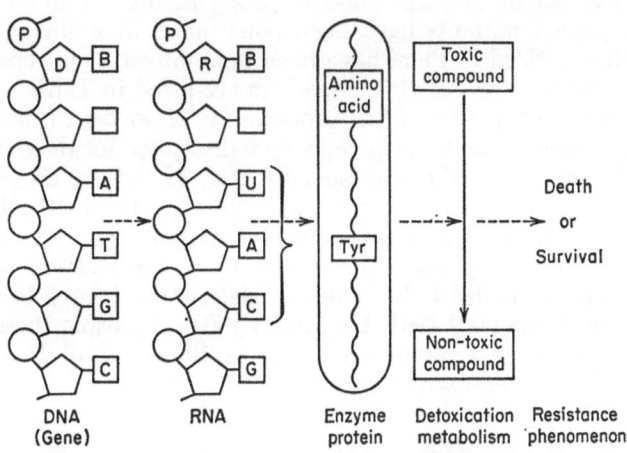

Fig. 1. Schematic view of genetic control of detoxication as a causal mechanism
of insecticide resistance. The order of genetic control is shown by broken
arrows. P = phosphate, D = deoxyribose, R = ribose, B = bases, A = adenine,
t = thymine, G = guanine, C = cytosine, U = uracil, and Tyr = tyrosine

tration for the order of genetic control (broken arrow) of resistance
via metabolism of an insecticide as a causal mechanism. Biochemists
usually deal with only middle parts of this scheme and, on the other
hand, geneticists deal with only the both ends. Relatively few works
have been done on this problem from a viewpoint of the biochemical
genetics (for recent reviews, see Agosin 1963, Georghiou 1965 and
1967, Oppenoorth 1965, etc.). Recent advances in molecular biology
are quite useful to understand that only a slight modification at a
coding pattern of the DNA molecule causes a different terminal expression.

II. Formal genetics of the housefly

Accumulation of knowledges on the formal genetics is remarkable
in insect pests of medical or agricultural importance, especially in the
housefly (Bodenstein 1939, Hiroyoshi 1960 and 1961, Hoyer 1966,
Milani 1954 and 1961, Tsukamoto et al. 1961). The use of visible
mutants in investigations of insecticide resistance problem is quite
effective not only for the genetic analysis itself but also for establishing special "purified" strains which are suitable for biochemical com-

parisons between strains or phenotypes. More than 100 mutations or aberrations of the housefly have been recorded during these decades, but some of them have later been proved to be allelic to each other and some of them have shown rather varied expressivity or incomplete penetrance of characters and/or poor viability. Up to date, about half of reported mutants have been determined their linkage group, and about one third of them have been determined their gene loci on the chromosome. Various visible markers are listed in Table I in order of the linkage group system first proposed by HIROYOSHI (1960).[1] The sex chromosomes seem to be genetically inert except for the determination of maleness by a Y chromosomal factor. Therefore, no sex-linked visible mutant nor resistance gene may be expected in the housefly. Reported sex-limited inheritance of visible mutant characters or DDT-resistance is now explained to be due to a translocation of the Y chromosome (or at least the maleness determiner locus) to the 2nd chromosome (SULLIVAN 1961, HIROYOSHI 1964). By using these visible mutants, genetic analyses of insecticide resistance or of biochemical properties have been carried out.

III. Crossing procedure

a) Estimation of mode of inheritance (reciprocal crosses)

Technically, reciprocal crosses between resistant and susceptible strains are the first step for further genetic analyses of resistance. By comparing toxicological responses to an insecticide between their F_1-hybrids, we can roughly assume some genetic properties of resistance, such as: dominant or recessive, cytoplasmic or chromosomal, sex-linked or autosomal, heterogeneous or homogeneous, etc. For these crosses, the use of visible markers is not essential, and subsequent interbreeding of these F_1-hybrids or backcross of the hybrid to one of parent strains may bring useful informations to estimate mode of inheritance of resistance. If the resistance to be analysed is monofactorial and discrimination of resistant and susceptible phenotypes is complete, some clear-cut segregations might be recognized by examining dosage-mortality curves. These genetic procedures can be applicable for any insect species in which no mutant marker is still available, but the interpretation of dosage-mortality curves is often difficult, vague or incorrect, especially in multifactorial inheritance systems. Most of the confusion or discrepancy of opinions on the mode of inheritance in earlier works are mainly due to such an incomplete genetic analysis without any marker gene. For a detailed discussion, see TSUKAMOTO (1963).

---

[1] More recently, WAGONER (1967) has succeeded in establishing the correspondence between PERJE's (1948) karyotype and confused number systems for the linkage group.

Table I. Linkage group of visible mutants of the housefly [a]

1	2	3	4	5	6
♂ (Male-determiner)	Bx (Beadex)	ocra (ocra)	ct (cut)	bu (brunette)	do (dark orange)
	N (Notch)	Lp (Loop)	rb (ruby)	ar (aristapedia)	acv (anterior-crossveinless)
	w (white)		abr (abrupt)	cm (carmine)	px (plexus-like)
	Sn (Singed)		ext (extended)	car (carnation)	bg (bridge)
	bwb (brown-body)			tw (twisted)	ho (hook)
	ge (green)			oc (ocelliless)	ac (ali curve)
	ro (rough)			ps (pigmented sternite)	sht (short)
	cp (clump)				bp (black puparium)
	dv (divergent)				ctc (countercoiled)
	bw (brown)				Rl (Rolled)
	pcv (posterior-crossveinless)				sb (subcostal-break)
					bkn (broken vein)
	Cw (Cleft wing)	cy (curly)	runt (runt)	atp (antennapedia)	it (interrupted)
	Iv (Irregular vein)	Iv (Irregular vein)	ye (yellow)	rd (reduced)	fs (fasciculate)
	cy-2 (curly-2)			claw (classic wing)	itu (interruptus)
	dov (dot vein)			stw (stubby wing)	ss (spineless)

[a] Loci for the mutants listed under the broken line are not yet determined.

b) Determination of linkage group (F_1-male backcross)

The second stage of genetic analyses of resistance is the F_1-male backcross to determine to which chromosome resistance factor (or factors) is linked. In the housefly, like drosophila, no crossing-over occurs in males. For this procedure, of course, the use of marker genes is essential, especially the use of multichromosomal marker strains is very effective not only to get precise informations on the linkage group qualitatively, but also to know quantitative informations on the relative contribution of each chromosome to resistance. For this purpose, the following multichromosomal mutant strains have been used in our laboratory:

$$bwb;ocra;ar;ac \quad (2;3;5;6) \quad \text{and} \quad ro;ext;cm;acv \quad (2;4;5;6).$$

They have visible markers for the indicated linkage groups, and hence the combination of these strains involve markers for all the autosomes. The design of crossing experiments and subsequent statistic treatments of data have been outlined in detail by TSUKAMOTO (1964).

c) Determination of gene locus (F_1-female backcross)

The third stage of the genetic analyses of resistance is to determine the locus of the resistance gene on a known linkage map. Females of the F_1-hybrid between a resistant strain and a marker strain have to be mated by males of the marker strain. Technically, at least two or more mutant markers are required for this purpose. Detailed methods for calculating recombination values between the resistance gene and marker genes have been described by TSUKAMOTO (1965). As shown in Table I, available marker mutants are still not sufficient in number, especially in linkage groups 3 and 4. Much more searches are awaited to find as many good mutants as possible.

IV. Genetic analysis of resistance

a) DDT-resistance

At least two major resistance factors are involved in kill-resistance to DDT in a Japanese resistant strain of the housefly. The 2nd chromosomal factor is rather incompletely recessive and the 5th chromosomal one is incompletely dominant over the susceptibility (TSUKAMOTO and SUZUKI 1964). Cooperative experiments with neurophysiologists have indicated that the 2nd chromosomal factor is responsible for low nerve sensitivity to DDT in the strain tested (TSUKAMOTO et al. 1965). No crossing experiment has been carried out to determine the gene locus for the 2nd chromosomal DDT-resistance factor nor the electrophysiological factor because of technical limitations, but it is likely to assume that the same single gene is responsible for DDT-resistance and low nerve sensitivity. Allelic relationships of this resistance gene to other 2nd chromosomal knockdown-resistance genes, *kdr* and *kdr-o* (MILANI and FRANCO 1959), are still unknown. Although the genetic control of DDT-dehydrochlorination was reported by LOVELL and KEARNS (1959) several years ago, the relation of this metabolism gene to two major resistance genes was unknown until recent genetic analyses have proved independently by *in vivo* experiments (TSUKAMOTO and SUZUKI 1964) and by *in vitro* experiments (OPPENOORTH 1965 b) that the 5th chromosomal resistance gene is responsible for the dehydrochlorination of DDT. The locus for this resistance gene has been located on the 5th chromosome, near to one of markers, *cm*. The use of DMC, which is known to inhibit the DDT-dehydrochlorinase activity, diminishes the effect of the 5th chromosomal resistance gene. More recently, OPPENOORTH (1967) has reported the presence of the third DDT-resistance gene which locates on the 3rd chromosome and its gene action is suppressible by the use of sesamex.

Further selections of an already developed DDT-resistant strain or population with a mixture of DDT and a synergist usually result in increases of resistance level to DDT. Several possible explanations can be applied to this phenomenon:

1) Selection of allelic genes which control production of a more efficient enzyme. In such case, the specific activity of DDT-dehydrochlorinase will become higher than that before selections.

2) Selection of allelic genes which control increased production of the same dehydrochlorinase protein. And hence, total enzyme activity may increase during the course of selections, but the specific activity of the enzyme will not change.

3) Selection of different metabolism genes which control other metabolic pathway than dehydrochlorination.

4) Selection of different genes which control other mechanisms than metabolism, such as penetration, storage, excretion, nerve sensitivity, etc.

It is probably not necessary to explain for the first possibility on the presence of qualitative variants of DDT-dehydrochlorinase, because different resistant strains have different dehydrochlorinase activities from highly resistant to susceptible strains. For the second possibility, no published evidence is available for metabolism of DDT, because most investigators describe only comparison of the *enzyme activity* but not of the *protein amounts* or *specific activity* of enzymes between resistant and susceptible strains. The third possibility should not be ignored because at least three other metabolic pathways of DDT are known in insects: hydroxylation of DDT to dicofol (TSUKAMOTO 1959), reductive dechlorination to TDE (HOOPER 1967), and formation of unknown metabolites probably conjugation via DDA

(PERRY *et al.* 1963). Indeed, even in the housefly, the presence of the hydroxylating enzyme system has been suggested *in vivo* (TSUKA-MOTO 1961) and has been reported *in vitro* (AGOSIN *et al.* 1961, TSUKAMOTO and CASIDA 1967 a). The presence of the 3rd chromosomal DDT-resistance gene which controls the sesamex-suppressible detoxication mechanism, probably further degradation of DDE to water soluble products, is also one of other metabolic mechanisms than the dehydrochlorination (OPPENOORTH 1967). The last possibility has been proved by the findings that the 2nd chromosome of resistant strains involves some genetic factors responsible for low nerve sensitivity, knockdown-resistance, resistance to *o*-chloro-DDT and Dilan (PLAPP and HOYER 1967), etc. Mechanisms of knockdown-resistance or low nerve sensitivity to DDT are still unknown, although some plausible explanations leave some room for consideration as follows: a) less excretion of a neurotoxin released by nerve cord excitations, b) rapid degradation of the neurotoxin, c) structural differences in lipoprotein nerve membranes, etc.

b) BHC-resistance

Earlier experiments on genetics of BHC-resistance failed to show any clear-cut segregation of resistant and susceptible phenotypes in progeny of crosses between resistant and susceptible strains, suggesting somewhat complicated polygenic inheritance system. Results of recent genetic analyses with visible markers have also confirmed the absence of any outstanding major gene for the resistance, namely, all the autosomes contribute more or less to total resistance level of resistant strains (TSUKAMOTO 1964 and unpublished data). In a highly resistant strain originated from a mixture of several field strains, ranking order of the contribution of each chromosomal factor to BHC-resistance is $3>5>2>6$ in heterozygous condition (dominant effect) and $2>3\cong5$ in homozygous condition (recessive effect). In a European diazinon-resistant strain which is also resistant to carbaryl, DDT, and BHC, an incomplete dominant factor on the 2nd chromosome seems to be major but no significant effect on BHC-resistance has been observed for the 4th and 5th linkage groups (KASAI and OGITA 1965). In contrary to these informations, OPPENOORTH and NASRAT (1966) have reported that a major resistance gene is on the 4th chromosome and an additional resistance gene is on the 5th chromosome in a resistant strain of Uruguayan origin.

In spite of numerous investigations and trials, metabolic pathways of BHC and related compounds are not yet completely clarified (CLARK *et al.* 1966). Rate of metabolism of BHC under *in vitro* conditions does not correlate with levels of resistance (SIMS and GROVER 1965, ISHIDA and DAHM 1965 a), thus metabolism of BHC might be a resistance mechanism of minor importance. Therefore, much more works should be done on biochemistry and physiology of BHC-resistance to correlate to such a complicated genetic system.

c) Dieldrin-resistance

Generally speaking, development of dieldrin-resistance is rather rapid in the housefly like other insect species, and the developed resistance levels are usually extremely high. In contrary to the complicated mode of inheritance of BHC, a monofactorial genetic system of dieldrin-resistance has been suggested by relatively clear-cut segregations of resistant and susceptible phenotypes in crossing experiments (GEORGHIOU et al. 1963, GUNEIDY and BUSVINE 1964, MILANI 1963, OPPENOORTH and NASRAT 1966). Results of genetic analyses with multichromosomal marker strains have confirmed results of previous experiments and indicate that dieldrin-resistance is mainly controlled by an incompletely dominant resistance gene on the 4th chromosome (TSUKAMOTO and SUZUKI, unpublished data). In addition, a recessive minor factor is often involved in the 2nd chromosome. In spite of its rather simple genetic system, mechanism of dieldrin-resistance of the housefly is almost entirely unknown. Any significant causal differences to account for the high level of resistance to cyclodiene compounds cannot be found out in absorption, metabolism, and excretion (WINTERINGHAM 1962). Therefore, prompt establishment of a new approaching method is awaited to this problem.

d) Organophosphorus-resistance

One of the most stimulating and attractive works in biochemical genetics of insecticide resistance is the finding by Dutch group (OPPENOORTH 1959, OPPENOORTH and VAN ASPEREN 1960) that parathion-, diazinon- and malathion-resistance in several strains examined are always associated with a low aliesterase activity in vitro and this property is controlled by a single locus. They assume that different alleles produce modified enzyme proteins of which aliesterase activities are low but this causes detoxication of organophosphorus (OP) insecticides. This hypothesis is very plausible. Available data, however, still insufficient to generalize this theory or to prove allelism among the resistances, metabolism of OP compounds, breakdown enzyme, and low aliesterase activity. For example, as will be discussed later, characterization of this breakdown enzyme is not yet clear. Results of crossing experiments by FRANCO and OPPENOORTH (1962) have suggested that the gene a for low aliesterase activity and OP-resistance is located on the 5th chromosome in a diazinon-resistant strain, and that crossing-overs between the a gene and marker genes ar or cm are either rare or absent. SAWICKI et al. (1966) have also reported that diazinon-resistance is due to non-recessive factors on the 3rd and 5th chromosomes.

Genetic analyses at the Osaka University, using several multi-chromosomal marker strains, have indicated that diazinon-resistance of three strains (of one Japanese and two European origins) is controlled by at least three major factors. The order of contribution of each chromosome to diazinon-resistance is 5>>3>2 as dominant effect, and 5>2>3 as recessive effect, respectively. The locus for the 5th chromosomal resistance gene has been determined to be almost at the terminal region of the left arm of the chromosome (TSUKAMOTO and SUZUKI 1966). If so, this diazinon-resistance gene is not an allele of the low aliesterase gene a, because our diazinon-resistance gene has about 30 percent or more recombination values from marker genes ar or cm.

A number of works have been reported on metabolism of OP insecticides *in vivo* and *in vitro* with insect materials. OPPENOORTH and VAN ASPEREN (1960 and 1961) demonstrated that OP-resistance of the housefly is due to higher enzyme activity in degrading paraoxon, diazoxon, or malaoxon in resistant strains. MATSUMURA and HOGENDIJK (1964 a and b) have partially purified OP-breakdown enzymes from susceptible and resistant strains. Although malathion is hydrolyzed by both phosphatase and carboxyesterase actions, it is not clear which enzyme is controlled by an allele of the modified aliesterase gene a. Interstrain comparisons of partially purified enzymes suggested that different strains have qualitatively different carboxyesterases in heat stability, pH-dependency and inhibition patterns by OP compounds. In the case of parathion-breakdown enzyme, both parathion-resistant and susceptible strains produce at least two groups of phosphatases, peak I and peak II of column chromatographic patterns. The parathion-resistant strain contains an increased activity of a parathion-degrading enzyme (peak I), which is very low in the susceptible strain, and a lowered peak II activity. However, these enzyme preparations hydrolyze parathion and diazinon but the enzyme activity for paraoxon degradation is rather low, suggesting that "thionase" and "oxonase" are not identical. These findings are not in accord with those by OPPENOORTH and VAN ASPEREN (1960 and 1961). In addition to these hydrolytic enzymes, more recent investigations have demonstrated the presence of other detoxication mechanisms in insect enzyme preparations: for instances, desmethylation of methyl parathion and Sumithion (FUKAMI and SHISHIDO 1966), oxidative degradation of parathion (NAKATSUGAWA and DAHM 1967), and microsomal oxidation of diazinon and glutathione-requiring degradation of diazinon by soluble enzyme fraction (FUKAMI, personal communication). Independently of these works, TSUKAMOTO and CASIDA (1967 a and unpublished data) have also found that diazinon and Imidan are metabolized by NADPH$_2$-requiring microsomal preparations of the housefly. The principal metabolite of diazinon has been identified to be a so-called "hydrolysis" product: 2-isopropyl-4-methyl-6-hydroxy-pyrimidine, and microsome-plus-soluble preparations from a diazinon-resistant strain have been more active in degrading diazinon than those from susceptible strains. No crossing experiment, however, has been tried to determine whether or not the 5th chromosomal diazinon-resistance gene is responsible for the microsomal oxidative breakdown of diazinon. Usually it is considered that a resistant strain selected with a methyl-type OP compound shows cross resistance to other methyl-type OP insecticides but not to ethyl-type OP insecticides (MARCH 1960). In this connection, recent papers on desmethylation of OP compounds by glutathione-requiring soluble enzyme system which is specific to methyl-type compounds (FUKAMI and SHISHIDO 1966) and desethylation of OP compounds by $NADPH_2$-requiring microsomal system in mammals (DONNINGER et al. 1967) are very interesting. Apart from metabolism, slower penetration of diazinon also seems to contribute to resistance (GWIAZDA and LORD 1967).

e) Carbamate-resistance

Depending upon the resistant strains tested, resistance to Isolan, to carbaryl, or to a combination of carbaryl + piperonyl butoxide (p.b.) seems to be controlled by a major gene on the 5th chromosome or by both 2nd and 5th chromosomal factors (GEORGHIOU 1966 and un-published data, HOYER et al. 1965, KASAI and OGITA 1965, PLAPP and HOYER 1967). Resistance to Baygon and Matacil in a diazinon-Baygon-resistant strain is due to three major factors on the 5th, 3rd, and 2nd linkage groups in decreasing order of dominant effects (TSUKAMOTO et al. 1968).

Metabolism of carbamate insecticides in resistant strains seems to be one of the most important mechanisms of carbamate-resistance. Homogenates or microsomal preparations can degrade various car-bamate chemicals in a variety of metabolic pathways. The enzyme activity is usually very low in vitro when the enzyme preparations are prepared from whole body of the housefly, but is very high when abdomens only are homogenized and albumin is added into the re-action system, because these techniques can remove the influence by natural inhibitary substances involved in the fly body (TSUKAMOTO and CASIDA 1967 a and b). As usual microsomal oxidations, this sys-tem requires oxygen and $NADPH_2$ and is inhibited by some methylene-dioxyphenyl synergists such as p.b. Rate of metabolism of carbamates is higher in carbamate-resistant strains than in susceptible strains both in vivo and in vitro. The major ether-extractable metabolites are 5-hydroxy, o-deisopropyl and N-hydroxymethyl derivatives from Bay-gon, and 4-methylamino, 4-amino, and N-hydroxymethyl derivatives from Matacil, respectively. Practically no hydrolysis product at the ester linkage has been detected even from carbaryl. It is still not known whether such a variety of metabolic pathways are catalyzed independently by each specific enzyme or are initiated as a hydroxy-lating step controlled by a single cooperative or non-specific enzyme system, whereas genetic analyses suggest that the 5th chromosomal factor is mainly responsible for over-all metabolism of these carbamate insecticides.

In contrary, recent results obtained by Riverside group are considerably interesting because they assume that the major carbamate-detoxifying enzymes might be phenolases or tyrosinases which are abundant in the housefly (ABD-EL-AZIZ 1966, METCALF et al. 1966). They have found that microsomal tyrosinase is not the principal detoxifying system but soluble tyrosinase of the housefly hydroxylates a variety of phenols and phenyl carbamates. The purified soluble tyrosinase requires NADP for the introduction of the first hydroxyl group. These data are quite strange for us because tyrosinases or phenolases are such the enzyme that catalyze the additional hydroxylation of already hydroxylated aromatic compounds but cannot catalyze the hydroxylation of unhydroxylated phenyl compounds. For example, the hydroxylation of phenylalanine to tyrosine is carried out by a pterin-$NADPH_2$-requiring enzyme system but not by NADP-requiring tyrosinase, and further oxidation of tyrosine is catalyzed by tyrosinase to 3,4-dihydroxyphenylalanine (dopa). Lack or decrease of tyrosinase activity in the housefly causes a decreased pigmentation of cuticle of brown-body mutant (bwb). Although this mutant gene for modified tyrosinase is linked with the 2nd chromosome, nothing is known about informations on the linkage group distribution for the gene which controls the carbamate-metabolizing tyrosinase. Such genetic approaches might be very important in connection with the evidence that resistance to carbamate insecticides in the resistant strains used by the Riverside group involves both 2nd and 5th chromosomal resistance genes (GEORGHIOU 1966 and 1967).

f) Pyrethrins-resistance

Results of genetic analyses of pyrethrins-resistance in a Swedish resistant strain with multichromosomal mutant strains indicate that an incompletely recessive factor on the 2nd chromosome is the most important one and additional resistance genes are associated with some other autosomes. The order of contribution of autosomes to the total kill-resistance to pyrethrins is $3 \geqq 2 = 5 > 4$ in heterozygous condition and $2 >> 5 > 3 > 4$ in homozygous condition. In the same type of genetic analysis for resistance to pyrethrins + p.b., a remarkable decrease has been noticed in the 3rd chromosomal dominant effect whereas any significant decrease has not occurred in other autosomal effect (TSUKAMOTO and SUZUKI, unpublished data). PLAPP and HOYER (1967) have also obtained similar results that resistance to pyrethrins + p.b. in a DDT-resistant strain is due to the 2nd chromosomal factor when F_2 progeny have been analysed with the 2nd and 5th chromosomal markers. For the mechanisms of pyrethroid-resistance, FINE et al. (1963) reported decreased absorption of pyrethrin I in the Swedish resistant strain, but reduced nerve sensitivity or detoxication of penetrated compounds might also be considered. It has long been interpreted that metabolism of pyrethroids is due to the hydrolysis of the ester linkage or initial detoxication step occurs on the keto-alcohol moiety. Recent studies with appropriately labeled C^{14}-pyrethroids have clearly indicated that the major metabolic pathway of pyrethrin I, allethrin and phthalthrin is not via hydrolysis but via oxidation of terminal isobutenyl carbon on the acid moiety (YAMAMOTO and CASIDA 1966). This detoxication process is also catalyzed by an $NADPH_2$-requiring microsomal preparations of the housefly. Relationships between resistance genes and these plausible mechanisms still remain unanalysed.

g) Resistance to other insecticides

According to PLAPP and HOYER (unpublished data), resistance to organotin compounds, such as tributyltin chloride and tributyltin acetate, is controlled by a single gene on the 2nd chromosome. This tin gene (or closely linked gene) also acts as an enhancer of DDT- and OP-resistance genes on the 5th chromosome. Organotin compounds are known to be an active inhibitor of ATPase and oxidative phosphorylation process of the fly mitochondria (PIEPER and CASIDA 1965). The mechanisms of organotin-resistance and of the gene action as an enhancer for the 5th chromosomal resistance genes are not yet known.

Hydroxylation of naphthalene is carried out by $NADPH_2$-requiring microsomal system of the housefly, and this seems to be an important mechanism of resistance (ARIAS and TERRIERE 1962, SCHONBROD et al. 1965) at least at the initial stages of metabolism (BOOSE and TERRIERE 1967). Genetics of naphthalene-resistance or the microsomal hydroxylation will bring valuable informations from a viewpoint of biochemical genetics because of its simplicity in chemical structure and metabolic pathways.

Resistance to fluoroacetate (FA) may also be another interesting subject for biochemical genetics, because the chemical structure of FA differs uniquely from those of usual synthetic insecticides and resistance level to FA in the resistant strain increased sharply after selections for more than 75 generations, suggesting the establishment of "true" resistance to FA by a newly occurred mutation. After conversion to toxic fluorocitrate (FC), FA inhibits aconitase of the Krebs cycle. Therefore citrate and its precursor pyruvate are accumulated in the housefly after feeding flies on FA. Citrate accumulation is less in the FA-resistant strain than in the susceptible strain (TAHORI 1966). One of some plausible mechanisms might be assumed as an action of FA-resistance gene, such as: a) a modified aconitase less sensitive to FC inhibition, b) a modified metabolic pathway of citrate, c) decreased conversion of FA to FC, etc.

V. General considerations

Linkage group distribution of major resistance factors of the house-fly outlined in the previous sections is summarized in Table II.

Visible expression of some mutant markers is based on biochemical aberrations. For example, green eye-color mutant (*ge*) cannot degrade tryptophan to formylkynurenine because of lack of the tryptophan pyrrolase activity; ocra eye-color mutant (*ocra*) cannot carry out the ring hydroxylation of kynurenine (Hiraga 1964); black puparium mutant (*bp*) forms pupal sheath protein in which β-alanine is not incorporated (Fukushi 1967, Fukushi and Seki 1965, Seki 1962); and so on. Insecticide resistance phenomena are also based on some biochemical or physiological mechanisms. Table III lists such known biochemical or physiological characters of the housefly. It is rather interesting to see that several hydrolytic enzymes are under the control of the 5th chromosomal genes, whereas distribution of genes for hydroxylating enzymes is not restricted into a special linkage group. Of course, such information may be largely changed after much extensive investigation along this line.

Usually, insecticide resistant strains show more or less resistance not only to the selected insecticide or analogs but also to structurally unrelated compounds. This so-called "cross resistance" phenomenon is very important from a practical viewpoint of pest control but relatively less attentions have been paid by biochemists. The term "cross resistance" has long been used rather in a broad sense and this involves both cross resistance *sensu lato* and *sensu stricto*. From a genetical standpoint, we may give the following terms and definitions:

Cross resistance (*s. l.*)
{
 Pleiotropic resistance (= cross resistance *s. str.*): resistance to other insecticides is due to a pleiotropic action of a single gene
 Co-existing resistance (= multiple resistance): resistance to other insecticides is due to co-existing different genes
}

or, if the term "cross resistance" should be used only in a narrow sense,

Co-resistance (= Cross resistance *s. l.*)
{
 Cross resistance (*s. str.*)
 Multiple resistance
}

Strictly speaking, no proved example for "pleiotropic resistance" is available yet except for resistance to some derivatives of the original insecticide. However, in *Drosophila melanogaster*, resistance genes to DDT, BHC (Tsukamoto and Ogaki 1954), parathion, carbaryl (Kikkawa 1961 and 1964) and negatively correlated susceptibility to phenylthiourea (PTU) (Ogita 1958) are concentrated into a very limited region of the 2nd chromosome, while resistance genes to nicotine sulfate (Tsukamoto and Hiroyoshi 1956), phenylurea (PU)

Table II. Linkage group distribution of major resistance factors in the housefly

1	2	3	4	5	6
	DDT DDT + DMC o-Chloro-DDT Dilan BHC	DDT BHC		DDT BHC	
	Diazinon Baygon Matacil	Diazinon Baygon Matacil	BHC Dieldrin	Parathion Malathion Diazinon Baygon Matacil Carbaryl	
	Carbaryl + p.b. Isolan Pyrethrins Pyrethrins + p.b. Organotin	Pyrethrins		Carbaryl + p.b. Isolan Pyrethrins Pyrethrins + p.b.	

Table III. Linkage distribution of known biochemical or physiological characters in the housefly

1	2	3	4	5	6
	Low nerve sensitivity to DDT P $\left\{\begin{array}{l}(r\text{-}DDT)\\(kdr)\\(kdr\text{-}o)\end{array}\right.$	Kynurenine hydroxylase (*ocra*)		DDT-dehydrochlorinase (*R-DDT*)	Absence of β-alanine in pupal sheath (*bp*)
	Late emerging males (*Lem*) Tryptophan pyrrolase (*ge*)	Amylases (*Amy-A*)		Metabolism of Baygon and Matacil Detoxication of OP compounds Aliesterases (*a*) Esterase bands 1, 2 + 4 Esterase-A (*Est-A*) Esterase-B (*Est-B*) Acid phosphomonoesterase (*Phos*)	
	Tyrosinase (*bwb*)				

(Ogita 1958), ether (Ogaki *et al.* 1967), and gamma-radiation (Og-AKI and Nakashima-Tanaka 1966) locate at separate loci on the 3rd chromosome (Fig. 2). Therefore, it is not unlikely that the 2nd

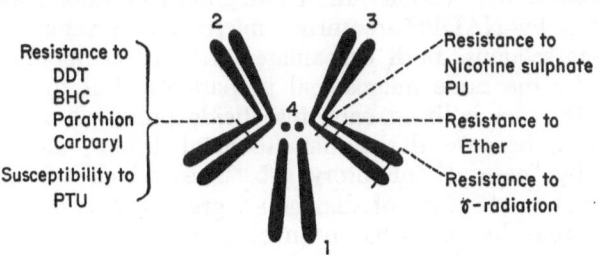

Fig. 2. Probable example for pleiotropic resistance and negatively correlated susceptibility controlled by a single locus on the 2nd chromosome of *Drosophila melanogaster*

chromosomal resistance characters are controlled by the pleiotropic action of a single resistance gene. The mechanism of pleitropism by this gene is still not known.

Resistance to carbamates and OP compounds is one of the interesting subjects in the housefly. Both insecticide groups are usually potential inhibitors of various esterases. In most cases, low aliesterase activity in resistant strains is inseparable from carbamate- or OP-resistance by repeated backcrosses. From such available data, Plapp *et al.* (1964) and Hoyer *et al.* (1965) have assumed that the same single gene locus is responsible for both carbamate- and OP-resistances and for the low aliesterase activity. Working with normal and OP-resistant strains and agar-gel electrophoretic techniques, van Asperen *et al.* (1965) have demonstrated a striking parallelism between the OP-breakdown enzyme activity and the esterase band 1 of the resistant strain, and the aliesterase activity and the band 1 of the susceptible strain, supporting their hypothesis that the OP-breakdown enzyme replaces the aliesterase in OP-resistant strains. Ogita and Kasai (1965 a and b) have determined the gene loci responsible for electrophoretic bands of esterases and acid phosphomonoesterase, but none of them was able to correlate to OP-resistance.

Pretreatments of mammals with some compounds which can be metabolized by microsomal oxidation often give rise to increases in the enzyme activity of microsomal oxidation system which is nonspecific not only to the inducer or analogs but also to other unrelated compounds (Gillette 1963). This phenomenon is suggestive of the presence of a common mechanism, at least in part, in the microsomal oxidations, and a similar mechanism might be present in insects too. Such an induction may not be hereditary directly to the next generation, but repeated selections with an insecticide which can be oxidized by microsomal enzyme system might help to concentrate the frequency of a gene or a group of genes which has some connection with the microsomal oxidation within a field population or a laboratory strain. Recently accumulated informations on insect biochemistry have indicated that various unrelated groups of insecticides can be metabolized by $NADPH_2$-requiring microsomal systems as outlined in previous sections. Both carbamates and OP compounds are metabolized by the same microsomal preparation. But the enzyme responsible for metabolizing each insecticide group does not seem to be identical, because the carbamate-metabolizing system is easily inhibited by "natural" inhibitory substances which are involved in fly body but in the case of diazinon-degrading microsomal enzyme, no appreciable difference has been observed between enzyme preparations from whole body and from abdomens only (TSUKAMOTO and CASIDA, unpublished data).

Another interesting subject to be mentioned is close relationship among resistance genes on the 2nd chromosome: *i.e.*, kdr, kdr-o, r-DDT, resistance genes to pyrethrins, Dilan, organotin, etc. Pyrethrins-resistant strains is almost always associated with DDT-(knockdown)-resistance as is reviewed by FINE (1963). One may assume some common detoxication mechanism of DDT and pyrethroids by microsomal oxidation. Hydroxylation of DDT *in vivo* is, however, an only minor metabolic pathway in usual DDT-resistant strains of the housefly, and such detoxication mechanism is probably important to kill-resistance but not to knockdown-resistance. If knockdown of flies is a reflection of some neurophysiological excitation mechanisms, knockdown-resistance might be caused by a low sensitivity in resistant strains. If much more imaginations are allowable here, some resistance genes concerning neurophysiological phenomena might be allelic to each other. Except for kdr gene, at present, no available data is reported to determine the loci for these 2nd chromosomal resistance genes.

According to ISHIDA and DAHM (1965 b), partially purified BHC-metabolizing enzyme preparations also act as the DDT-dehydrochlorinase. This dehydrochlorination of DDT is inhibited by an addition of γ-BHC, and vice versa. Responses to various inhibitors are also same in both DDT- and BHC-metabolisms. If BHC-metabolism involves the same mechanism for DDT-dehydrochlorination, we can explain some parts of cross resistance to DDT in the BHC-resistant strains.

Adoption of a synergist into genetic analysis of insecticide resistance seems to be a very useful technique to identify a special resistance mechanism when this synergist is known to inhibit a particular metabolic pathway. Results of genetic analysis of DDT-resistance have clearly indicated that DMC suppresses the 5th chromosomal resistance effect specifically (TSUKAMOTO and SUZUKI 1964). Similar experiments have also been done by using pyrethrins + p.b., where only the 3rd chromosomal dominant effect has decreased remarkably. Therefore, the microsomal oxidation system involved in pyrethrins-resistance may be controlled by the 3rd chromosomal gene, if p.b. inhibits only microsomal oxidation process (TSUKAMOTO and SUZUKI, unpublished data). In the case of Baygon-resistance, noticeable decreases in resistance effect have observed in both 3rd and 5th chromosomal genes (TSUKAMOTO and CASIDA 1968). OPPENOORTH (1967) has also reported the example of sesamex-suppressible resistance to DDT and diazinon under the control of the 3rd chromosome, suggesting that DDT-diazinon-resistance in a Danish strain is caused by one and the same mechanism probably via oxidative detoxication.

Such adoption methods of an inhibitor in genetic analyses are useful to distinguish a detoxication gene from other non-metabolism genes, but when a series of similar metabolic reactions is involved in the over-all detoxication process, interpretations of experimental results might not be easy. For instance, if the reaction velocity v_1 for the initiatory hydroxylation step of an insecticide A to its metabolite B is much slower than the velocity v_2 for further conversion of B to C (Fig 3), the gene (Gene$_B$) which controls the reaction from B

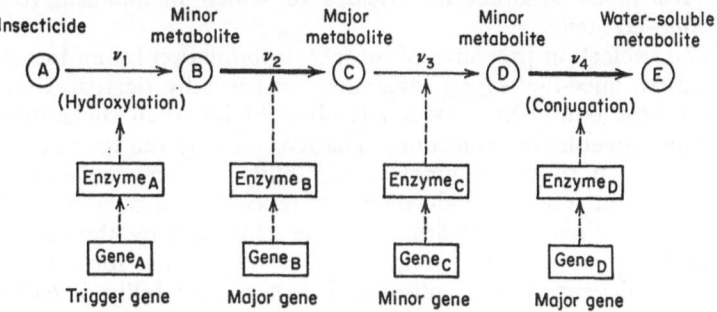

Fig. 3. Schematic diagram showing genetic control of insecticide metabolism initiating from hydroxylation and terminating to conjugation. Reaction velocity of each metabolic step is shown by v. Solid arrows represent the sequence of the reactions and broken arrows indicate the order of genetic control

to C might be considered as a major resistance gene responsible for the oxidative metabolism of the insecticide A, whereas the true trigger gene (Gene$_A$) is such a one which controls the initiation of subsequent detoxication processes. Under such a circumstance, genetic analysis of resistance to a combination of insecticide + synergist might indicate remarkable decreases in relative effect of both major and minor resistance genes.

The majority of foreign compounds are known to be excreted in water-soluble conjugate forms. Therefore, the role of conjugating enzyme systems also should not be overlooked. In the housefly, formation of water-soluble metabolites, mostly in conjugate forms, are known from DDT (OPPENOORTH 1967, TERRIERE and SCHONBROD 1955), from BHC (BRADBURY and STANDEN 1959 and 1960, COHEN et al. 1964, ISHIDA and DAHM 1965 b), from carbamates (SHRIVASTAVA 1967), and from naphthalene (BOOSE and TERRIERE 1967, TERRIERE et al. 1961). In the scheme of Figure 3, the metabolite E represents such a water-soluble conjugate. If the reaction velocity v_4 is very rapid, the gene which controls this conjugating process might be greatly important for the over-all metabolism and hence for resistance too. Substrate specificity of conjugating systems involved in insecticide metabolisms will be an interesting subject for discussions on cross resistance (s. l.) in the future.

In a similar way, genetic control of the availability of co-factors in the detoxication process is also undoubtedly important for resistance to insecticides. It is now a common knowledge for biochemists that a variety of metabolic reactions often require a single and the same co-factor such as $NADPH_2$ or glutathione. If availability of a common co-factor or activator for some detoxifying reactions is controlled by a gene or genes, selection of an insect population with an insecticide will give rise to so-called vigor tolerance or low level of cross resistance to other insecticides of which metabolism requires the same co-factor.

When selection pressures of an insect population by an insecticide are ceased, once-developed resistance level usually decreases until to reach a new equilibrium, which is often higher than the initial susceptibility level before selections. The reasons why the resistance level does not drop to the original susceptibility or why such an insect population can rapidly re-develop resistance to the once-selected insecticide or other insecticides might be explained by the concentration of the gene or gene group which is responsible for a relatively common biochemical mechanism such as the availability of co-factors or conjugating systems.

During earlier stages of investigations on insecticide resistance, the majority of examined interstrain differences in morphological, ecological, and biochemical characters have been discarded as causal mechanisms of resistance because of inconsistency of these characters in different strains examined. However, it should be kept in mind that resistance to even one insecticide is frequently due to different mechanisms in different strains. In this sense, once given-up characters as causal mechanisms, such as increased length of the larval period, respiratory enzyme systems, quantitative or qualitative differences of tissue lipids or fats, have to be re-examined from a viewpoint of advanced biochemical concepts such as hydroxylation steps involved in growth and/or moulting hormon systems, electron transporting systems, availability of co-factors, and so on.

Slight changes of the DNA configuration at a single locus of a chromosome may result in appearance of different alleles of the same gene. The biochemical activity of abnormal proteins or modified proteins controlled by these alleles may differ from that of the normal enzyme. Usually we can investigate normal phenomena through the comparison with abnormal cases. For example, DDT-dehydrochlorinase may be produced by a mutant allele of a normal gene. But, at present, we do not know what kind of the normal enzyme in a susceptible strain corresponds to DDT-dehydrochlorinase in a resistant strain, what is the normal substrate for such the normal enzyme, and what kinds of chemical or physical changes have occurred at the DNA or protein levels during the course of evolution. To answer these questions is more than our present knowledges, but it should not be forgotten that much more extensive investigations of insect pests based on general comparative biochemistry and formal genetics will finally help to bring invaluable information on the mechanism of insecticide resistance.

Summary

Housefly resistance to insecticides has been reviewed from a standpoint of biochemical genetics. In most cases, major resistance genes are linked to 2nd and/or 5th chromosomes, except for the 4th chromosomal control of dieldrin-resistance. It is interesting that detoxications of DDT, organophosphorus, and carbamate insecticides are under the control of the 5th chromosomal dominant genes but incompletely recessive genes responsible for knockdown-resistance or low nerve sensitivity are involved in the 2nd chromosome. Resistance and cross resistance to unrelated compounds should be re-examined on the basis of recent biochemical information and more extensive genetic analyses, especially of precise gene loci, for both resistance and plausible biochemical mechanisms.

Résumé *

Génétique biochimique de la résistance aux insecticides chez la mouche domestique

La résistance de la mouche domestique aux insecticides a été examinée du point de vue de la génétique biochimique. Dans la plupart des cas, les principaux gènes de la résistance sont liés au deuxième et (ou) au cinquième chromosome, à l'exception de la résistance de la dieldrine, liée au quatrième chromosome. Il est intéressant de constater que la détoxication du DDT, des insecticides organo-phosphorés et des carbamates sont sous le contrôle des gènes dominants du cinquième chromosome, tandis que les gènes incom-

---

* Traduit par S. DORMAL-VAN DEN BRUEL.

plètement récessifs, responsables de la résistance au "knockdown" ou d'une faible sensibilité nerveuse, sont liés au deuxième chromosome. La résistance et la résistance croisée à des composés non apparentés devraient être examinées à nouveau sur la base d'informations biochimiques récentes et d'un plus grand nombre d'analyses génétiques approfondies portant, en particulier, sur la localisation précise des gènes, tant pour la résistance que pour les mécanismes plausibles.

Zusammenfassung *

Biochemische Genetik der Insektizidresistenz in der Hausfliege

Die Resistenz der Hausfliegen gegenüber Insektiziden wird vom Standpunkt der biochemischen Genetik überprüft. In den meisten Fällen sind die Hauptresistenzgene mit zweiten und/oder fünften Chromosomen verbunden, ausser der Dieldrinresistenz, die durch die vierten Chromosomen kontrolliert wird. Es ist interessant, dass Detoxifizierungen von DDT, Organophosphor- und Carbamatinsektiziden unter der Kontrolle der an den fünften Chromosomen vorherrschenden Gene sind, jedoch unvollständig rezessive Gene, welche verantwortlich für die "knockdown"—Resistenz oder niedrige Nervensensibilität sind, sind im zweiten Chromosom enthalten. Resistenz und Co-Resistenz zu nicht verwandten Verbindungen sollten nachgeprüft werden—sowohl für die Resistenz als auch für annehmbare biochemische Mechanismen—auf der Basis jüngster biochemischer Information und ausgedehnteren genetischen Analysen, besonders der genauen Genelokalisierung.

References

ABD-EL-AZIZ, S. A.: Housefly phenolase and detoxication of carbamate insecticide. Abstr. Pacific Branch Meeting, Entomol. Soc. Amer, p. 13 (1966).

AGOSIN, M.: Present status of biochemical research on the insecticide resistance problem. Bull. World Health Org. 29 (Suppl.), 69 (1963).

——, D. MICHAELI, R. MISKUS, S. NAGASAWA, and W. M. HOSKINS: A new DDT-metabolizing enzyme in the German cockroach. J. Econ. Entomol. 54, 340 (1961).

ARIAS, R. O., and L. C. TERRIERE: The hydroxylation of naphthalene-1-C[14] by house fly microsomes. J. Econ. Entomol. 55, 925 (1962).

BODENSTEIN, G.: Die Auslösung von Modifikationen und Mutationen bei Musca domestica L. Roux' Archiv für Entwicklungsmechanik 140, 614 (1939).

BOOSE, R. B., and L. C. TERRIERE: Quantitative aspects of the detoxication of naphthalene by resistant and susceptible house flies. J. Econ. Entomol. 60, 580 (1967).

BRADBURY, F. R., and H. STANDEN: Metabolism of benzene hexachloride by resistant houseflies, Musca domestica. Nature 183, 983 (1959).

---

* Übersetzt von A. SCHUMANN.

—— —— Mechanism of insect resistance to the chlorohydrocarbon insecticides. J. Sci. Food Agr. 1960, 92 (1960).

CLARK, A. G., M. HITCHCOCK, and J. N. SMITH: Metabolism of gammexane in flies, ticks and locusts. Nature 209, 103 (1966).

COHEN, A. J., J. N. SMITH, and H. TURBERT: Comparative detoxication. 10. The enzyme conjugation of chloro compounds with glutathione in locusts and other insects. Biochem. J. 90, 457 (1964).

DONNINGER, C., D. H. HUTSON, and B. A. PICKERING: Oxidative cleavage of phosphoric acid triesters to diesters. Biochem. J. 102, 26 (1967).

FINE, B. C.: The present status of resistance to pyrethroid insecticides. Pyrethrum Post 7, 18 (1963).

——, P. J. GODIN, and E. M. THAIN: Penetration of pyrethrin I labelled with carbon-14 into susceptible and pyrethroid resistant houseflies. Nature 199, 927 (1963).

FRANCO, M. G., and F. J. OPPENOORTH: Genetical experiments on the gene for low aliesterase activity and organophosphate resistance in Musca domestica L. Entomol Expt. Appl. 5, 119 (1962).

FUKAMI, J., and T. SHISHIDO: Nature of a soluble, glutathione-dependent enzyme system active in cleavage of methyl parathion to desmethyl parathion. J. Econ. Entomol. 59, 1338 (1966).

FUKUSHI, Y.: Genetic and biochemical studies on amino acid compositions and color manifestation in pupal sheaths of insects. Japan. J. Genet. 42, 11 (1967).

——, and T. SEKI: Differences in amino acid compositions of pupal sheaths between wild and black pupa strains in some species of insects. Japan. J. Genet. 40, 203 (1965).

GEORGHIOU, G. P.: Genetic studies on insecticide resistance. Adv. Pest Control Research 6, 171 (1965).

—— Genetic analysis of carbamate-resistance in Musca domestica L. Abstr. Pacific Branch Meeting, Entomol. Soc. Amer., p. 67 (1966).

—— Genetics of resistance to insecticides in insects of medical importance. Inform. Circ. Insecticide Resistance 60, 38 (1967).

——, R. B. MARCH, and G. E. PRINTY: A study of the genetics of dieldrin-resistance in the housefly (Musca domestica L.). Bull. World Health Org. 29, 155 (1963).

GILLETTE, J. R.: Metabolism of drugs and other foreign compounds by enzymatic mechanisms. Progress in Drug Research 6, 11 (1963).

GUNEIDY, A. M., and J. R. BUSVINE: Genetical studies on dieldrin-resistance in Musca domestica L. and Lucilia cuprina (Wied.). Bull. Entomol. Research 55, 499 (1964).

GWIAZDA, M., and K. A. LORD: Factors affecting the toxicity of diazinon to Musca domestica L. Ann. Applied Biol. 59, 221 (1967).

HIRAGA, S.: Tryptophan metabolism in eye-color mutants of the housefly. Japan. J. Genet. 39, 240 (1964).

HIROYOSHI, T.: Some new mutants and linkage groups of the house fly. J. Econ. Entomol. 53, 985 (1960).

—— The linkage map of the house fly, Musca domestica L. Genetics 46, 1373 (1961).

—— Sex-limited inheritance and abnormal sex ratio in strains of the housefly. Genetics 50, 373 (1964).

HOOPER, G. H. S.: A new metabolite of DDT in Culex pipiens fatigans Wied. Proc. Roy. Soc. Queensland 79, 9 (1967).

HOYER, R. F.: Some new mutants of the house fly, Musca domestica, with notations of related phenomena. J. Econ. Entomol. 59, 133 (1966).

——, F. W. PLAPP, JR., and R. D. ORCHARD: Linkage relationships of several insecticide resistance factors in the house fly (Musca domestica L.). Entomol. Expt. Appl. 8, 65 (1965).

Ishida, M., and P. A. Dahm: Metabolism of benzene hexachloride isomers and related compounds in vitro. I. Properties and distribution of the enzyme. J. Econ. Entomol. **58**, 383 (1965 a).

—— —— Metabolism of benzene hexachloride isomers and related compounds in vitro. II. Purification and stereospecificity of house fly enzymes. J. Econ. Entomol. **58**, 602 (1965 b).

Kasai, T., and Z. Ogita: A genetic study on Sevin-resistance and joint toxic action of Sevin with γ-BHC against house flies. Botyu-Kagaku **30**, 12 (1965).

Kikkawa, H.: Genetical studies on the resistance to parathion in *Drosophila melanogaster*. I. Gene analyses. Ann. Rept. Sci. Works, Faculty Sci., Osaka Univ. **9**, 1 (1961).

—— The genetic study on the resistance to Sevin in *Drosophila melanogaster*. Botyu-Kagaku **29**, 42 (1964).

Lovell, J. B., and C. W. Kearns: Inheritance of DDT-dehydrochlorinase in the house fly. J. Econ. Entomol. **52**, 931 (1959).

Matsumura, F., and C. J. Hogendijk: The enzymatic degradation of malathion in organophosphate resistant and susceptible strains of *Musca domestica*. Entomol. Expt. Appl. **7**, 179 (1964 a).

—— —— The enzymatic degradation of parathion in organophosphate-susceptible and -resistant houseflies. J. Agr. Food Chem. **12**, 447 (1964 b).

Metcalf, R. L., T. R. Fukuto, C. Wilkinson, M. H. Fahmy, S. Abd-el-Aziz, and E. R. Metcalf: Mode of action of carbamate synergists J. Agr. Food Chem. **14**, 555 (1966).

Milani, R.: The genetics of the house-fly: Preliminary note. Atti IX Congress Intern. Genetica, 1954, Caryologia, Suppl., p. 791 (1954).

—— Results of genetic research on *Musca domestica* L. Atti A. G. I. **6**, 427 (1961).

——, and M. G. Franco: Comportamento ereditario della resistenza al DDT in incroci tra il ceppo *Orlando-R* e ceppi *kdr* e *kdr+* di *Musca domestica* L. Symp. Genet. **6**, 269 (1959).

Nakatsugawa, T., and P. A. Dahm: Microsomal metabolism of parathion. Biochem. Pharmacol. **16**, 25 (1967).

Ogaki, M., and E. Nakashima-Tanaka: Inheritance of radioresistance in *Drosophila*. I. Mutation Research **3**, 438 (1966).

—— ——, and S. Murakami: Inheritance of ether resistance in *Drosophila melanogaster*. Japan. J. Genet. **42**, 387 (1967).

Ogita, Z.: The genetical relation between resistance to insecticides in general and that to phenylthiourea (PTU) and phenylurea (PU) in *Drosophila melanogaster*. Botyu-Kagaku **23**, 188 (1958).

——, and T. Kasai: Genetic control of multiple esterases in *Musca domestica*. Japan. J. Genet. **40**, 1 (1965 a).

—— —— Genetic control of multiple molecular forms of the acid phosphomonoesterases in the house fly, *Musca domestica*. Japan. J. Genet. **40**, 185 (1965 b).

Oppenoorth, F. J.: Genetics of resistance to organophosphorus compounds and low ali-esterase activity in the housefly. Entomol. Expt. Appl. **2**, 304 (1950).

—— Biochemical genetics of insecticide resistance. Ann. Rev. Entomol. **10**, 185 (1965 a).

—— DDT-resistance in the housefly dependent on different mechanisms and the action of synergists. Medl. Landbouwhogesch. Opzoekingsstat. Gent **30**, 1390 (1965 b).

—— Two types of sesamex-suppressible resistance in the housefly. Entomol. Expt. Appl. **10**, 75 (1967).

——, and G. E. Nasrat: Genetics of dieldrin and γ-BHC resistance in the housefly. Entomol. Expt. Appl. **9**, 223 (1966).

——, and K. VAN ASPEREN: Allelic genes in the housefly produces modified enzymes that cause organophosphate resistance. Science 132, 298 (1960).

—— —— The detoxication enzymes causing organophosphate resistance in the housefly; properties, inhibition, and the action of inhibitors as synergists. Entomol. Expt. Appl. 4, 311 (1961).

PERJE, A. M.: Studies on the spermatogenesis in Musca domestica. Hereditas 34, 209 (1948).

PERRY, A. S., S. MILLER, and A. J. BUCKNER: The enzymatic in vitro degradation of DDT by susceptible and DDT-resistant body lice. J. Agr. Food Chem. 11, 457 (1963).

PIEPER, G. R., and J. E. CASIDA: House fly adenosine triphosphatases and their inhibition by insecticidal organotin compounds. J. Econ. Entomol. 58, 392 (1965).

PLAPP, F. W., JR., G. A. CHAPMAN, and W. S. BIGLEY: A mechanism of resistance to Isolan in the house fly. J. Econ. Entomol. 57, 692 (1964).

——, and R. F. HOYER: Insecticide resistance in the house fly: resistance spectra and prelininary genetics of resistance in eight strains. J. Econ. Entomol. 60, 768 (1967).

SAWICKI, R. M., M. G. FRANCO, and R. MILANI: Genetic analysis of non-recessive factors of resistance to diazinon in the SKA strain of the housefly (Musca domestica L.). Bull. World Health Org. 35, 893 (1966).

SCHONBROD, R. D., W. W. PHILLEO, and L. C. TERRIERE: Hydroxylation as a factor in resistance in house flies and blow flies. J. Econ. Entomol. 58, 74 (1965).

SEKI, T.: Absence of beta-alanine in hydrolyzate of the pupal sheaths of ebony mutant of Drosophila virilis. Drosophila Inform. Serv. 36, 115 (1962).

SHRIVASTAVA, S. P.: Metabolism of o-isopropoxyphenyl methylcarbamate (Baygon) in house flies. Ph.D. Dissertation, Univ. Calif., Berkeley, 101 p. (1967).

SIMS, P., and P. L. GROVER: Conjugations with glutathione: The enzymic conjugation of some chlorocyclohexenes. Biochem. J. 95, 156 (1965).

SULLIVAN, R. L.: Linkage and sex limitation of several loci in the housefly. J. Heredity 52, 282 (1961).

TAHORI, A. S.: Changes in the resistance pattern of a fluoroacetate-resistant fly strain. J. Econ. Entomol. 59, 462 (1966).

TERRIERE, L. C., R. B. BOOSE, and W. T. ROUBAL: The metabolism of naphthalene and 1-naphthol by houseflies and rats. Biochem. J. 79, 620 (1961).

——, and R. D. SCHONBROD: The excretion of a radioactive metabolite by house flies treated with carbon-14 labelled DDT. J. Econ. Entomol. 48, 736 (1955).

TSUKAMOTO, M.: Metabolic fate of DDT in Drosophila melanogaster. I. Identification of a non-DDE metabolite. Botyu-Kagaku 24, 141 (1959).

—— Metabolic fate of DDT in Drosophila melanogaster. III. Comparative studies. Botyu-Kagaku 26, 74 (1961).

—— The log dosage-probit mortality curve in genetic researches of insect resistance to insecticides. Botyu-Kagaku 28, 91 (1963).

—— Methods for the linkage-group determination of insecticide-resistance factors in the housefly. Botyu-Kagaku 29, 51 (1964).

—— The estimation of recombination values in backcross data when penetrance is incomplete, with a special reference to its application to genetic analysis of insecticide-resistance. Japan. J. Genet. 40, 159 (1965).

——, Y. BABA, and S. HIRAGA: Mutations and linkage groups in Japanese stains of the housefly. Japan. J. Genet. 36, 168 (1961).

——, and J. E. CASIDA: Metabolism of methylcarbamate insecticides by the NADPH$_2$-requiring enzyme system from houseflies. Nature 213, 49 (1967 a).

—— —— Albumin enhancement of oxidative metabolism of methylcarbamate insecticide chemicals by the house fly microsome-NADPH$_2$ system. J. Econ. Entomol. **60**, 617 (1967 b).

——, and T. Hiroyoshi: Further studies on the mode of inheritance of resistance to nicotine sulfate in *Drosophila melanogaster*. Botyu-Kagaku **21**, 71 (1956).

——, T. Narahashi, and T. Yamasaki: Genetic control of low nerve sensitivity to DDT in insecticide-resistant houseflies. Botyu-Kagaku **30**, 128 (1965).

——, and M. Ogaki: Gene analysis of resistance to DDT and BHC in *Drosophila melanogaster*. Botyu-Kagaku **19**, 25 (1954).

——, S. P. Shrivastava, and J. E. Casida: Biochemical genetics of house fly resistance to carbamate insecticide chemicals. J. Econ. Entomol. **61**, 50 (1968).

——, and R. Suzuki: Genetic analyses of DDT-resistance in two strains of the housefly, *Musca domestica* L. Botyu-Kagaku **29**, 76 (1964).

—— —— Genetic analyses of diazinon-resistance in the house fly. Botyu-Kagaku **31**, 1 (1966).

van Asperen, K., M. van Mazijk, and F. J. Oppenoorth: Relation between electrophoretic esterase patterns and organophosphate resistance in *Musca domestica*. Entomol. Expt. Appl. **8**, 163 (1965).

Wagoner, D. E.: Linkage group-karyotype correlation in the housefly determined by cytological analysis of X-ray induced translocations. Genetics **57**, 729 (1967).

Winteringham, F. P. W.: Action and inaction of insecticides. J. Royal Soc. Arts, London **110**, 719 (1962).

Yamamoto, I., and J. E. Casida: *O*-Demethyl pyrethrin II analogs from oxidation of pyrethrin I, allethrin, dimethrin, and phthalthrin by a house fly enzyme system. J. Econ. Entomol. **59**, 1542 (1966).

Metabolism of strychnine nitrate
applied for the control of the bear

by
Tetsuo Inukai *

Contents

I. The problem

Damage by the bear, *Ursus arctos yesoensis* Lydekker, which inhabits Hokkaido, the northernmost island of Japan, is very serious. It attacks cattle, horses, sheep, pigs, and chickens, and eats corn, buckwheat, oats, and squash. Recently, damage has been reported in orchards of apples and pears. The worst of it is that a few people are killed by the bear every year. Therefore, the development of Hokkaido has been disturbed greatly by the existence of these bears.

The prefectural government encourages hunters to kill the bears by offering a bounty. However, there appears no sign of any decrease of the bear population, even though around 500 bears have been killed every year for the past 60 years or so. Extermination of the bears is almost impossible under present conditions. The chief reasons are the following: First, the bear hibernates in a den under the snow during the winter, which can be found only with extreme difficulty. Second, it is a dweller of the deep forest in summer, which is almost impenetrable on account of thick undergrowth. Third, the bear is nocturnal in its activity.

From the above it is understandable why resort to the shotgun or rifle will not succeed in killing off the bear population. When the

---

* Department of Agriculture, Hokkaido University, Sapporo, Japan.

bear attacks livestock, for example cattle, it customarily eats a part of the body and leaves the rest of the carcass, concealing it with grass and earth until the next night. At this point, the farmer can use poison to kill the bear, putting it in the carcass. This is naturally illegal, being against Japanese game laws, but the serious havoc caused by the bear has obliged us to carry out experiments in order to determine how to apply poison without any hazard to man and wildlife.

II. Strychnine nitrate as a "bearicide"

The farmers in Hokkaido have secretly used strychnine nitrate for poisoning bears. Poisoning with this chemical has been proved very effective, and fortunately, no trouble has been so far reported, even though sometimes the meat of a poisoned bear has been afterwards eaten by the farmer. However, there has been no fundamental scientific study of this matter. So we have done a series of experiments concerning the metabolism of strychnine nitrate taken by the bear.

First, the lethal dose of strychnine nitrate for the bear was determined by the following experiments:

Case I.   One mg. of strychnine nitrate/kg. of body weight was given *per os*.

Result: The bear died after two hours and 30 minutes.

Case II.   0.5 mg. of strychnine nitrate/kg. of body weight was given *per os*.

Result: The bear died after 13 hours.

Case III. 0.25 mg. of strychnine nitrate/kg. of body weight was given *per os*.

Result: No harmful effects occurred.

Next, in order to see if any part of the body of the poisoned bear contained poison sufficient to harm other animals when eaten, we used dogs, 40 in number, as test animals, having kept them hungry for 24 hours. We gave them muscle from various parts of the body, the heart, the blood, the liver, the intestine, and the esophagus of the dead bear. However, no part of the body of the poisoned bears from Case I or Case II harmed the dogs.

Two foxes and one raccoon caught in the wild were also fed muscle and liver from the above poisoned bears. They showed no signs of poisoning from strychnine nitrate. Chemical analysis of the body parts of the dead bears showed no traces of strychnine nitrate in them, except in the stomach of Case I, in which there was 1.1 mg. of strychnine nitrate/100 g. of sample. Such a small amount of the drug does no harm to other animals.

III. Conclusion

We believe that strychnine nitrate can be used for the control of the bear population in the wilds of Hokkaido under strict supervision by a specialist.

Summary

In order to control the bear population, which causes very serious damage in Hokkaido, Japan, poisoning by strychnine nitrate may prove advisable, since shooting them has been ineffective to diminish their number. A lethal dose of the drug was determined to be 0.5 mg./kg. of body weight. No part of the body of a poisoned bear contains drug concentration sufficient to harm dogs, foxes, or raccoons, if ingested by them. This means that there is no hazard to man or other wildlife in using strychnine nitrate as a poison against the wild bear population of Hokkaido, provided that the lethal dose is not exceeded.

Résumé *

Métabolisme du nitrate de strychnine appliqué dans la lutte contre l'ours

L'empoisonnement par le nitrate de strychnine peut être judicieux pour détruire la population des ours, qui causent de graves dégâts dans l'Hokkaido, Japon, car leur destruction par les armes à feu s'est révélée inefficace pour en réduire le nombre. La dose létale du poison a été déterminée comme étant de 0.5 mg./kg. de poids corporel. Aucune partie du corps de l'ours empoisonné ne contient le poison en quantités suffisantes pour nuire aux chiens, aux renards ou aux ratons laveurs, qui en absorberaient. Ceci signifie que l'utilisation du nitrate de strychnine pour empoisonner la population des ours sauvages de l'Hokkaido ne présente pas de danger pour l'homme ou le gibier, à condition de ne pas dépasser la dose létale.

Zusammenfassung **

Metabolismus von Strychninnitrat, angewandt zur Kontrolle des Bären

Um die Bärenbevölkerung, welche ernsthaften Schaden in Hokkaido, Japan, verursacht, zu kontrollieren, mag die Vergiftung mit Strychninnitrat sich als ratsam erweisen, da Jagen unwirksam ge-

---

* Traduit par S. DORMAL-VAN DEN BRUEL.
** Übersetzt von A. SCHUMANN.

wesen ist, um ihre Zahl zu verringern. Als lethale Dosis der Droge wurde 0.5 mg./kg. Körpergewicht ermittelt. Kein Teil des vergifteten Bärenkörpers enthält Drogenkonzentrationen, die ausreichend wären, um Hunde, Füchse oder Waschbären zu schädigen, falls er von diesen als Nahrung eingenommen wird. Das bedeutet, dass keine Gefahr für Menschen oder anderes Wild im Gebrauch von Strychninnitrat als Gift gegen die Wildbärenbevölkerung von Hokkaido besteht vorausgesetzt, dass die lethale Dosis nicht überschritten wird.

The correlation between physiological activity
and physicochemical property of the substituted phenols

by
Toshio Fujita * and Minoru Nakajima *

Contents

I. Introduction

Recently, a method for correlating the physiological activity of a series of variously substituted compounds using substituent constants such as the Hammett σ constant and a hydrophobicity constant, π, has been developed by Hansch and coworkers (Hansch et al. 1963, Hansch and Fujita 1964, Hansch and Steward 1964, Hansch et al. 1965, Hansch and Anderson 1967, Fujita and Hansch 1967). According to this method, the contributions of the electronic and hydrophobic characters of substituent to a specific biological activity of a series of substituted compounds can be analyzed by equation 1.

$$\log\frac{1}{C} = a\pi - b\pi^2 + \rho\sigma + c \qquad (1)$$

In this equation, C is the equieffective molar concentration such as CD_{50}, ED_{50}, LD_{50}, minimum inhibitory concentration, etc., and a, b (≥ 0), ρ and c are constants which are determined by the method of least squares. This method has been proved to delineate nicely a number of structure-activity problems in the field of pesticides (Hansch and Deutsch 1966 a and b, Bracha and O'Brien 1966, Ish-

---

* Department of Agricultural Chemistry, Kyoto University, Kyoto, Japan.

IDA 1966, Büchel et al. 1966, Johnson et al. 1967, Muir et al. 1968) as well as of medicinal drugs (Hansch and Fujita 1964, Hansch and Anderson 1967, Pratesi et al. 1966, Miller and Hansch 1967, Hansch and Lien 1968), antibiotics (Hansch et al. 1963, Hansch and Steward 1964, Hansch and Deutsch 1965), and enzyme reactions (Hansch et al. 1965, Hansch and Deutsch 1966 b, McMahon 1965, Mitsuda et al. 1967).

More recently, we have applied this procedure to compounds such as a series of substituted phenols where there is considerable variance in dissociation under physiological conditions (Fujita 1966). For these compounds, the biological activity can be expressed by either equation 2 or equation 3 regardless of whether the sites of action are located inside or outside the cell. In equations 2 and 3, [H+] is the hydrogen ion concentration of the extracellular phase and K_A is the

$$\log\frac{1}{C} + \log\frac{K_A + [H^+]}{[H^+]} = a\pi - b\pi^2 + \rho\sigma + c \qquad (2)$$

$$\log\frac{1}{C} + \log\frac{K_A + [H^+]}{K_A} = a\pi - b\pi^2 + \rho'\sigma + c' \qquad (3)$$

dissociation constant. ρ' and c' are constants for the ionized form. Equation 2 describes the structure-activity correlation when the action of a series of compounds is solely due to the neutral molecule. Likewise, equation 3 holds when the ionic form is regarded as the active form. However, if the structure-activity correlation is considered on the basis of the activity data obtained at a single extracellular pH, equations 2 and 3 are interrelated by equations 4 and 5. Therefore, whether the active form is the neutral or ionized form or both, equations 2 and 3 should hold simultaneously.

$$\rho - \rho' = \rho_A \qquad (4)$$

$$c' = c + pK_{Ao} - pH \qquad (5)$$

In these equations (4 and 5), ρ_A is the Hammett reaction constant for the ionization of the dissociable group in the molecule and pK_{Ao} is the pK_A of a standard compound (in most cases, the unsubstituted compound). When $\Delta pK_A = (\log K_{Ax} - \log K_{Ao})$ is used for the analysis in place of the Hammett σ constant, σ_A becomes one in equation 4.

Here, we wish to present the application of this approach to the physiological activity of the substituted phenols such as toxicity to plants.

II. Phytotoxic activity

Blackman and his coworkers tried to analyze the relationship between the activity of phenols causing chlorosis in Lemna minor and their structure (BLACKMAN *et al.* 1955). They tried to correlate the activity with pK_A and solubility of the phenols and found that, in mono-, di-, and some of the tri-substituted phenols, $\log\dfrac{1}{C} + \log\dfrac{K_A + [H^+]}{[H^+]}$ + log solubility is linearly correlated with their pK_A values, where C is the molar concentration required to produce 50 percent chlorotic fronds. However, with the other phenols—in particular those substituted at *ortho*-positions—this kind of correlation is rather poor. They postulated that combined steric and electronic effects of the *ortho*-substituent on the hydrogen bond formation between hydroxyl-group and biosurface might be operative in these *ortho*-substituted compounds.

Using the method of least squares with 25 substituted phenols of their work, the following equations were derived in terms of the equieffective concentration of the neutral form obtained at pH = 5.1, where n is the number of points used in the regression, s is the standard deviation, and r is the correlation coefficient. Equation 8 is an application of equation 2 with the π^2 term deleted, and equations 6 and 7 are those of equation 2 with two terms each deleted. The addition of a π^2 term does not improve the correlation. The π values for polysubstituted phenols which have not been determined experimentally are those obtained by summing up π values of the individual substituents (FUJITA *et al.* 1964). Comparison of equations 6, 7, and

$$\log\frac{1}{C} + \log\frac{K_A + [H^+]}{[H^+]} = 1.007\pi + 1.531 \qquad \begin{array}{ccc} n & s & r \\ 25 & 0.275 & 0.972 \end{array} \quad (6)$$

$$\log\frac{1}{C} + \log\frac{K_A + [H^+]}{[H^+]} = 0.546\Delta pK_A + 2.902 \qquad 25 \;\; 0.572 \;\; 0.873 \quad (7)$$

$$\log\frac{1}{C} + \log\frac{K_A + [H^+]}{[H^+]} = 0.146\Delta pK_A + 0.865\pi + 1.758 \quad 25 \;\; 0.230 \;\; 0.981 \quad (8)$$

8 would indicate that the role of lipohydrophilic (or hydrophobic bonding) character of the substituent is very important, whereas that of the electronic effect is only of subsidary significance. An F-test indicates, however, that both ΔpK_A and π terms in equation 8 are justified at better than 0.995 confidence level when compared with equations 6 and 7, respectively (for ΔpK_A term: $F_{1,22} = 10.79$, for π term: $F_{1,22} = 119.87$; $F_{1,22,0.005} = 9.72$). The calculated values for $\log\dfrac{1}{C} + \log\dfrac{K_A + [H^+]}{[H^+]}$ in Table I were obtained with equation 8.

The same action of the substituted phenols is also examined in terms of the concentration of the phenoxide ion at pH = 5.1 using the value $\log\dfrac{1}{C} + \log\dfrac{[H^+] + K_A}{K_A}$ for the 25 compounds. Equations 9, 10,

Table I. Activity of phenols causing chlorosis in Lemna minor

Substituent	pK_A [a]	ΔpK_A	π	$\log \frac{1}{C} + \log \frac{K_A+[H^+]}{[H^+]}$ [c]		$\log \frac{1}{C} + \log \frac{K_A+[H^+]}{K_A}$ [c]	
				Obs.	Calc. [b]	Obs.	Calc. [d]
H	9.9	0	0	1.8	1.76	6.6	6.56
4-Cl	9.4	0.5	0.9	2.7	2.61	7.0	6.91
2,3-Cl$_2$	7.8	2.1	1.6	3.4	3.45	6.1	6.15
2,4,6-Cl$_3$	6.1	3.8	2.3	4.5	4.30	5.5	5.30
2,4,5-Cl$_3$	7.0	2.9	2.7	5.1	4.52	7.0	6.42
2,3,4,6-Cl$_4$	5.3	4.6	3.4	5.6	5.37	5.8	5.57
Cl$_5$	4.8	5.1	4.4	6.2	6.31	5.9	6.01
2-Me-4-Cl	9.6	0.3	1.4	3.2	3.01	7.7	7.51
2-Me-6-Cl	8.7	1.2	1.2	2.4	2.97	6.0	6.57
3-Me-4-Cl	9.4	0.5	1.5	3.2	3.13	7.5	7.43
4-Me-2,6-Cl$_2$	7.6	2.3	1.9	3.4	3.74	5.9	6.24
3-Me-2,4,6-Cl$_3$	6.6	3.3	2.9	4.7	4.75	6.2	6.25
3-Me-2,4,5,6-Cl$_4$	5.9	4.0	4.0	5.4	5.80	6.2	6.60
2-Me	10.0	-0.1	0.5	2.2	2.18	7.1	7.08
2,6-Me$_2$	10.6	-0.7	1.0	2.4	2.52	7.9	8.02
2,4-Me$_2$	10.4	-0.5	1.0	2.6	2.55	7.9	7.85
2,5-Me$_2$	10.3	-0.4	1.1	2.6	2.65	7.8	7.85
3,5-Me$_2$	10.1	-0.2	1.1	2.7	2.68	7.7	7.68
2,4,6-Me$_3$	10.9	-1.0	1.5	2.8	2.91	8.6	8.71
2,3,5-Me$_3$	10.6	-0.7	1.6	3.0	3.04	8.5	8.54
3-Me-5-Et	10.1	-0.2	1.5	3.1	3.03	8.1	8.03
3,5-Me$_2$-4-Cl	9.6	0.3	2.0	3.6	3.53	8.1	8.03
2,5-Me$_2$-4-Cl	9.7	0.2	2.0	3.6	3.52	8.2	8.12
2,6-Me$_2$-4-Cl	9.9	0	1.9	3.4	3.40	8.2	8.20
3-Me-5-Et-4-Cl	9.6	0.3	2.4	4.0	3.88	8.5	8.38

[a] Taken from Table IV and Fig. 6 of BLACKMAN et al. (1955).   [c] $[H^+] = 10^{-5.1}$.
[b] Calculated by equation 8.   [d] Calculated by equation 11.

and 11 were obtained by the method of least squares. The difference between coefficients of ΔpK_A term of equations 8 and 11 is equal to one and that between constant terms is equal to $pK_{A_0} - pH = 9.9 - 5.1 = 4.8$ as theoretically expected by equations 4 and 5.

$$\log\frac{1}{C} + \log\frac{[H^+] + K_A}{K_A} = \qquad\qquad \begin{array}{ccc} n & s & r \\ 25 & 0.947 & 0.385 \end{array} \quad (9)$$
$$-0.374\pi + 7.885$$

$$\log\frac{1}{C} + \log\frac{[H^+] + K_A}{K_A} = \qquad\qquad 25 \ \ 0.572 \ \ 0.830 \quad (10)$$
$$-0.454\Delta pK_A + 7.702$$

$$\log\frac{1}{C} + \log\frac{[H^+] + K_A}{K_A} = \qquad\qquad 25 \ \ 0.230 \ \ 0.976 \quad (11)$$
$$-0.854\Delta pK_A + 0.865\pi + 6.558$$

The plus sign of the coefficient of π term in these equations suggests that the higher the π value the higher is the potency of the phenols as weed killers. However, the dissociation of the phenols does not necessarily correlate with the apparent potency. The most favorable dissociation constant for the apparent toxicity is calculated by equation 12 or 13. For this series of phenols, equation 12 is de-

$$\frac{\partial \log\dfrac{1}{C}}{\partial \log K_A} = \rho - \frac{K_A}{K_A + [H^+]} = 0 \qquad (12)$$

$$K_A = -\frac{\rho}{\rho'}[H^+] \qquad (13)$$

rived from equation 8 or 11, i.e., other factors being equal, a maximum contribution from the electronic factor to the apparent potency, $\log\frac{1}{C}$, would be made when $K_A = \dfrac{0.146}{0.854}[H^+]$, i.e., $pK_A = 5.9$.

The above analyses indicate an important role of the hydrophobicity or oil-water partitioning rather than the solubility on the process of the toxic action. The result that the π^2-term is not significant for the phenols where π value ranges from 0 to 4, i.e., the partition coefficient changes over 10,000-fold, would suggest that the toxic action is exerted by one step partitioning, e.g., the absorption of the molecule to the cell surface. The fact that phenols with and without ortho-substituent(s) are equally well accomodated by equation 8 or 11 would indicate that steric effects or any other proximity effects of the ortho-substituent on the hydroxyl group are unimportant.

In order to draw a definite conclusion on the active form, it is necessary to obtain the activity data at several different medium pH values. Nevertheless, our procedure is able to describe the pK_A dependence of the apparent toxicity. The optimal pK_A value for the apparent potency is located near the value of experimental pH. This has been shown to be also true for the antibacterial activity of the sulfanilamides (Fujita and Hansch 1967). Thus, our overall results would suggest that, other factors being constant, the pK_A value should be as close to the plant tissue pH as possible in order to obtain a maximal herbicidal activity.

III. Concluding remarks

We have also shown that the same procedure can be nicely applied to the uncoupling activity with the oxidative phosphorylation and antibacterial activity of the substituted phenols (Fujita 1966) as well as the antibacterial activity of the sufanilamide drugs (Fujita and Hansch 1967) and the anticholinesterase and insecticidal activity of the nicotine analogs (Fujita et al. 1968). In this way, the electronic and hydrophobic demands for a specific biological action of a series of dissociable compounds can be analyzed so that the physiochemical characteristics of the site of action could be predicted in certain cases.

Summary

A structure-activity relationship of substituted phenols has been analyzed using substituent constants such as the ΔpK_A value and the hydrophobicity constant π. Separating the effect of ionization from other electronic effects of the substituents, the toxicity of phenols is nicely delineated by means of π together with ΔpK_A, wherever the site of action may be located outside or inside the cell. The hydrophobic property of the molecule is found to play a definite role on various toxic actions of the phenols.

Résumé *

Relation entre l'activité physiologique et propriété physicochimique des phénols substitués

Une relation structure-activité des phénols substitués a été analysée en utilisant des constantes de substituants telles que la valeur du ΔpK_A et la constante d'hydrophobicité π. En séparant l'effet de l'ionisation des autres effets électroniques des substituants, la toxicité des phénols est correctement interprétée au moyen de π et de ΔpK_A, que le site d'action soit localisé à l'extérieur ou à l'intérieur de la

---

* Traduit par R. Mestres.

cellule. La propriété hydrophobe de la molécule joue un rôle défini sur divers effets toxiques des phénols.

Zusammenfassung *

Die Beziehungen zwischen physiologischer Aktivität und physiko-chemischen Eigenschaften der substituierten Phenole

Eine Struktur-Aktivitätsbeziehung von substituierten Phenolen wurde analysiert, indem Substituentenkonstanten wie der ΔpK_A-Wert und die Hydrophobkonstante π benutzt wurden. Wenn man die Wirkung der Ionisierung von andern Elektronenwirkungen der Substituenten trennt, ist die Toxizität der Phenole ausgezeichnet mittels π zusammen mit ΔpK_A dargestellt, wo immer der Sitz der Aktion auch sein mag, ausserhalb oder innerhalb der Zelle. Es wurde gefunden, dass die hydrophoben Eigenschaften der Moleküle eine genau festgesetzte Rolle in den verschiedenen toxischen Aktionen der Phenole spielen.

References

BLACKMAN, G. E., M. H. PARKE, and G. GARTON: The physiological activity of substituted phenols. Arch. Biochem. Biophys. 54, 45 and 55 (1955).

BRACHA, P., and R. D. O'BRIEN: The relation between physical properties and uptake of insecticides by eggs of the large milkweed bug. J. Econ. Entomol. 59, 1255 (1966).

BÜCHEL, K. H., W. DRABER, A. TREBST, and E. PISTORIUS: Zur Hemmung photosynthetischer Reaktionen in isolierten Chloroplasten durch Herbizide des Benzimidazol-Typs und deren Struktur-Aktivitat-Beziehung unter Berücksichtigung des Verteilungskoeffizienten und des pK_A-Wertes. Z. Naturforsch. 21b, 243 (1966).

FUJITA, T.: The analysis of physiological activity of substituted phenols with substituent constants. J. Med. Chem. 9, 797 (1966).

——, and C. HANSCH: Analysis of the structure-activity relationship of the sulfonamide drugs using substituent constants. J. Med. Chem. 10, 991 (1967).

——, J. IWASA, and C. HANSCH: A new substituent constant, pi, derived from partition coefficients. J. Amer. Chem. Soc. 86, 5175 (1964).

——, Y. SOEDA, I. YAMAMOTO, and M. NAKAJIMA: Analysis of physiological activity of nicotine analogs with substituent constants. (Unpublished, 1968).

HANSCH, C., and S. M. ANDERSON: The structure-activity relationship in barbiturates and its similarity to that in other narcotics. J. Med. Chem. 10, 745 (1967).

——, and E. W. DEUTSCH: The structure-activity relationship in penicillins. J. Med. Chem. 8, 705 (1965).

—— —— The structure-activity relationship in amides inhibiting photosynthesis. Biochim. Biophys. Acta 112, 381 (1966).

—— —— The use of substituent constants in the study of structure-activity relationships in cholinesterase inhibitors. Biochim. Biophys. Acta 126, 117 (1966).

---

* Übersetzt von A. SCHUMANN.

—— ——, and R. N. Smith: The use of substituent constants and regression analysis in the study of enzymatic reaction mechanisms. J. Amer. Chem. Soc. 87, 2738 (1965).

——, and T. Fujita: RHO-SIGMA-PI analysis. A method for the correlation of biological activity and chemical structure. J. Amer. Chem. Soc. 86, 1616 (1964).

——, and E. J. Lien: An analysis of the structure-activity relationship in the adrenergic blocking activity of the β-haloalkylamines. Biochem. Pharm. 17, in press (1968).

——, R. M. Muir, T. Fujita, P. P. Maloney, F. Geiger, and M. Streich: The correlation of biological activity of plant growth regulators and chloromycetin derivatives with Hammett constants and partition coefficients. J. Amer. Chem. Soc. 85, 2817 (1963).

——, and A. R. Steward: The use of substituent constants in the analysis of the structure-activity relationship in penicillin derivatives. J. Med. Chem. 7, 691 (1964).

Ishida, S.: Study on the correlation of chemical structure and ovicidal activities of 2-bromoethylthiobenzenes. Part 1. Agr. Biol. Chem. (Tokyo) 30, 800 (1966).

Johnson, H. L., W. A. Skinner, H. I. Maibach, and T. R. Pearson: Repellant activity and physical properties of ring-substituted N,N-diethylbenzamides. J. Econ. Entomol. 60, 173 (1967).

McMahon, R. E.: The enzymatic reduction of aromatic ketones. Abstracts of papers, 150th meeting, Amer. Chem. Soc. p. 74S (1965).

Miller, E., and C. Hansch: A structure-activity analysis of tetrahydrofolate analogs using substituent constants and regression analysis. J. Pharm. Sci. 56, 92 (1967).

Mitsuda, H., K. Yasumoto, and A. Yamamoto: Inhibition of lipoxygenase by saturated monohydric alcohols through hydrophobic bondings. Arch. Biochem. Biophys. 118, 664 (1967).

Muir, R. M., T. Fujita, and C. Hansch: Structure-activity relationship in the auxin activity of mono-substituted phenylacetic acids. Plant Physiol. in press (1968).

Pratesi, P., L. Villa, and E. Grana: Physico-chemical requirements for β-sympatholytic action. Il Pharmaco 21, 409 (1966).

Physico-organic chemical approach
to the mode of action of organophosphorus insecticides

by

T. Roy Fukuto *

Contents

I. Introduction

The toxicity of organophosphorus compounds to insects and mammals generally is associated with the inactivation of the cholinesterase enzymes. The biochemistry of these enzymes, including inhibition by organophosphorus esters, has been the subject of numerous review articles (COHEN and OOSTERBAAN 1963, HEATH 1961, O'BRIEN 1960). Briefly, the inhibition process may be described as proceeding through the following sequence of reactions. For simplicity the model postulated by KRUPKA (1964) for the active site of the cholinesterase enzyme is used.

In this scheme, B is a basic group (histidine imidazole moiety), $HO-$ is a serine hydroxyl, $HA-$ is an acidic group (tyrosine hydroxyl), S is the anionic site, and X is the leaving group in the phosphorus ester. According to this scheme the enzyme and organophosphorus ester combine to form a complex with subsequent phosphorylation of the serine hydroxyl. In certain cases X may interact with the anionic site and aid (or hinder) the inhibition process. The phosphorylated enzyme thus obtained is unable to carry out its normal function, *i.e.*, the hydrolysis of acetylcholine, a reaction which occurs by a similar mechanism. By utilizing different kinetic methods the values for k_e (bi-

* Department of Entomology and Department of Chemistry, University of California, Riverside, California.

molecular inhibition constant), K_i (affinity constant), and k_p (phosphorylation constant) may be determined (ALDRIDGE and DAVISON 1952, MAIN 1964).

The relationship between chemical structure and anticholinesterase activity of organophosphorus compounds has been studied in great detail. These studies have shown that the anticholinesterase activity of phosphorus esters, including phosphates, phosphonates, phosphinates, and phosphoramidates depends largely on the reactivity of the ester (FUKUTO 1957). According to the mechanism of inhibition given above, phosphorylation of cholinesterase takes place by a serine hydroxyl attack on the phosphorus atom, the reaction being aided by the acidic and basic moieties present in the esteratic site. Thus, organophosphorus esters containing a readily displaceable group X are found to be good inhibitors. Good correlation between anticholinesterase activity and reactivity of the phosphorus ester as determined by such parameters as hydrolysis and solvolysis rates, Hammett's sigma constants, and shifts in infrared absorption have been reported (ALDRIDGE and DAVISON 1952, FUKUTO and METCALF 1956). Examples for some of these correlations are given below.

II. Established correlations

A linear correlation between log. k_e and log. $k_{hyd.}$ (logarithm of the first-order hydrolysis constant in phosphate buffer) for the inhibition of erythrocyte cholinesterase by a series of diethyl substituted-phenyl phosphates was first demonstrated by ALDRIDGE and DAVISON 1952. This relationship showed unequivocally that the rate of enzyme inhibition is directly related to the reactivity of the P-O-phenyl bond. Further investigations in our laboratory have shown that other physicoorganic parameters may be used to estimate anticholinesterase activity of these esters. An example is given below in Figure 1 which

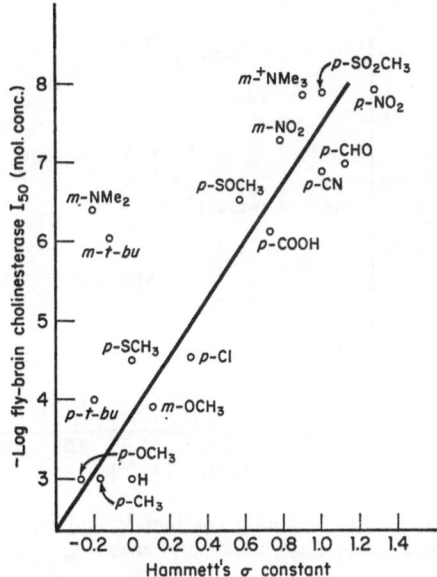

Fig. 1. Relation between log fly-brain cholinesterase I_{50} and Hammett's σ constants for *meta*- and *para*-substituted phenyl diethyl phosphates

shows the correlation between Hammett's sigma constants and the negative logarithm of the I_{50} value (molar concentration required to produce 50 percent inhibition of a given amount of enzyme after 15 minutes incubation).

Hammett's sigma (σ) constants determine the electron-withdrawing or donating properties of the substituent in the phenyl nucleus and the linear relationship found in Figure 1 shows that this parameter may be used to estimate the reactivity of the ester toward fly-head cholinesterase. Similar correlation was obtained between enzyme inhibition and shift in infrared absorption stretching frequency of the P-O-phenyl bond.

Steric factors also play a significant role in the inhibition process. Although this factor has not been examined in depth, the existing evidence indicates that the affinity constant K_i may be affected by steric interaction between the phosphorus ester and the enzyme. Examples of the importance of steric effects in cholinesterase inhibition are given below.

In earlier investigations in our laboratory (FUKUTO and METCALF 1959 a) with a series of ethyl p-nitrophenyl alkylphosphonates the following relationship (Fig. 2) was demonstrated when log. k_e was plotted against log. $k_{hyd.}$ (solvolysis in phosphate buffer). Figure 2 shows that although there is a general trend, the relationship is not linear. Examination of the data shows that in over half of the cases

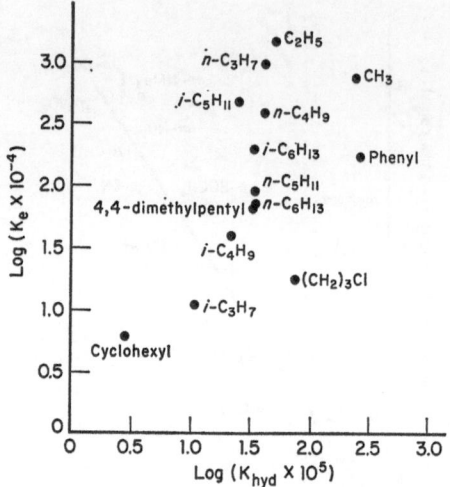

Fig. 2. Relation between log inhibition constant k_e and log. hydrolysis constant $k_{hyd.}$ for ethyl p-nitrophenyl alkylphosphonates

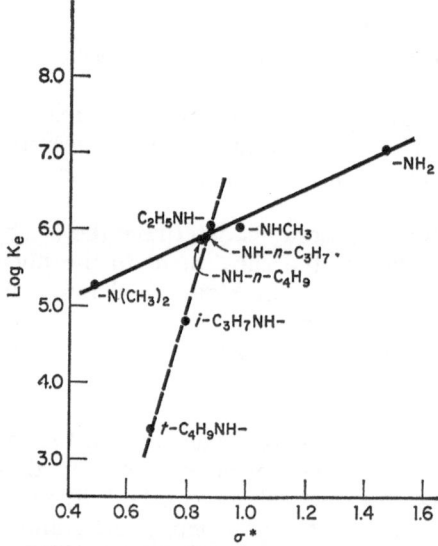

Fig. 3. Relation between log inhibition constant k_e and σ^* values for methyl 2,4,5-trichlorophenyl N-alkylphosphoramidates

the change in anticholinesterase activity is not reflected in the hydrolysis constants. More recently (HANSCH and DEUTSCH 1966) have re-examined these data and have found good correlation between K_e and Taft's steric constant E_s (TAFT 1956) according to the following equation.

$$\log k_e = 3.738 \; E_s + 7.539$$

This relationship gave a correlation coefficient of 0.901 and suggests that the longer chain, hence bulkier alkyl groups, interfere with the overall phosphorylation process as determined by k_e owing to steric effects. Although not strictly proven, steric effects probably are more important in interfering with the formation of the enzyme-inhibitor complex (K_i) than with the actual phosphorylation step (k_p).

Another example of this type is found in results obtained with a series of methyl 2,4,5-trichlorophenyl N-methylphosphoramidates (FU-KUTO et al. 1963). Figure 3 gives the relation between log. k_e and Taft's polar substituent constant σ^*. There is a linear correlation between this parameter for the compounds with straight chain alkyls (solid line). The isopropylamido and t-butylamido compounds are considerably lower in anticholinesterase activity than predicted by σ^* values, suggesting the operation of steric factors. HANSCH's treatment of the data gave the following equation which provides an improved correlation between log k_e and Taft's E_s and σ^* constants

$$\log k_e = 2.359 \; E_s - 3.913 \; \sigma^* + 4.948$$

(correlation coefficient of 0.939). In this case the additional use of the steric substituent constant E_s brought better fit to the data as shown in Figure 4, again demonstrating the importance of steric effects.

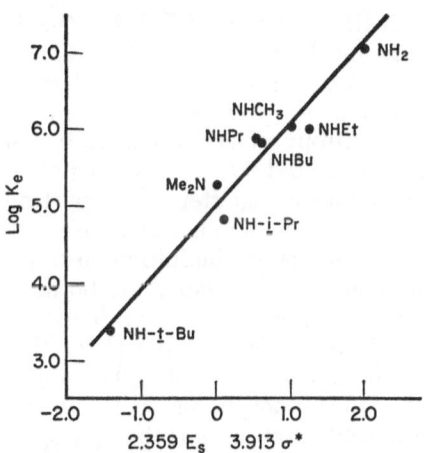

Fig. 4. Relation between log inhibition constant k_e and $(2.359E_s - 3.913 \; \sigma^*)$ for methyl 2,4,5-trichlorophenyl N-alkylphosphoramidates

Other examples which show the importance of steric factors may be seen in differences in anticholinesterase inhibition shown by geometric and optical isomers of certain organophosphorus esters. Mevinphos [methyl 3-(dimethoxyphosphinyloxy)-crotonate] exists as the two isomeric forms depicted below.

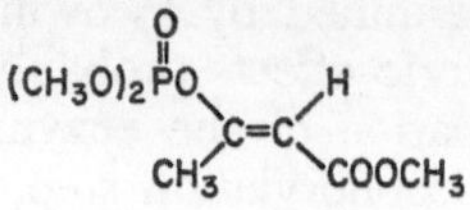

I. Cis-crotonate 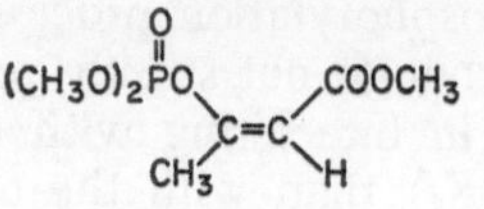   II. Trans-crotonate

I inactivates fly-head cholinesterase at a rate approximately 10 times faster than II (FUKUTO et al. 1961). The rate constants (k_e) for the reaction with the enzyme at three different temperatures are given below.

Table I. *Bimolecular rate constants for the inhibition of fly-head cholinesterase by cis and trans isomers of mevinphos*

Temp. (° C.)	k_e (l. moles^{-1} min.$^{-1}$)	
	I	II
29.0	3.9×10^4	8.7×10^3
33.0	1.0×10^5	1.4×10^4
37.5	2.8×10^5	2.2×10^4

The calculated activation energies (Ea) for the inhibition reaction by I (41.6 kcal.) is much higher than that of II (19.9 kcal.), indicating that the phosphate ester linkage in II is more reactive. The higher rate of inhibition of I over II must then be attributed to its greater PZ factor or entropy of activation, $\Delta S^{\pm}$. The values for log. PZ and $\Delta S^{\pm}$ for I are 32.9 and 90.3 e.u. compared to 14.7 and 6.6 e.u. for II. These values indicate that steric factors strongly influence the reactivity of the isomers with the cholinesterase enzymes.

Differences in cholinesterase inhibition also have been demonstrated with enantiomorphs of organophosphorus esters which contain an asymmetric phosphorus atom. Evidence for this is seen in the biological data given below in Table II for the *d*- and *l*-isomers of O-ethyl S-2-(ethylthio)ethyl ethylphosphonothiolate (FUKUTO and METCALF 1959 b). Since phosphorus is tetrahedral, four different groups bonded to it lead to an asymmetric phosphorus atom.

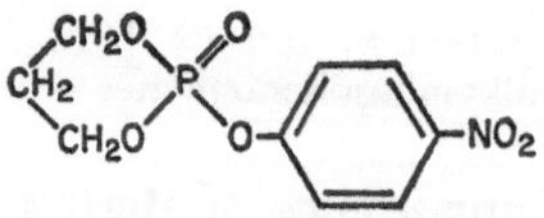

III 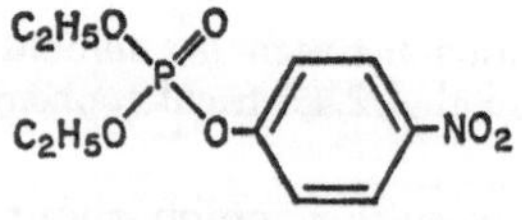   IV. Paraoxon

Table II. k_e and toxicity values for enantiomers of O-ethyl
S-2-(ethylthio)ethyl ethylphosphonothiolate

| Isomer | k_e (l. mole^{-1} min.$^{-1}$) | LD$_{50}$ (μg./g.) | | LC$_{50}$ Mosquito larvae (p.p.m.) |
		Housefly	Honey Bee	
d	6.7 x 10^4	30.0	6.6	0.80
l	7.6 x 10^5	5.0	0.68	0.08
d,l	4.4 x 10^5	9.5		0.15

The k_e values given above show that the l-isomer reacts 11 times faster with fly-head cholinesterase than the d-isomer and 1.8 times faster than the d,l-mixture. These values are reflected in the toxicity data which show that the l-isomer is 6, 10, and 9.7 times more toxic to the house fly, mosquito larva, and honey bee, respectively, than the d-isomer. These results add further dimension to steric effects in inhibition, *i.e.*, stereospecificity in cholinesterase inactivation associated with an asymmeric center.

A very striking example illustrating the importance of steric effects in cholinesterase inhibition is found in results obtained from investigations with a series of 2-p-nitrophenoxy-1,3,2-dioxaphosphorinane 2-oxides exemplified by compound III below (FUKUTO and METCALF 1965).

III is slightly more reactive to aqueous sodium hydroxide than the open chain analog IV or paraoxon. The bimolecular rate constant k_{OH^-} (l. mole^{-1} min.$^{-1}$) is 1.56 for III compared to 0.94 for paraoxon. III, however, is practically devoid of anticholinesterase activity ($I_{50} > 1.3 \times 10^{-3}$ M), while paraoxon is an extremely potent inhibitor with an I_{50} value 2.6×10^{-8} M for fly-head cholinesterase. Since organophosphorus esters of the same order of reactivity as paraoxon are generally inhibitors of cholinesterase, the low activity of the six-membered ring ester is surprising and its inactivity must be attributed to steric effects.

Recent investigations in our laboratory (HOLLINGWORTH et al. 1967) have shown that the selectivity of Sumithion and other 3-alkyl analogs to insects, *i.e.*, high toxicity to insects but low toxicity to mammals, is due partially to differences in the inhibition of insect and mammalian cholinesterase as illustrated by the information given below in Table III.

Since the toxic action of thionate phosphorus esters occurs through the corresponding phosphate ester the anticholinesterase activity of the thionate esters are not given. The data show that there is a gradual increase in fly-head anticholinesterase activity with increasing size of the 3-alkyl substituent compared to a strong decrease in mouse brain activity. Although there is a slight decrease in housefly toxicity

Table III. *Toxicity and antiChE activity of Sumithion and analogs*

$$(CH_3O)\overset{\overset{\text{X}}{\|}}{P}\text{-O-}\underset{}{\bigcirc}\overset{\text{Y}}{-NO_2}$$

No.	X	Y	LD$_{50}$ (mg./kg.) House-fly	LD$_{50}$ (mg./kg.) White mouse	ChE I$_{50}$ (M) Fly-head	ChE I$_{50}$ (M) Mouse brain
V	S	H	1.2	23	—	—
VI	S	CH$_3$	3.1	1250	—	—
VII	S	i-C$_3$H$_7$	6.3	880	—	—
VIII	O	H	2.5	21	1.0×10^{-7}	3.3×10^{-7}
IX	O	CH$_3$	4.3	120	5.6×10^{-8}	1.9×10^{-6}
X	O	i-C$_3$H$_7$	6.5	>500	1.6×10^{-8}	1.2×10^{-5}

with the approximately six-fold increase in fly-head inhibition in going from VIII to X, there is significant correlation between mouse toxicity and mouse-brain anticholinesterase activity.

A critical analysis of fly-head and mouse-brain inhibition by VIII, IX, and X according to MAIN (1964) showed that the change in the overall rate of inhibition (k_e) of both enzymes is due primarily to change in the value for k_i or the affinity constant. The values of K_i, k_e, k_p, and $k_{hyd.}$ are given in Table IV.

There is a slight decrease in k_p value in going from H to i-C$_3$H$_7$ and this is due to the electron donating tendency of the alkyl group plus the inhibition of resonance of the nitro moiety by steric interference. Comparison of k_p values with rate constants for solvolysis ($k_{hyd.}$) shows that the phosphorylating ability of these compounds is directly related to $k_{hyd.}$ values and decreases with increase in size of the 3-alkyl group. This linear correlation between k_p *and* $k_{hyd.}$ gives excellent confirmation that phosphorylation of the enzyme depends on the factors which also determine purely chemical hydrolytic rates. The substantial changes in K_i values for the three compounds suggest that enzyme-inhibitor complex formation with housefly ChE is aided by interaction of the 3-alkyl substituent with the anionic site but is hindered in the case of mouse ChE. The data suggest that a significant difference exists between the structures of insect and mammalian enzymes and these differences may explain in part the selectivity of Sumithion.

Summary

The toxicity of organophosphorus compounds to insects and mammals generally is associated with the inactivation of the cholinesterase enzymes. Briefly, the inhibition process takes place by nucleophilic attack of a serine hydroxyl, situated in the esteratic site of the enzyme, on the phosphorus atom of the organophosphorus ester. This reaction,

Table IV. *Bimolecular, affinity, phosphorylation, and hydrolysis constants for the inhibition of fly-head and bovine erythrocyte ChE*

$$(CH_3O)_2P\text{-}O\text{-}\langle\bigcirc\rangle^{Y}\text{-}NO_2 \quad O{=}$$

No.	Y	$k_e \times 10^{-5}$ $(M^{-1}\,min.^{-1})$	$K_i \times 10^5$ (M)	k_p $(min.^{-1})$	$k_{hyd.} \times 10^{-4}$ $(min.^{-1})$
		Fly-head			
VIII	H	2.9	3.7	10.6	5.25
IX	CH_3	7.6	1.1	8.3	3.50
X	$i\text{-}C_3H_7$	22.6	0.33	7.5	2.89
		Bovine erythrocyte			
VIII	H	5.2	1.3	6.6	
IX	CH_3	0.73	6.7	5.0	
X	$i\text{-}C_3H_7$	0.22	15.8	3.5	

which results in a phosphorylated enzyme, probably is catalyzed by basic (imidazole nitrogen) and acid (tyrosine hydroxyl) groups also present in the esteratic site, the mechanism of catalysis being similar to that proposed for the hydrolysis of acetyl choline. The evidence indicates that the enzyme and organophosphorus ester combine to form a complex with subsequent phosphorylation of the serine hydroxyl. By utilizing different kinetic methods, the values for the overall bimolecular inhibition constant for the reaction between inhibitor and enzyme, the affinity constant for complex formation, and the phosphorylation constant may be determined.

The relationship between chemical structure and anticholinesterase activity of organophosphorus compounds has been studied in great detail. These studies have shown that the anticholinesterase activity of phosphorus esters, including phosphates, phosphonates, and phosphinates, depends largely on the reactivity of the phosphorus atom. Thus, organophosphorus esters of the type $(RO)_2P(O)X$ in which X is a readily displaceable group are found to be good inhibitors. Good correlation between anticholinesterase activity and reactivity of the phosphorus ester as determined by such parameters as hydrolysis and solvolysis rates, Hammett's sigma constants, Taft's polar and steric substituent constants, and shifts in infrared absorption has been obtained.

Steric factors occasionally play a significant role in the inhibition process. Although this factor has not been examined in depth, the existing evidence indicates that the affinity constant for enzyme-ester complex formation may be affected by steric interaction between the phosphorus ester and the enzyme. Thus, differences in anticholinesterase activity for geometric and optical isomers, and examples demonstrating interaction between the anionic site of the cholinesterase enzyme and substituents in the phosphorus ester have been reported. Steric factors also are believed to be responsible for the poor anticholinesterase activity of the cyclic trimethylene p-nitrophenyl phosphate, a compound of similar hydrolytic reactivity as paraoxon, a potent inhibitor. The difference in anticholinesterase activity of Sumioxon and other dimethyl 3-alkyl-4-nitrophenyl phosphates between insect and mammalian cholinesterase has been attributed to steric effects.

Résumé *

Approche physicochimique du mode d'action des insecticides organophosphorés

La toxicité des composés organo-phosphorés pour les insectes et les mammifères est généralement associée à l'inactivation des cholinestérases. Brièvement, le processus d'inhibition a lieu par attaque nu-

* Traduit par R. Mestres

cléophile d'un hydroxyle de la sérine, situé dans le site "fonction ester" de l'enzyme, sur l'atome de phosphore de l'ester oganophosphoré. Cette réaction qui produit un enzyme phosphorylé est probablement catalysée par des groupes basiques (azote d'imidazole) et acide (hydroxyle de la tyrosine) également présents dans le site "fonction ester", le mécanisme de catalyse étant semblable à celui propose pour l'hydrolyse de l'acétylcholine. De toute évidence, l'enzyme et l'ester organophosphoré se combinent pour former un complexe suivi d'une phosphorylation de l'hydroxyle de la sérine. En utilisant différentes méthodes cinétiques, les valeurs de la constante de l'inhibition bimoléculaire globale pour la réaction entre l'inhibiteur et l'enzyme, la constante d'affinité pour la formation du complexe et la constante de phosphorylation peuvent être déterminées.

La relation entre la structure chimique et l'activité anticholinestérasique des composés organo-phosphorés a été étudiée en détail. Ces études ont montré que l'activité anticholinestérasique des esters phosphorés, comprenant les phosphates, les phosphonates et les phosphinates, dépend grandement de la réactivité de l'atome de phosphore. Ainsi, les esters organophosphorés du type $(RO)_2P(O)X$ dans lesquels X est un groupe aisément déplacable sont de bons inhibiteurs. Une bonne corrélation a été obtenue entre l'activité anticholinestérasique et la réactivité de l'ester phosphoré, valeurs déterminées par des paramètres tels que les vitesses d'hydrolyse et de solvolyse, les constantes sigma d'Hammett, les constantes des substituants polaires et stériques de Taft et les déplacements de l'absorption dans l'infra-rouge.

Les facteurs stériques jouent occasionnellement un rôle important dans le processus d'inhibition. Bien que ce facteur n'ait pas été examiné profondément, il apparaît nettement que la constante d'affinité pour la formation du complexe enzyme-ester peut être affectée par une intéraction stérique entre l'ester phosphorique et l'enzyme. Ainsi des différences dans l'activité anticholinestérasique des isomères géométriques et optiques, et des exemples démontrant l'intéraction entre le site anionique de la cholinestérase et les substitutions dans l'ester phosphorique ont été signalées. Des facteurs stériques sont aussi jugés responsables de la faible activité anticholinestérasique du cyclo triméthylène p-nitro phényl phosphate, composé à réactivité hydrolytique similaire à celle du paraoxon, puissant inhibiteur. La différence dans l'activité anticholinestéraque du sumioxon et d'autres diméthyl 3-alkyl-4-nitrophényl phosphates entre les cholinestérases des insectes et des mammifères a été attribuée à des effets stériques.

Zusammenfassung *

Physiko-organisch-chemische Betrachtung zur Wirkungsweise von Organophosphorinsektiziden

Die Toxizität von Organophosphorverbindungen gegenüber Insek-

---

* Übersetzt von A. Schumann.

ten und Säugetieren ist im allgemeinen mit der Inaktivierung der Cholinesteraseenzyme verbunden. Kurz gesagt, der Hemmungsprozess findet durch nukleophilen Angriff eines Serinhydroxyl, welches sich an der veresterten Seite des Enzyms befindet, auf das Phosphoratom des Organophosphorsäureesters statt. Diese Reaktion, die ein phosphoryliertes Enzym ergibt, wird wahrscheinlich durch basische (Imidazolstickstoff) und saure (Tyrosinhydroxyl) Gruppen katalysiert, welche ebenfalls an der Esterseite vorhanden sind, der Mechanismus der Katalyse ist ähnlich wie der für die Hydrolyse von Acetylcholin vorgeschlagene. Das Beweismaterial zeigt an, dass das Enzym und der Organophosphorsäureester sich verbinden und einen Komplex bilden mit nachfolgender Phosphorylierung des Serinhydroxyls. Unter Benutzung von verschiedenen kinetischen Methoden können die Werte für die gesamt-bimolekulare Hemmungskonstante für die Reaktion zwischen Hemmer und Enzym, die Affinitätskonstante für die Komplexbildung und die Phosphorylierungskonstante bestimmt werden.

Die Beziehung zwischen chemischer Struktur und der Anticholinesteraseaktivität von Organophosphorverbindungen ist sehr ausführlich untersucht worden. Diese Studien haben gezeigt, dass die Anticholinesteraseaktivität von Phosphorestern, einschliesslich Phosphaten, Phosphonaten und Phosphinaten, stark von der Reaktionsfähigkeit des Phosphoratoms abhängt. Auf diese Weise werden Organophosphorsäureester des Typs $(RO)_2P(O)X$, in welchen X eine leicht zu ersetzende Gruppe ist, als gute Hemmer erachtet. Gute Wechselbeziehung zwischen Anticholinesteraseaktivität und Reaktionsfähigkeit des Phosphorsäureesters, bestimmt mit Parametern wie Hydrolyse- und Solvolyseraten, Hammetts Sigma-Konstante, Tafts Polar- und sterische Substituentenkonstante und Verschiebungen in der Infrarotabsorption sind erhalten worden.

Sterische Faktoren spielen gelegentlich eine bedeutsame Rolle im Hemmungsprozess. Obwohl dieser Faktor noch nicht in aller Tiefe untersucht worden ist, zeigt das vorhandene Beweismaterial an, dass die Affinitätskonstante für die Enzym-Ester-Komplexbildung durch sterische Wechselwirkunk zwischen dem Phosphorsäureester und dem Enzym beeinflusst sein kann. Daher sind Unterschiede in der Anticholinesteraseaktivität für geometrische und optische Isomere und Beispiele, welche die Wechselwirkung zwischen der anionischen Seite des Cholinesteraseezyms und den Substituenten im Phosphorsäureester demonstrieren, berichtet worden. Auch nimmt man an, dass sterische Faktoren für die geringe Anticholinesteraseaktivität des cyklischen Trimethylen-(p-nitrophenyl)-phosphat verantwortlich sind, einer Verbindung von ähnlich hydrolytischer Reaktionsfähigkeit wie Paraoxon, eines potentiellen Hemmers. Der Unterschied in der Anticholinesteraseaktivität von Sumioxon und anderen Dimethyl-(3-alkyl-4-nitrophenyl)-phosphaten zwischen Insekten- und Säugetiercholinesterase wird sterischen Wirkungen zugeschrieben.

References

ALDRIDGE, W. N., and A. N. DAVISON: The inhibition of erythrocyte cholinesterase by triesters of phosphoric acid. Biochem. J. 51, 62 (1952).

COHEN, J. A., and R. A. OOSTERBAAN: The active site of acetylcholinesterase and related esterases and its reactivity towards substrates and inhibitors. In: Handbuch der experimentellen Pharmakologie, Ergänzungswerk XV. Berlin-Göttingen-Heidelberg: Springer-Verlag (1963).

FUKUTO, T. R.: The chemistry and action of organic phosphorus insecticides. In: Advances in pest control research, Vol. I. New York: Interscience (1957).

——, E. O. HORNIG, R. L. METCALF, and M. Y. WINTON: Configuration of the α- and β-isomers of methyl 3-(dimethoxyphosphinyloxy)-crotonate (Phosdrin). J. Org. Chem. 26, 4620 (1961).

——, and R. L. METCALF: Structure and insecticidal activity of some diethyl substituted phenyl phosphates. J. Agr. Food Chem. 4, 930 (1956).

—— —— The effect of structure on the reactivity of alkylphosphonate esters. J. Amer. Chem. Soc. 81, 372 (1959 a).

—— —— Insecticidal activity of the enantiomorphs of O-ethyl S-2-(ethylthioethyl)-ethyl ethylphosphonothiolate. J. Econ. Entomol. 52, 739 (1959 b).

—— —— Reactivity of some 2-p-nitrophenoxy-1,3,2-dioxaphospholane 2-oxides and -dioxaphosphorinane 2-oxides. J. Med. Chem. 8, 759 (1965).

—— ——, M. Y. WINTON, and R. B. MARCH: Structure and insecticidal activity of alkyl 2,4,5-trichlorophenyl N-alkylphosphoramidates. J. Econ. Entomol. 56, 808 (1963).

HANSCH, C., and E. A. DEUTSCH: The use of substituent constants in the study of structure-activity relationships in cholinesterase inhibitors. Biochem. Biophys. Acta 126, 117 (1966).

HEATH, D. F.: Organophosphorus poisons. London: Pergamon Press Ltd. (1961).

HOLLINGWORTH, R. M., T. R. FUKUTO, and R. L. METCALF: Selectivity of Sumithion compared with methyl parathion. Influence of structure on anticholinesterase activity. J. Agr. Food Chem. 15, 235 (1967).

KRUPKA, R. M.: Acetylcholinesterase. Can. J. Biochem. 42, 677 (1964).

MAIN, A. R.: Affinity and phosphorylation constants for the inhibition of esterases by organophosphates. Science 144, 992 (1964).

O'BRIEN, R. D.: Toxic phosphorus esters. New York: Academic Press (1960).

TAFT, R. W., JR.: Separation of polar, steric, and resonance effects in reactivity. In: Steric effects in organic chemistry. New York: Wiley (1956).

Physico-chemical studies on the absorption of pesticides by the insect cuticle and penetration to the insect body

by
MASANA SUWANAI *

Contents

I. Introduction

Physico-chemical studies were conducted on the mode of action of several organic insecticides such as γ-BHC, ethyl parathion, and methyl parathion in gaseous state and in aqueous solutions on adults of the adzuki-bean weevil, *Callosobruchus chinensis*. These insecticides are absorbed at first by the surface of the insect body and then moved inwards for their action. The first study was, therefore, made on the absorption of these insecticides by the surface of insects based on the kinetic theory of molecules (SUWANAI 1957). The movement of insecticide chemicals into the insect body was then studied based on the diffusion theory.

II. Vapor action of insecticides

The vapor action of γ-BHC, ethyl parathion, and methyl parathion was observed in rubber-stoppered glass tubes (3×50 cm.) as shown in Figure I. In each tube, filter paper impregnated with one of these

---

* Tokyo University of Agriculture and Technology, Fuchu, Tokyo, Japan.

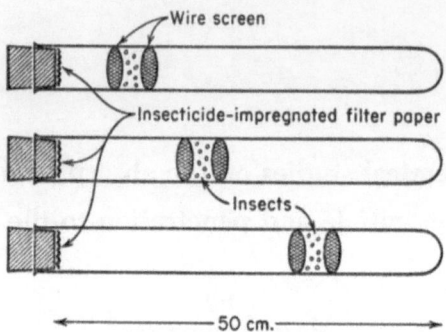

Fig. 1. Glass tubes used for the observation of vapour action

insecticides was placed at an end attached to the bottom of the rubber stopper and 20 adult weevils were placed between two pieces of wire screen at a distance from the filter paper. Undoubtedly, the insecticide must have acted on the insects as gaseous molecules. The first experiment was carried out with ethyl parathion. The lethal time varied with the distance between the filter paper and the insects. Thus, the shorter the distance was, the shorter was the lethal time. Figure 2 gives the relationship between the distance and the lethal

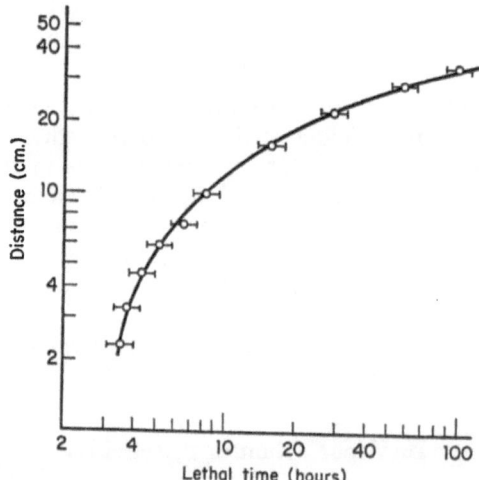

Fig. 2. Relation between the distance and the lethal time for ethyl parathion at 30° C.

time. It can be seen from this figure that 99 percent of 20 insects were killed within four hours when placed three cm. from the filter paper, while none of the insects was killed during an exposure period of 100

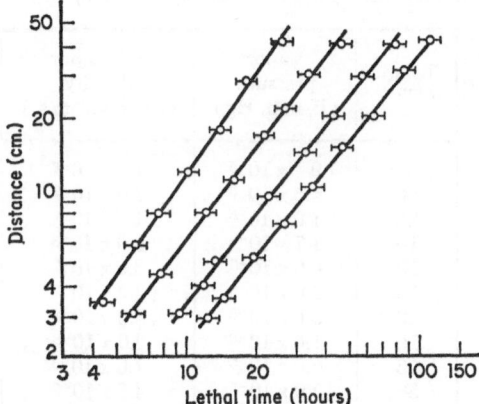

Fig. 3. Relation between the distance and the lethal time for γ-BHC at 20°, 23°, 30°, and 35° C.

hours when placed 50 cm. from the paper. Figure 3 illustrates the result of a similar experiment with γ-BHC. The relations between the distance and the lethal time show a common tendency at various temperatures between 20° and 35° C.

In order to discuss these experimental results, it seemed necessary to introduce the kinetic theory of molecules and to determine lethal doses for these insecticides toward the weevils. Measurements and calculations were, therefore, made of physical constants and LD_{99} for these insecticides. Saturated vapor densities were calculated from the ideal gas law which is given by the following equation:

$$C_0 = \frac{PM}{RT} \qquad \text{(Equation 1)}$$

in which R, T, P, and M are the gas constant, the absolute temperature, the vapor pressure, and the molecular weight, respectively. Diffusion coefficients of insecticides in air were calculated from Gilliland's semi-theoretical equation given as follows (ARNOLD 1930, BALSON 1948, GILLILAND 1943, HIRSCHELDER 1934, LEBASS 1906):

$$D_{1\cdot2} = \frac{0.0043T^{2/3}}{P(V_1^{1/3}+V_2^{1/3})^2} \sqrt{\frac{1}{M_1}+\frac{1}{M_2}}$$

$$\text{(Equation 2)}$$

where T is the absolute temperature, P is the total pressure, M_1 and M_2 are molecular weights of the two gases, and V_1 and V_2 are the molecular volumes of the two gases. These constants and coefficients calculated are given in Table I together with the vapor pressures.

Table I. *Physical constants and coefficients of insecticides*

Insecticide	Temp. (C°.)	Vapor pressure (P, mm. Hg)	Sat. vapor density (Co, mg./cm³.)	Diffusion coefficient (D, cm²./hr.)
γ-BHC	35	9 x 10⁻⁵	1.7 x 10⁻⁶	1.85 x 10²
	30	4.5 x 10⁻⁵	7.0 x 10⁻⁷	1.8 x 10²
	20	1.0 x 10⁻⁵	1.5 x 10⁻⁷	1.7 x 10²
	10	1.7 x 10⁻⁶	3.0 x 10⁻⁸	1.6 x 10²
Ethyl parathion	35	4.0 x 10⁻⁵	8.0 x 10⁻⁷	1.6 x 10²
	30	2.0 x 10⁻⁵	3.2 x 10⁻⁷	1.55 x 10²
	20	5.0 x 10⁻⁶	8.0 x 10⁻⁸	1.45 x 10²
	10	1.0 x 10⁻⁶	2.0 x 10⁻⁸	1.4 x 10²
Methyl parathion	35	6.0 x 10⁻⁵	1.0 x 10⁻⁶	1.70 x 10²
	30	3.0 x 10⁻⁵	4.5 x 10⁻⁷	1.66 x 10²
	20	7.0 x 10⁻⁶	1.0 x 10⁻⁷	1.55 x 10²
	10	1.8 x 10⁻⁶	2.0 x 10⁻⁸	1.5 x 10²

On the other hand, LD_{99} values for several insecticides toward 20 adults of the adzuki-bean weevil at a 24-hour exposure period were estimated by topical application and dry film methods. The values for γ-BHC, ethyl parathion, and methyl parathion were 0.7, 0.6, and 0.5 microgram (μg.)/20 adults, respectively. The volume of each glass tube used for the observation of the vapor action was about 300 ml. When the tube is saturated with the vapor of ethyl parathion at 20°C., the amount of gaseous ethyl parathion in the tube should be as follow:

$$C_o \times (\text{vol. of tube}) = 8 \times 10^{-8} \text{ mg./ml.} \times 300 \text{ ml.}$$
$$= 0.02 \ \mu\text{g.}$$

This amount is only one-thirtieth of the LD_{99} for ethyl parathion to 20 weevils. It means that 20 weevils would not be killed by the vapor of the insecticide saturated in the tube alone. This marked difference may be explained by absorption phenomena.

As the insect body absorbs the vapor of insecticide, the vapor density in the tube decreases. To make up for the absorbed vapor, the insecticide vaporizes from the filter paper to which the insecticide has been applied. Upon reaching the insects, the gaseous insecticide molecules are quickly absorbed by the surfaces of the insects. The insects are killed when the total amount of absorbed insecticide reaches the lethal dose.

If the vapor density around the insects is expressed as C', the velocity of vaporization, V, is given by Fick's diffusion equation as follows:

$$V = \frac{A}{H} D(C_o - C') \qquad \text{(Equation 3)}$$

where A is the cross-sectional area of the tube and D is the diffusion coefficient. If the insects absorb the insecticide vapor rapidly, the vapor density, C', would tend to zero. When C' is zero, Equation 3 can be modified to Equation 3′ (it can be proved experimentally that C' is zero) (see Fig. 4).

$$V = \frac{A}{H} DC_0 \qquad \text{(Equation 3′)}$$

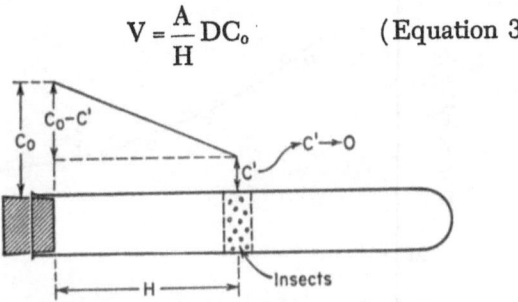

Fig. 4. Density gradient of gaseous ethyl parathion in the glass tube

If vaporization goes on at a constant velocity, the following equation is given:

$$M = \frac{A}{H} DC_0 T \qquad \text{(Equation 4)}$$

where M is the amount of the insecticide which vaporizes from the filter paper during T hours (SUWANAI 1957).

Since the amount of the insecticide vapor remaining in the glass tube is very small, most of the insecticide which vaporizes from the filter paper is assumed to be absorbed by the insects. When the amount of insecticide absorbed by the insects is expressed as W, the following relation is obtained:

$$M = W$$

If all of these 20 insects are killed during T hours, the amount of the insecticide which vaporizes during T hours is nearly equivalent to the lethal dose toward 20 weevils. By substituting H and T in Equation 4 with the values obtained from Figure 2, the amount of ethyl parathion absorbed during T hours can be calculated. Figure 5 shows the relation between the amount of ethyl parathion absorbed by 20 weevils and the lethal time toward 99 percent of 20 insects.

III. Insecticides absorbed by insects from aqueous solutions

A small volume of aqueous ethyl parathion solution was pipetted into a small test tube and test insects were dipped in this solution about 30 seconds. The insects were then transferred from the solution into another vessel for the estimation of the lethal time. On the other hand, the residual amount of the insecticide in the test tube was determined chemically. The actual amount of insecticide deposited on the insects was then calculated from the following equation:

$$W = CL - R \qquad \text{(Equation 5)}$$

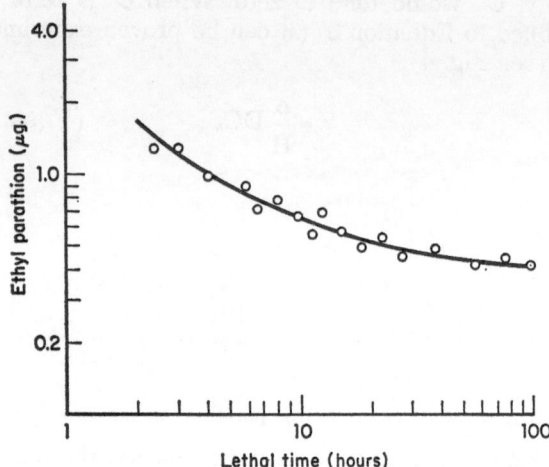

Fig. 5. Relation between the amount of ethyl parathion absorbed by 20 Weevils
from vapour phase and the lethal time for the insecticide toward 99% of
the insects

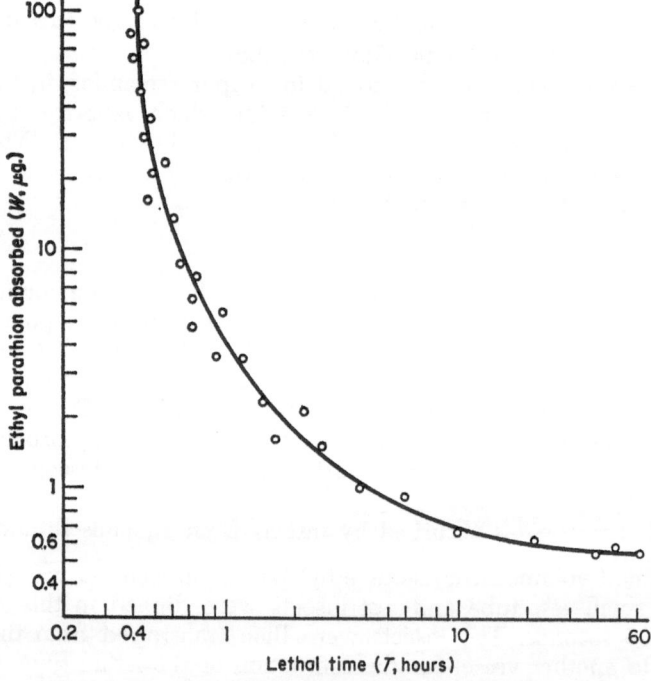

Fig. 6. Relation between the amount of ethyl parathion absorbed by 20 Weevils
from aqueous solution and the lethal time for the insecticide toward 20
Weevils

in which W is the actual amount of the insecticide deposited on the insects, C is the concentration of the insecticide in the solution, L is the volume of the solution, and R is the residual amount of insecticide in the test tube.

Figure 6 shows the relation between the lethal time toward 99 percent of 20 weevils and the actual amount of ethyl parathion deposited on the insects calculated from Equation 5.

Values of W obtained from two experiments (vapor action and the dipping method) were plotted against T on the same scales, as shown in Figure 7. It is clear that the $W - T$ relations obtained from

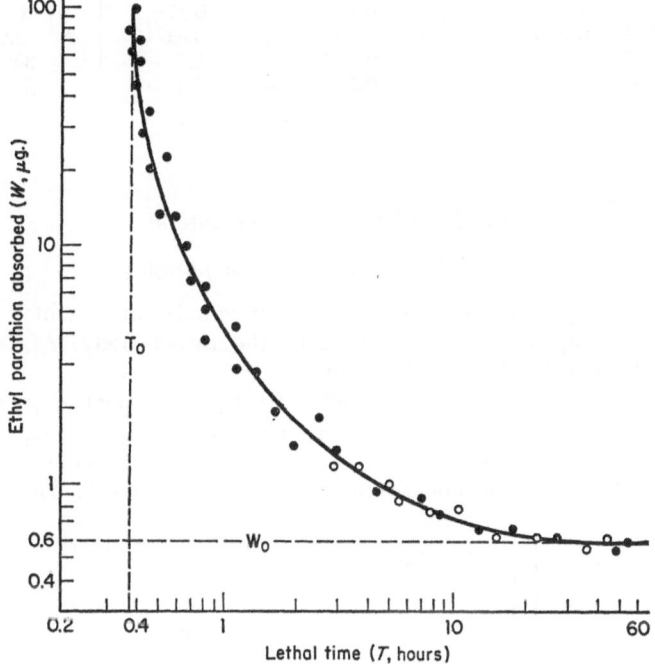

Fig. 7. Relation between the amount of ethyl parathion absorbed by 20 weevils from vapour phase or aqueous solution and the lethal time for the insecticide toward 20 weevils: ● = from aqueous solution and ○ = from vapor phase

two experiments give a single curve. The value T tends to a limit of T_o with the increase in the value of W, while the value of W tends to a limit of W_o with the increase in the value of T. The products of $(W - W_o)$ and $(T - T_o)$ calculated had an approximately constant value as given in Equation 6:

$$(W - W_o)\ (T - T_o) = K \qquad \text{(Equation 6)}$$

This equation shows that the curve obtained in Figure 7 is an equilateral hyperbola. Similar curves were obtained with three other insecticides. The values of W_o, T_o, and K for ethyl parathion and these three insecticides obtained with the adzuki-bean weevil are listed in Table II.

Table II. *Values of* T_o, W_o, *and* K *for four insecticides toward 20 adults of the adzuki-bean weevil* [a]

Insecticide	W_o (mg.)	T_o (hr.)	K (g.hr.)
Methyl parathion	0.18	0.17	0.56
Ethyl parathion	0.30	0.31	2.9
γ-BHC	0.4	1.0	60
EPN	0.5	5.0	80

[a] 86 mg.

IV. Paraffin from insect cuticle

a) Extraction of the insect paraffin

It was supposed that there must be some substances which absorb insecticides rapidly on the surface of the insect body. An attempt was made to extract these substances.

Adults of the azuki-bean weevil (10 g.) were packed in a glass column and percolated with ether. The eluate was collected into 3-ml. fractions. Each fraction was evaporated to dryness and the residue was weighed. The result is shown in Figure 8. The residue

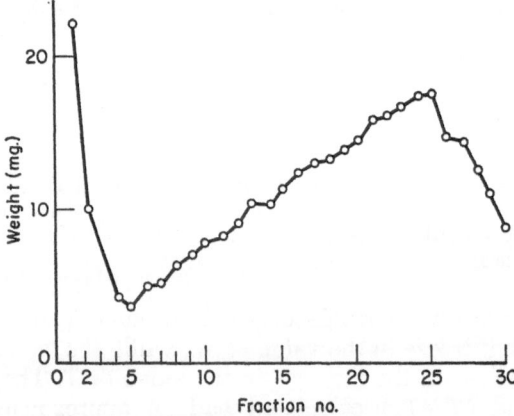

Fig. 8. Weight of substance eluted into 3-ml. fractions from adzuki-bean weevils (10g.) with ether

of each fraction was then analyzed by infrared spectroscopy. It was found that residues in early eluted fractions mainly consisted of paraffin. If the surface of the insect body is covered with substances very soluble in ether these substances would have been eluted into early fractions. The paraffin obtained in this experiment may, therefore, be considered as the main substance on the surface of the insect body.

Based on this result, 20 kg. of adult weevils were washed with ether for a short time. In this way, about 20 g. of the insect paraffin was obtained and refined. The result of elementary analysis showed that this refined paraffin has a molecular formula of C_nH_{2n+2} ($n = 24$ to 26).

b) Partition of insecticides between water and the insect paraffin

Partition coefficients of ethyl parathion and methyl parathion were measured between distilled water and the paraffin from the adzuki-bean weevil. The results are given in Figure 9.

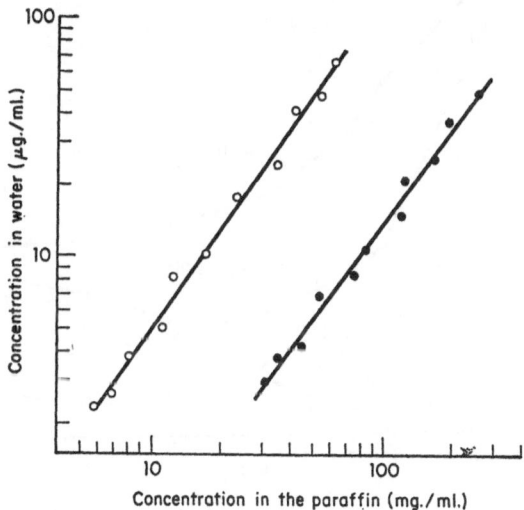

Fig. 9. Partition of ethyl parathion (●) and methyl parathion (○) between water and the insect paraffin

c) Diffusion of insecticides through the insect paraffin layer

The velocity of diffusion through the insect paraffin layer from an aqueous solution to distilled water was measured for ethyl parathion and methyl parathion with an H-shaped glass apparatus shown in Figure 10. The insecticide concentration in chamber A decreased

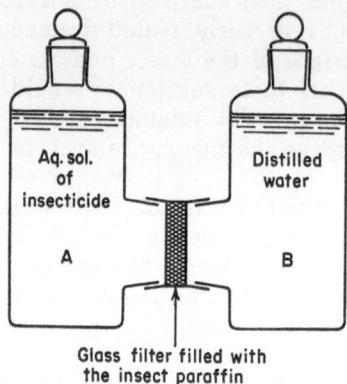

Fig. 10. H-shaped glass apparatus used for the observation of insecticide diffusion through the paraffin layer

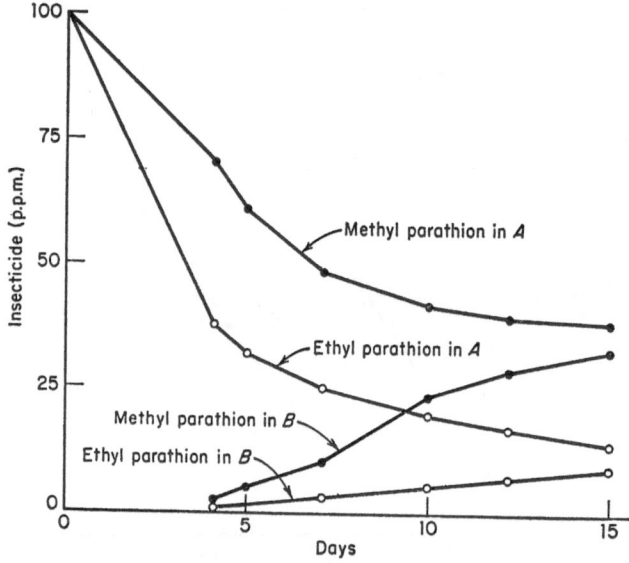

Fig. 11. Changes in the concentration of insecticides in water held in two chambers separated by the paraffin layer

gradually, while that in chamber B increased as shown in Figure 11. The density gradient in the paraffin layer was supposed to be dependent upon the thickness of the layer as illustrated in Figure 12.

By Fick's diffusion equation, the diffusion velocity of an insecticide in the insect paraffin is given as follows:

$$V = \frac{S}{H} D(C_A^P - C_B^P) \qquad \text{(Equation 7)}$$

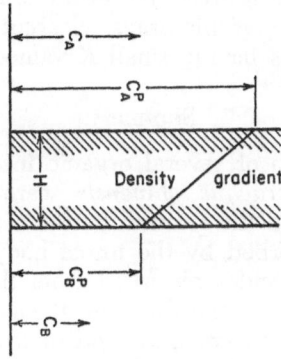

Fig. 12. Density gradient of ethyl parathion in the insect paraffin: C_B = concentration of ethyl parathion in water held in chamber B, C_A = concentration of ethyl parathion in water held in chamber B

where V is the diffusion velocity of the insecticide, S is the cross-sectional area of the glass filter, H is the thickness of the glass filter, and C_A^P and C_B^P are the densities of the insecticide in the paraffin at the interface with chamber A and chamber B, respectively. As is known, S/H in Equation 7 is a diaphragm constant. The diffusion coefficients of ethyl parathion and methyl parathion, calculated from Equation 7, are given below

Ethyl parathion: $D = 2.9 \times 10^{-7}$ cm./hr. ($30°$ C.)
Methyl parathion: $D = 9.2 \times 10^{-7}$ cm./hr. ($30°$ C.)

These two figures indicate that the diffusion coefficient of methyl parathion is about three times larger than that of ethyl parathion. It means that methyl parathion passes through the surface paraffin of the adzuki-bean weevil faster than ethyl parathion. It can, therefore, be said that methyl parathion has a more advantageous physical property than ethyl parathion.

V. Conclusions

Physico-chemical approaches are very important for understanding the mode of action of insecticides in details. In this paper, some parts of the author's studies on the physicochemical properties of insecticides have been described in relation to the insecticidal action of these substances. The most interesting result to be emphasized is the relation given by the following equation:

$$(W - W_0)\ (T - T_0) = K$$

It is not too much to say that insecticides having small K values have high efficiency. This formula indicates that, when T is selected arbitrarily, a value of W is given, because W_o, T_o, and K are constants. Moreover, in carrying out bioassays, clearcut results would be obtained with insecticides having small K values.

Summary

The mode of action of several organic insecticides on the adzki-bean weevil, *Callosobruchus chinensis* were studied. Experiments were carried out to find the relation between the amounts of insecticides which were absorbed by the insect body and the lethal times. The amounts of insecticides absorbed from the gaseous phase were calculated on the basis of the diffusion theory of the insecticides to air. Partition coefficients were measured of insecticides between paraffin collected from the surface of the insects and water, and diffusion coefficients of the insecticides in the paraffin were calculated.

Résumé *

Etudes physico-chimiques sur l'absorption des pesticides par la cuticule des insectes et la pénétration dans l'organisme des insectes

Le mode d'action de plusieurs insecticides organiques sur le charançon du haricot Adzki, *Callosobruchus chinensis*, a été étudié. Les expériences ont été menées en vue de trouver la relation entre les quantités d'insecticides absorbées par l'organisme de l'insecte et le temps requis pour obtenir l'effet létal. Les quantités d'insecticides absorbées de la phase gezeuse ont été calculées en se basant sur la théorie de la diffusion des insecticides dans l'air. Les coefficients de partition des insecticides entre la paraffine récoltée de la surface des insectes et l'eau ont été measurés; les coefficients de diffusion des insecticides dans la paraffine ont été calculés.

Zusammenfassung **

Physiko-chemische Studien über die Absorption von Pestiziden durch die Insekten-Kuticula und ihr Eindringen in den Insektenkörper

Die Wirkungsweise von verschiedenen organischen Insektiziden auf den "Adzki"—Bohnenkäfer, *Callosobruchus chinensis*, wurde untersucht. Es wurden Versuche durchgeführt, um die Beziehung zwischen den Insektizidmengen, welche durch den Insektenkörper absorbiert werden, und den lethalen Zeiten zu finden. Die Insektizidmengen, die von der Gasphase absorbiert werden, wurden auf der

---

* Traduit par S. DORMAL-VAN DEN BRUEL.
** Übersetzt von A. SCHUMANN.

Grundlage der Diffusionstheorie der Insektizide in Luft kalkuliert. Verteilungskoeffizienten für Insektizide wurden zwischen Paraffin, welches von der Oberfläche der Insekten gesammelt worden war, und Wasser gemessen, und Diffusionskoeffizienten von Insektiziden in Paraffin wurden berechnet.

References

ARNOLD, J. H.: Studies in difusion (1). Estimation of diffusivities in gaseous systems. Ind. Eng. Chem. **22**, 1091 (1930).

BALSON, E. W.: An effusion manometer sensitive to DDT and other slightly volatile substances. Trans. Faraday Soc. **43**, 48 (1947).

GILLILAND, J. E. R.: Diffusion coefficients in gaseous systems. Ind. Eng. Chem. **26**, 681 (1934).

HIRSCHELDER, J. H., R. B. BIRD, and E. L. SPATZ: Transport properties of gases and gaseous mixtures. Chem. Rev. **44**, 205 (1949).

LEBASS, G.: A relation between the volumes of the atoms of certain organic compounds at the melting points and their valencies. Proc. Chem. Soc. **22**, 322 (1906).

SUWANAI, M.: Physico-chemical studies on the mode of action of several organic insecticides (1). Studies on the mode of action and the molecular diffusion of gases. Bull. Nat. Inst. Agr. Sci. (Japan) Series c **7**, 113 (1957).

References

Subject Index